THIRD EDITION

FIRE AND EMERGENCY SERVICES ADMINISTRATION

MANAGEMENT AND LEADERSHIP PRACTICES

L. Charles Smeby, Jr.

JONES & BARTLETT
LEARNING

World Headquarters
Jones & Bartlett Learning
5 Wall Street
Burlington, MA 01803
978-443-5000
info@jblearning.com
www.jblearning.com
www.psglearning.com

Jones & Bartlett Learning books and products are available through most bookstores and online booksellers. To contact the Jones & Bartlett Learning Public Safety Group directly, call 800-832-0034, fax 978-443-8000, or visit our website, www.psglearning.com.

21540-3

Production Credits
VP, Product Development: Christine Emerton
Senior Managing Editor: Donna Gridley
Director Product Management: Bill Larkin
Content Strategist: Andrew Labelle
Content Coordinator: Paula-Yuan Gregory
Project Manager: Kristen Rogers
Digital Project Specialist: Rachel DiMaggio
Director of Marketing Operations: Brian Rooney
Production Services Manager: Colleen Lamy
VP, Manufacturing and Inventory Control: Therese Connell
Composition: S4Carlisle Publishing Services
Project Management: S4Carlisle Publishing Services
Cover Design: Kristin Parker

Text Design: Scott Moden
Media Development Editor: Faith Brosnan
Rights & Permissions Manager: John Rusk
Rights Specialist: Liz Kincaid
Cover Image (Title Page, Chapter Opener): © Karl R. Martin/ShutterStock; © Volodymyr Baleha/Shutterstock; © Jones & Bartlett Learning. Photographed by Glen E. Ellman; © Everett Collection/Shutterstock; © Michael Vi/Shutterstock; © Stevecoleimages/E+/Getty Images; © rarrarorro/Shutterstock; © kali9/E+/Getty Images; © Filo/Vetta/Getty Images; © Rubberball/Mike Kemp/Getty Images; © Enisaksoy/DigitalVision Vectors/Getty Images; © iQoncept/Shutterstock; © Rawpixel.com/Shutterstock; © Vladyslav Lehir/Shutterstock; © Halfpoint/Shutterstock
Printing and Binding: LSC Communications

Library of Congress Cataloging-in-Publication Data
Names: Smeby, L. Charles., author.
Title: Fire and emergency services administration : management and leadership practices / L. Charles Smeby, Jr., Adjunct Professor (retired), University of Florida, Fire & Emergency Services Distance Learning.
Description: Third edition. | Burlington, Massachusetts : Jones & Bartlett Learning, [2022] | Includes bibliographical references and index.
Identifiers: LCCN 2020043783 | ISBN 9781284180213 (paperback)
Subjects: LCSH: Fire departments--Management. | Personnel management. | Leadership.
Classification: LCC TH9158 .S64 2022 | DDC 363.37068/4--dc23
LC record available at https://lccn.loc.gov/2020043783

Printed in the United States of America
25 24 23 22 21   10 9 8 7 6 5 4 3 2 1

Brief Contents

Contents

About the Author

Chief (Retired) Smeby recently retired after 16 years as an adjunct professor at the University of Florida where he developed, created, and presented fire science courses. In addition, he administered the Fire and Emergency Service Bachelor's distance learning program at the University.

Previously, he was a Senior Fire Service Specialist at the National Fire Protection Association (NFPA) in the Public Fire Protection Division. He coordinated several NFPA fire service technical committees and provided guidance to the implementation, understanding, and intent of NFPA fire service standards for professional qualifications; training; protective clothing; deployment of emergency resources; and safety, health, and wellness.

Before joining the NFPA, Mr. Smeby had a 20-year career with the Prince George's County, Maryland, Fire Department (PGFD), where he retired as a Battalion Chief. At PGFD he served as a commander of a very busy seven-station battalion for several years. Subsequently, he was assigned as the first safety chief for this metro department that protected over 700,000 people in 530 square miles from 47 fire stations. He helped the department implement policies to comply with OSHA's Fire Brigade Standard and the NFPA 1500 *Standard on Fire Department Occupational Safety, Health, and Wellness Program.*

In addition, he served on the Fire Department Accreditation Task Force for the International Association of Fire Chiefs. Presently, he is a member of the NFPA 1710 *Standard for the Organization and Deployment of Fire Suppression Operations, Emergency Medical Operations, and Special Operations to the Public by Career Fire Departments* technical committee. He is a Life Member of the International Association of Fire Chiefs and the National Fire Protection Association.

He has also achieved the following professional certifications:

Fire Officer IV, Maryland Fire Service Personnel Qualifications Board

Fire Instructor III, State of Florida, Bureau of Fire Standards and Training

Fire Inspector, State of Florida, Bureau of Fire Standards and Training.

Prior to his salaried PGFD career, Mr. Smeby began his service as a volunteer with the Bethesda-Chevy Chase Rescue Squad in Bethesda, Maryland and the College Park Volunteer Fire Department in College Park, Maryland. He has also lectured in the fire science programs at Prince George's Community College and the University of Maryland. His education includes a Bachelor's degree in Fire Protection Engineering and a Master's of Public Policy from the University of Maryland.

Reviewers

Terrence Blackmer
Assistant Fire Chief
Delta College Fire Science Training Coordinator
University Center MI
Mayville Fire Department
Mayville, MI

David S. Blevins, Director of EMS Education
Roane State Community College
Knoxville, TN

Brian K. Blomstrom, MPA, CFO, CTO, PEM
Staff Sergeant
Greenville Department of Public Safety
Greenville, MI

Bradley B. Brooks, LT/AEMT
Pooler Fire-Rescue
Pooler, GA

Robert DiPietro, Deputy Fire Chief (retired)
Central Arizona Valley Institute of Technology
Coolidge, AZ

Vance L. Duncan III, EMT-P, MPA, MiFireE
Deputy Chief, Training Division
City of Erie, Bureau of Fire
Erie, PA

Christopher L. Gilbert, Ph.D.
Training Officer, Tactical Medic, Hazardous Materials Officer
Alachua County
Gainesville, FL

Tony Hargett, Senior Developer/ Instructor
Aqua 7 Rescue
Surfside Beach, SC

Gary R. Johnson, MBA, MIFireE
Program Director/Associate Professor
Fire Science Technology/Technical Rescue
Central Ohio Technical College
Newark, OH

Jeffrey LeBlanc, Captain
Meriden Fire Department
Meriden, NH

Geoffrey Low, Fire Chief
Orono Fire Department
Orono, ME

Christopher Maeder, BA, EMT-P
Chief of Department
Fairview Fire District
Poughkeepsie, NY

Matthew Marlow, Fire Chief
Medway-Grapeville Fire District
Climax, NY

Fred J. McAnn, MS, CFO, MIFireE
Assistant Fire Chief
Haughton Fire Department
Haughton, LA

Larry Murphy, Asst. Chief
HFD/Lakeside EMS
Hurst, IL / Effingham, IL

Douglas Rohn, Assistant Chief
Verona Fire Department
Verona, WI

Leigh H. Shapiro
Deputy Chief/Senior Tour Commander (Ret.)
Hartford Fire Department
Hartford, CT

Brian J. Ward
Global Safety and Health: Emergency Preparedness
    and Response Leader
Georgia Pacific
Atlanta, GA

Tim Whitham, MS, EFO, CFO
Fire Chief
Edwardsville Fire Department
Edwardsville, KS

Warren Whitley, Assistant Chief (ret)
Prince William County Department of Fire & Rescue
Prince William, VA

Jason Zangara, MPH, NREMT, AIFireE
Chief Executive Officer
Fire & Emergency Services Consultants
Manville, NJ

Foreword

This text is focused on all levels of chief officers who administer, manage, or lead fire, emergency medical services (EMS), rescue, and/or emergency management organizations. After reading and studying this text, chief officers (who, when acting in their various roles, are also called "managers," "leaders," and "administrators") will have the knowledge to make policy decisions and the skills needed to lead the organization through progressive change. These same abilities will help anyone in a top position thrive and survive—despite any political pressures that are present in the public or for-profit emergency services agency.

Many fire departments and EMS organizations were propelled into a new level of consciousness after the September 11, 2001 terrorist attack; these organizations are entering a new era of expanded and professional emergency services, including EMS, hazardous materials response, fire prevention, and disaster planning. In addition to these services, however, traditional fires still occur. Although fewer fires occur now than in the past, these incidents remain a challenge, perhaps even more so now because firefighters have less real-world experience. For example, a 2013 *Boston Globe* story reported that "…city records show that major fires are becoming vanishingly rare. In 1975, there were 417 of them. Last year, there were 40. That's a decline of more than 90 percent. A city that was once a tinderbox of wooden houses has become—thanks to better building codes, automatic sprinkler systems, and more careful behavior—a much less vulnerable place [for fire threat]."*

However, today's fires can be hotter and faster spreading. Newer buildings use structural components that do not have the same fire resistance possessed by older construction, often resulting in earlier structural collapse during a fire. The contents of modern buildings have higher rates of heat release and smoke, resulting in shorter times to flashover.

The fire and emergency services (FES) profession has made many progressive changes in equipment, training, and funding, but the job of making improvements is never complete. This text provides the knowledge needed by chief officers to identify and implement progressive change that will help keep first responders safe while providing the best public fire and emergency services.

The Challenge of Gaining New Knowledge

Remember the firehouse pronouncement "200 years of tradition unhampered by progress"? (It is now actually closer to 285 years since the first volunteer fire department was created in 1736!) If you are reading this text, you are probably open-minded about the future and new knowledge; but remember, many people remain fearful of change. This fear can result in strong opposition to new ideas. Expect to hear "we tried that before and it did not work," "it is too expensive," "the union/elected officials/management will never approve," etc. Keep in mind that this resistance is not personal. It can, however, be overcome by persistence and truthful knowledge.

The advice in this text will challenge accepted wisdom in FES' custom, culture, and traditions. Be a risk taker and consider new ideas, but remain cautious. Ask yourself if the idea is based on hard scientific data or if it is just someone's opinion. Seek out other open-minded individuals in your agency with whom you can speak privately. You may scare some members of your organization if you openly express your newfound ideas. Turn on the light of knowledge slowly for these members. Perseverance is the key. Never give up and be prepared for a long, slow journey. Your successes will be worth the wait.

---

* Plenty of firefighters, but where are the fires? As 'emergency' changes its meaning, some critics are arguing it's time to revisit a century-old system, Leon Neyfakh, September 8, 2013, © 2020 Boston Globe Media Partners, LLC. Retrieved from https://www3.bostonglobe.com/ideas/2013/09/07/plenty-firefighters-but-where-are-fires/isCPrIPauX078UKHdixu0K/story.html?arc404=true

CHAPTER

Historical Foundations of Fire and Emergency Services

LEARNING OBJECTIVES

After studying this chapter, the student will be able to:

- Understand the history of fire and emergency services (FES) and its impact on contemporary organizations (p 3).
- Examine the effect that the insurance industry has had on building and fire prevention codes (p 4).
- Examine the influence of the National Fire Protection Association (NFPA) on the fire service (p 5).
- Explore federal involvement in regulation and funding of state and local FES (pp 6–10).
- Examine challenges in FES (pp 10–12).
- Examine progressive trends in FES (pp 12).

Case Study

Slow Progress in Fire and Emergency Services

In 1994, a state fire marshal was in the process of reviewing plans for a new casino ship that would dock in his state and sail on the Mississippi River when he decided to make inquiries to find out if there was a fire safety standard that covered casino ships. He found that the NFPA did not have a standard at that time (subsequently, at the request of the U.S. Coast Guard, the NFPA did create a fire safety standard for these ships).

The fire marshal then noticed that the permit applicant was the previous owner of the Beverly Hills Supper Club in Southgate, Kentucky. This nightclub was the location of a tragic fire in 1977 that took the lives of 165 victims. The marshal asked himself how this person, who was at least partially responsible for many fire deaths, could be allowed to own and operate a ship that needed to be safe enough to protect the lives of thousands.

As he began to think about this situation, he remembered a more recent tragic fire that occurred. In 1991, a fire broke out at a meat processing plant in Hamlet, North Carolina; the fire claimed the lives of 25 workers. The owner was convicted of 25 counts of manslaughter, lost all his wealth in civil court action, and spent many years in prison. This sentence was in sharp contrast to the consequences encountered by the owner of the Beverly Hills Supper Club some 14 years prior.

As this case exemplifies, there has been a great deal of progress in the FES, but because it is relatively slow and not always obvious to the casual observer, some might believe that these tradition-based services are unaffected by change. This is not true; the following examples demonstrate relatively recent progressive changes in the FES:

- Smoke alarms
- Seat belts and air bags
- Hurricane- and earthquake-resistant construction
- Paramedic services
- Professional firefighter and officer competency standards
- Federal grants for FES
- Compressed air foam
- National standards for emergency medical training and certification
- Higher education opportunities
- United States Fire Administration (USFA) and the National Fire Academy
- National incident management system
- Residential fire sprinklers
- Safety standards for firefighters and fire suppression operations
- Defibrillators
- Federal support for disaster mitigation
- NFPA standard for the deployment and staffing of fire companies
- Thermal imaging cameras
- Hydraulic rescue tools
- National focus of emergency response to major disaster events
- Chem-bio detectors

Discussion Questions

1. Identify and list three examples of progress for the U.S. Wildfire problem.

2. Name Issues with future progress hampered by lack of nationwide standards or regulations (for example: Occupational Health and Safety Administration [OSHA], emergency medical service professionals, NFPA, or insurance industry).

3. Identify and list protocols or techniques in need of progress for contagious diseases.

Prologue to the Future

FES administrators should always be thinking about the future and looking ahead. However, it is important to begin with a basic understanding of how we arrived at where we are today. Like any journey, you need to know where the trip began and where you are at present to plan the route to the future, including all the points along the way. The history of a profession is important because it explains many current policies and points to issues that remain unresolved. By examining the history of FES specifically, you can gain insight into the foundation of these organizations, avoid mistakes, and identify what worked in the past to serve as a guide toward a brighter future. Ultimately, your role in leadership, administration, and management can be enhanced with knowledge of the history and current status of the FES profession.

Local Beginnings

The first fire protection service on record in North America started in 1648 in the town of New Amsterdam, now part of New York City. Like other early fire departments that soon followed, it was an all-volunteer organization. However, it was not long before the public began to demand fire safety codes and paid fire departments to combat the massive fires that struck many large cities across the United States as a result of combustible construction and careless fire safety behaviors. In 1853, the nation's oldest fully paid fire department was organized in Cincinnati, Ohio **FIGURE 1-1**.

As demand for fire protection grew, other towns and cities developed their own independent fire departments. However, the need for communication and standardization across departments soon became clear. When a large conflagration consumed Baltimore in 1904, the mayor called on fire departments in Washington, DC, Philadelphia, and New York to assist. These agencies responded but were dismayed to find that their hose couplings would not fit the Baltimore fire hydrants, and the city continued to burn.

These local origins also explain why there are now more than 29,000 fire departments across the United States. Most of these departments protect a population of fewer than 2,500 citizens, and only 16 departments protect more than one million people each (Evarts & Stein, 2020). Interestingly, the largest

FIGURE 1-1 Early American firefighters.
© Alpha Historica/Alamy Stock Photo.

FIGURE 1-2 Vintage Ambulance 1934. At the beginning of EMS, patients were just transported to hospitals without emergency medical treatment.
© Margo Harrison/ShutterStock.

department—New York City Fire Department—is actually a consolidation of five counties (New York City is made up of five boroughs, each of which is also a county in New York State.)

Emergency medical response also started out on a small scale. The first prehospital medical response system on record served with Napoleon's army in 1797. This system was designed to triage and transport the injured from the battlefield to aid stations. More modern field care and transport began after the first year of the Civil War. Civilian ambulance services in the United States began transporting patients to hospitals in Cincinnati in 1865 and in New York City in 1869. Later, funeral home hearses, which often doubled as ambulances, were eventually replaced by fire department, municipal, and private ambulances **FIGURE 1-2**.

National Development Efforts

Development of the Modern Fire Service

As of 2017, the United States had an estimated 29,819 fire departments. Because there are relatively few federal regulations for fire services in the United States, there could be 29,819 different ways to provide public fire protection. For example, NFPA Fire Service standards are voluntary unless adopted as law by a government. And for FES, they are written more as an outline and do not have specific educational, training, or testing criteria.

Volunteer firefighters staff most of the U.S. fire departments, about 65%, and serve primarily small populations of 2,500 or less. Presently, all-paid or mostly paid departments protect 69% of the U.S. population, all-volunteer departments protect 19.0%, and the rest are protected by combination departments (Evarts and Stein, 2020). Even with the lack of federal regulation, the equipment and procedures of these departments are relatively uniform because of past initiatives, primarily the Insurance Service Office (ISO) Grading Schedule.

Insurance Service Office

Most of the uniformity among fire service practices and equipment that exists today is a direct result of the *ISO Grading Schedule for Municipal Fire Protection* (ISO, 2012). This document was created in the early 1900s by the National Board of Fire Underwriters, an organization whose members come from the insurance industry. The effort was undertaken after several insurance providers went bankrupt trying to pay claims from citywide fires that resulted in financial panic in the insurance industry.

The insurance industry wisely realized that it needed to encourage fire departments to prepare for large-scale fires by giving them the ability to provide mutual aid to other fire departments that could not handle major fires on their own. The grading schedule surveyed public fire protection departments in urban centers throughout the United States and aimed to create uniformity. This generally resulted in increases in the number of fire stations and staffing of fire companies. Municipalities that did not adopt the ISO's recommendations faced the prospect of higher insurance costs.

As a result of improvements in public fire services, lower insurance rates primarily benefited commercial properties. For example, Highly Protected Risk buildings have their own insurance rates that are lower than the general class rates. These properties are primarily those protected by engineered automatic supervised sprinkler systems and are found in buildings owned by large corporations.

Today, municipalities are graded with a numerical rating from Class 1 to 10, with 1 being the best rating. A community that has a Class 1 rating has lower insurance premiums than a community with a higher rating. Many fire chiefs (administrators) have justified budgetary increases for items recommended by the ISO to improve a community's rating, thus lowering insurance costs. Unfortunately, this national standardization initiative has not been accepted completely by all departments because the numerous ratings have several levels of compliance.

Some fire officials and municipal administrators have been openly critical of the grading schedule. To receive a high rating, a fire department has to invest heavily in many fire companies, including increased staffing levels. However, as the probability of large-scale fires has decreased, public officials have questioned the need for large crews (up to six firefighters per company) and close spacing for fire stations (sometimes as close as three-quarters of a mile apart).

In addition, there have been complaints regarding the relevancy of the grading schedule. For example, in June 2007, a tragic fire in Charleston, South Carolina resulted in the death of nine firefighters. At the time, Charleston Fire Department was one of 38 departments in the country with an ISO Class 1 rating. In a December 2008 article in *Fire Chief* magazine, Charles Jennings, Ph.D., MIFireE, CFO commented: "Numerous studies have questioned whether compliance with the [ISO Grading] schedule is correlated with better fire services or lower losses. In the wake of the deaths in the Class 1 Charleston (SC) Fire Department, an embarrassed insurance industry trotted out proposed revisions to the schedule, which hasn't been updated since 1980" (Jennings, 2008). Some of the potential revisions include increased reference to NFPA standards, additional emphasis on firefighter safety and training, and credit for adoption and enforcement of model building and fire prevention codes.

Still, although complaints regarding the validity of this grading scale may be well-founded, it has helped fire departments from different areas fight fires together, and many departments rely on the grading schedule to justify and plan improvements to their departments.

Commission on Fire Accreditation International

As an alternative to ISO for planning, the International Association of Fire Chiefs (IAFC) and the International

City/County Management Association have created a new accreditation program for fire and emergency rescue agencies. The Commission on Fire Accreditation International self-assessment process can be used to measure fire department capabilities, response times, and compliance with nationally recognized standards. Some criticism about this process indicates that departments receive accreditation even though they do not meet all of NFPA standards, especially staffing and response time criteria.

As of August 2018, 247 fire departments have been accredited. The program is a comprehensive self-analysis of each agency's ability to provide professional FES to the public. A committee of public fire protection experts appointed by the IAFC and International City/County Management Association evaluates the competencies using professional standards as guidelines. Although this effort is a step in the right direction, stricter requirements are still needed for accreditation and compliance with standards on firefighter safety, deployment, and staffing.

Facts and Figures

Subzone Rating Factors

When State Farm insurance company abandoned the ISO grading schedule in 2001 for homeowner's single family dwellings insurance rates, it was uncertain what budgetary planning tool would be used as a replacement for funding local fire departments. State Farm changed to a system that determines insurance rates based solely on the loss experience in *subzone rating factors*. These factors include fire, wind, hail, water damage, theft, and liability. (This is the primary statistical tool U.S. state regulator agencies use to set insurance premiums.) For example, if the company pays out $20 million during a year, it wants to charge rates that recover the payouts along with a percentage profit. This loss data is collected at the state level since each state regulates insurance independently. State Farm has said that 70% of claims paid under its homeowner's program are nonfire losses, although it has not produced hard data to support this claim.

Therefore, in most jurisdictions across the country a better ISO rating will not provide lower insurance premiums for homeowners. Again, only commercial buildings not protected by automatic sprinklers receive the substantial benefit.

National Fire Protection Association

In the early 1900s, national and regional building and fire codes were developed in response to the loss of life and property occurring because of catastrophic fires plaguing many U.S. cities. The NFPA, which was formed in 1896 by insurance companies, was and still is the leader in national consensus fire, safety, and electrical codes. The first NFPA standard addressed uniform automatic fire sprinkler installation. Around 1913, in response to several fatal fires, the NFPA added a standard for Safety to Life in an effort to prevent fire-related deaths.

To this day, NFPA continues to create, update, and adopt codes and standards using a consensus method from committees that represent the fire protection community and the general public. The NFPA uses subject matter experts, including local, state, and national officials and public input to create consensus standards for firefighter training and education, equipment, and emergency operations with an emphasis on safety and efficient emergency service. These standards are updated on a schedule of every 3-5 years. The last step in approving a code or standard is a vote of its membership at a national meeting of NFPA. In addition, there is an appeal process to correct or add any justifiable changes.

For a full explanation of this process, the NFPA provides a comprehensive description in detail at its web page (https://www.nfpa.org/Codes-and-Standards/Standards-development-process/How-the-process-works). Anyone can participate, including the general public, either by attending the committee meetings in person or submitting in writing proposals and comments.

America Burning Report

In 1968, the Fire Research and Safety Act was created. This Act, which established the National Commission on Fire Prevention and Control, stated, "The [U.S.] Congress finds and declares that the growing problem of the loss of life and property from fire is a matter of grave national concern" (USFA, 1973). In 1973, the national effort to professionalize the fire service started with the Commission's America Burning Report.

The report influenced the development of several NFPA standards for training, safety, and professional qualifications of fire service members and hazardous materials response personnel. In total, the Commission formally listed 90 recommendations, including:

- Establish a national fire data system.
- Increase the research and medical facilities for burn treatment.

- Make fire prevention at least equal to suppression in planning priorities.
- Use women for fire service duties.
- Increase the use of automatic aid.
- Recognize advanced and specialized education in hiring and promotions.
- Provide federal financial help for planning, training, equipment, research, and fire safety education.
- Provide ambulance, paramedical, and rescue services where they are not provided by other agencies.
- Create the USFA and the National Fire Academy.
- Enforce adequate building and fire prevention codes.
- Require smoke alarms to protect sleeping areas and automatic extinguishing systems for high-rise buildings.
- Implement fire safety education in schools throughout the school year.
- Annually inspect homes for fire safety.
- Develop technology for automatic extinguishing systems for all kinds of dwellings, including single family.

Even though many of its recommendations have not been fully implemented, the report has been the basis for many of the changes in progressive FES organizations and continues to influence the modern fire service. Because of its continued relevance, the USFA updated and reprinted the report in 1989.

Development of Modern Emergency Medical Services

Modern emergency medical services (EMS) started with the 1966 publication of the paper **Accidental Death and Disability: The Neglected Disease of Modern Society,** by the National Academy of Sciences' National Research Council (National Academy of Sciences, 1966). This document pointed out an injury epidemic that was the leading cause of death among persons between the ages of 1 and 37, and served as a blueprint for a national effort to improve emergency medical care. These deaths were primarily the result of motor vehicle and work accidents that were the incentive for U.S. OSHA and the U.S. Department of Transportation (DOT) National Safety Act.

The report revealed that many ambulances and emergency medical treatment facilities were inappropriately designed, ill-equipped, and staffed with

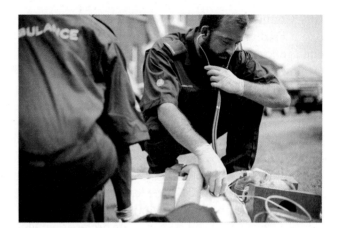

FIGURE 1-3 Paramedics in action.
© Rawpixel.com/ShutterStock.

inadequately trained personnel. When the Highway Safety Act of 1966 established the Emergency Medical Services Program in the DOT, the DOT was given the authority to improve EMS through program implementation and development of standards for provider training **FIGURE 1-3**. States were required to develop regional EMS systems, and some of the costs of these systems were funded by the **National Highway Traffic Safety Administration (NHTSA)**.

Through this federal agency, funding became available for training, equipment, and planning. At about the same time, a number of military medics were returning from the Vietnam War. They had experience with specialized emergency medical training, organization, and equipment that had never before been seen in any other military operation. These medics brought that experience, which included the use of medical air transport for rapid evacuation, to the field of prehospital care.

In addition, the war provided opportunity for trauma research, more effective tourniquets, chest seals, and bandages with clotting material. Even synthetic blood and intravenous solutions that can carry oxygen have now been field tested because of war experience. Prehospital care has benefitted from this improved equipment and the resulting procedures.

In 1971, another phenomenon changed the general public's attitude concerning the fire service and emergency medical care. When the television show *Emergency!* aired for the first time, there were only six regions in the country providing paramedic service: Seattle, Miami, Jacksonville, Los Angeles City and County, and Columbus. By 1978, a mere 7 years later, there were more than 300 regions providing EMS. The show *Emergency!* generated an increased interest in providing EMS and paramedic service by local governments and effectively launched the careers of thousands of fire and EMS personnel throughout the country. Television has a strong influence on the general public, including politicians, resulting in demand and funding for paramedic programs. This was especially true for this popular TV show.

In 1999—more than 20 years after the show's debut—Project 51, a website dedicated to the show and its cast, stated, "Who would have guessed that the humble beginnings of the Mobile Intensive Care Unit would evolve to one of today's most recognized emergency based services, and one that is used most routinely by millions of people" (Project 51, 1999). The show had more to do with the fast adoption of advanced and basic EMS by the fire service, hospitals, and ambulance services than anything that had come before.

However, although *Emergency!* was visionary, it was not a completely accurate portrayal of the state of EMS at the time. For example, it suggested that paramedics existed everywhere, which was not true in the 1970s. It also portrayed paramedics as having frequent success saving lives, which may have led to unrealistic expectations. Even today, portrayals of EMS in television, movies, and other media may create unrealistic perceptions and expectations regarding EMS, such as high success rates for heart attacks, etc. This causes problems with depression of EMS providers who expect to save every patient and with the public that expects emergency medical miracles on every call.

FES Professional Standards

To develop a professional service organization—one whose members are trained and qualified—FES professionals must rely on recognized national or international fire and emergency standards of competency. In the fire service, this effort is led by the NFPA.

Fire Standards

NFPA publishes voluntary codes and standards that cover fire department professional benchmarks for firefighter training, safety, and emergency operations, including planning tools for fire station location and response times. These codes and standards to be legally enforced must be adopted by a state or local government. In many cases, they are implemented locally by voluntary compliance. For example, in 2001, an NFPA committee finished work on a new standard—1710, *Standard for the Organization and Deployment of Fire Suppression Operations, Emergency Medical Operations, and Special Operations to the Public by Career Fire Departments*—which measures fire and EMS deployment, response times, and staffing (NFPA, 2016). This standard is commonly used as a planning tool to help improve the professional and

CASE STUDY

The Need for National Standards

In a tragic example of not learning the lessons of history, 25 people died and nearly 3,500 homes were destroyed in a large wildfire in Oakland, California, in 1991. Surrounding jurisdictions that responded to the mutual aid call were unable to connect their hoses to the Oakland fire hydrants, severely handicapping firefighting capabilities. Even today, many FES cannot communicate with one another because of different radio channels and systems of communication. Progress in this critical area has begun with funding and radio spectrum allocation for a new nationwide system called FirstNet. ATT is the partner and contractor for the implementation. Further efforts are needed to standardize fire service equipment, training, and communication. Adherence to national standards is necessary to keep fire departments up-to-date with personnel, equipment, and training.

effective operation of a fire department that includes safety of firefighters.

Expert fire protection consultants can also be hired to study and make recommendations on any aspect of a FES organization; this practice is discussed in more detail in the chapter *Public Policy Analysis*. This can be a necessary step in policy analysis to implement these standards for a local jurisdiction. In most cases these changes would result in employee and officers altering existing policies, causing resistance. In addition, many need additional funding that can be justified by an outside expert consultant.

EMS Standards

Within the field of EMS, the path toward national unifying standards began with the nationally recognized EMT-A curriculum created by the National Highway Traffic Safety Administration (NHTSA) in 1969. The field of EMS quickly grew with the Statewide EMS Technical Assessment program initiated by the NHTSA in 1988, along with the Trauma Care Systems Planning and Development Act of 1990. These efforts provided guidelines for a trauma system to address the needs of injured patients and match them to available resources. Local EMS authorities were vested with the responsibility of establishing trauma systems and designating trauma centers in their areas, and states were encouraged and given funds to develop inclusive trauma systems.

In 1991, the Commission on Accreditation of Ambulance Service began administering an EMS accreditation program that is still used today. This program is a comprehensive self-evaluation analysis with the main objective being to provide quality and timely EMS to the community.

In 1994, the National Standard Curriculum, a national training model developed by the DOT, was developed and included examinations, skills sheets, and evaluation tools from the National Registry of EMTs, the entity that conducts testing for national EMS certification. This is still a work in progress as some states have their own EMS certification systems and do not recognize the national certifications or those of other states.

Federal Involvement

Federal agencies, such as the Federal Emergency Management Agency (FEMA), along with the Departments of Defense and Homeland Security and the Federal Bureau of Investigation, provide training, equipment, and resources to prepare local FES organizations for terrorist activities, weapons of mass destruction, and other extremist violence. The Environmental Protection Agency, a major regulator of hazardous substances, has created regulations for storage, use, and transportation of hazardous materials. OSHA has developed rules to protect fire and EMS personnel against the hazardous materials that they come into contact with in their jobs. The USFA and its National Fire Academy also provide valuable training and education for the fire service, although their impact has been limited as a result of insufficient federal budget support.

However, the federal effort continues to expand as a result of the September 11, 2001 terrorist attacks on the United States, which led to increased national attention and prestige for FES. These federal programs continue to play an important role in creating a path to national uniformity by stressing curriculum, ideas, management, and leadership practices that prove to be successful, including providing a path to the interoperability between jurisdictions.

Melding of Fire Service and EMS

Over time, the role of the fire service has expanded greatly beyond fire suppression to include medical response, fire prevention, natural disasters, and public education. Citizens now depend on the fire department to protect them against the dangers of fire, entrapment, hazardous materials, emergency

medical incidents, and explosion, and to come to their aid in any emergency event that might occur in the community. Domestic acts of terrorism, such as those in Oklahoma City, Orlando, El Paso, Parkland, San Bernardino, and New York City, have added a new mission of safe and effective response to terrorist events for the fire service. As such, the job of the chief officer is becoming more complicated and challenging every day.

With the recognition that firefighters are dedicated to saving lives and are strategically positioned to deliver timely response, fire service–based EMS have gained popularity. Currently, four levels of EMS providers exist: (1) emergency medical responder, (2) emergency medical technician, (3) advanced emergency medical technician, and (4) paramedic. Firefighters may or may not be trained to an EMS level of certification, and fire departments may or may not provide ambulance transport. This varies considerably throughout the nation and is usually based on local policy.

On a national level, nearly 60% of fire departments provide some EMS. Those that do not are mostly all-volunteer departments serving areas with small populations that do not have sufficient resources to provide EMS. In many areas, volunteer fire departments have trouble recruiting and retaining members, so there is concern that the additional demands of a volunteer's time to train for and provide EMS might discourage membership.

Unified Federal Emergency Response

After a series of devastating fires in New Hampshire in 1803, the U.S. Congress passed a measure that provided financial relief to the affected New Hampshire residents. This measure is considered the first piece of legislation passed by the federal government that provided relief after a disaster. Over the years, the federal government created various other programs to fund recovery from disasters, including flood control by the U.S. Army Corps of Engineers.

During the 1960s and 1970s, such major disasters as hurricanes and earthquakes brought focus to this fragmented approach to federal disaster assistance **FIGURE 1-4**. In 1979, FEMA was established to consolidate the federal disaster response, which at the time included the USFA and the Civil Preparedness Agency in the Defense Department. With the end of the Cold War, resources once allocated to civil defense were redirected to support disaster relief, recovery, and mitigation programs. FEMA started programs to prevent and reduce the risk before a disaster struck.

FIGURE 1-4 Hurricane damage.
© Leonard Zhukovsky/ShutterStock.

Currently, the agency is focused on preparing and delivering training to mitigate and recover from terrorist incidents.

The U.S. Forest Service, under the Department of Agriculture, also played a crucial role in developing the ability to provide a unified emergency response through the help of two key initiatives: the development of the National Interagency Incident Management System—now called the National Incident Management System (NIMS)—and the red card system. The Forest Service developed the National Interagency Incident Management System in the early 1970s to ensure that fire protection agencies were prepared and organized to be able to fight destructive wildfires in southern California. In the red card program, firefighters who are assigned to a wildland fire managed by a federal agency (U.S. Forest Service, Bureau of Land Management, National Park Service, Bureau of Indian Affairs, or U.S. Fish and Wildlife) or by many state agencies are required to have a red card, which documents the current wildfire qualifications of an individual. In a sense, it is similar to a driver's license. The credentials specify levels of competency, including firefighting and incident command, achieved by successfully completing training and testing to a national standard.

This credentialing system is remarkably effective and prepares firefighters from all over the country to work together, efficiently and collaboratively, on large wildfire firefighting efforts. For example, a wildland firefighter from North Carolina can travel to Wyoming and be assigned to a team made up of individuals from all over the country because there is a national system of competency credentialing.

On November 22, 2002, the federal government took the effort for emergency response a step further with the establishment of the U.S. Department of Homeland Security (DHS). At the time, threats against the United States were growing in magnitude, and the DHS was created to protect American citizens from terrorist attacks. In 2003, FEMA became part of the DHS and continues to be the nation's incident manager for all hazards and major disasters. This command system provides a direct line of authority and communications from the President of the United States—by way of the Secretary of Homeland Security—down to the local level.

The number one priority of the DHS is to prevent the occurrence of a terrorist attack. The same goal is also very important to fire and EMS providers, because preventing problems has always been more effective than responding to an emergency in progress. Traditionally, however, prevention has not been a major priority with most emergency services. In December 2008, the DHS released its NIMS. This system supersedes all previous emergency management systems when the federal government is involved and is the default system in the United States. In responding to future disasters, a plan that details mutual aid across jurisdictions and state lines using the incident management command system will be a top priority.

In addition, there are now more mutual aid agreements being signed to prepare for major emergency incidents that could result from terrorist actions and catastrophic natural disasters. National credentialing efforts for first responders are being developed with the support of the DHS.

Although DHS has not implemented a national credentialing system at this time, several state, regional, and local government credentialing programs have been created to provide real world case studies. When large scale emergencies occur and first responders from distant jurisdictions respond to assist, the resources and staffing have to be identified, including skill and competency levels. This is especially true for a terrorist or major weather incidence.

As successful as these efforts have been, greater standardization among fire departments is still necessary to ensure departments and individual responders are able to work together safely and effectively. The NFPA and DHS are continuing to develop standards and initiatives to improve the fire service's ability to present a unified emergency response. These include NFPA 1500, *Standard on Fire Department Occupational Safety and Health Program* (NFPA, 2018) and NFPA 1710, *Standard for the Organization and Deployment of Fire Suppression Operations, Emergency Medical Operations, and Special Operations to the Public by Career Fire Departments* (NFPA, 2016).

When many different emergency services departments come together to respond to national events, effective management is critically important. According to *NFPA Fire Protection Handbook*, "The absence of incident command at an incident scene puts firefighters at great risk and is one of five leading contributing factors of firefighter fatalities, as reported by the National Institute of Safety and Health" (NFPA, 2008). Therefore, incident scene management is critical to the successful mitigation of large- and small-scale emergencies and the health and safety of emergency responders.

Critical to incident scene management is the ability to communicate to other responders and their dispatch centers using mobile two-way radios. Currently, there are around 10,000 different and incompatible mobile radio emergency services networks in the United States. A new effort by the Federal government, FirstNet, is the nationwide public safety communications system dedicated to America's first responders. Congress allotted $7 billion and 20 MHz of valuable radio spectrum to build the network. This effort is a work in progress and does not incorporate voice communications at this time.

Current Problems Facing Fire Service and EMS

Although the fire service in the United States has come a long way from its local beginnings to create a system of national standards and unified emergency response, there are still many areas where further professional progress is needed. The modern fire service is continuing to work toward measurable outcomes for firefighter and fire officer training, certification, education, and physical fitness standards nationwide.

There are still many challenges to be faced, including breaking down traditional ways of operating and getting officials to help set emergency management priorities. A few specific challenges are mentioned in the following sections.

U.S. Fire Experience and Trends

Trends on a yearly basis may be misleading so it is always helpful to look at a multi-year analysis such as:

		Trend from 2008 to 2017
FIRES	1,319,500	−6.2%
DEATHS	3,400	+9.6%
INJURIES	14,670	−15.8%
$LOSS*	$23.0 billion	+12.0%

* Adjusted to 2017 dollars; includes $10 billion in losses from Northern California wildfires.
Data from National Fire Protection Association for more statistics on U.S. fire loss.

See: National Fire Protection Association for more statistics on U.S. fire loss.

After looking at these U.S. statistics you may ask the question, "Why are fire deaths up 9.6% over this 10-year period?" This is a great observation and is something that should be investigated by NFPA and other national fire research organizations to determine a cause(s). Once a cause (not always just one) is identified, then appropriate fire prevention measures would be suggested. At the local level the fire administrator may want to strengthen or implement programs such as smoke detector battery replacement or installations, for example. This may also indicate that modern furnishings are more flammable and that residential sprinklers should be considered.

We have solid statistical evidence that these smoke detectors have been very effective at lowering fire deaths historically. From 1980 to 2016 U.S. fire statistics documented a 63% drop in fire deaths. Smoke detectors are now installed in over 90% of homes. The other major contributor to this reduction was a sharp decline in fires started by smoking materials. This was the result of two factors: fewer smokers and, more recently, fire safe cigarettes. This trend is also supported by the present statistic that 51.6% of residential fires were caused by "cooking" in 2017 and that was the biggest cause. A more in-depth discussion of the use of trends and statistics can be found in Chapter 12, *Public Policy Analysis*.

Challenges of the Modern Fire Service

In 2016, the NFPA conducted research on fire-related trends and statistics and noted a number of issues relating to smoke alarms. To begin with, 38% of all home fire deaths reported occurred in households without smoke alarms. What is even more concerning is that, although most households in the United States have at least one smoke alarm, these alarms were found to operate correctly in only about half (40%) of reported home fires (NFPA, 2015). This is a trend that the fire service will have to continue to battle.

A more recent NFPA study conducted in 2016 found the following additional concerns regarding the U.S. Fire Service's needs and response capabilities. A few highlights are listed below:

- Two out of five (43%) fire stations are at least 40 years old, up from 32% in 2001.
- There is an increasing need for additional stations in both the largest and smallest communities.
- Only one quarter (27%) of fire departments have a basic firefighter fitness and health program.

- Twenty percent of fire departments did not have anyone conducting fire code inspections in the community, down from 24% in 2010.

- Nine percent of departments reported that they could not communicate via two way radio with their response mutual aid partners at an incident scene, lower than 13% in 2010.

- Forty-nine percent of all fire departments in 2015 have not formally trained all of their personnel involved in structural firefighting, up from 46% in 2010. However, 67% of these formally untrained departments serve under 25,000 residents.

- Except for cities protecting populations of greater than 500,000, most cities do not assign at least four firefighters to an engine or pumper company and are probably not in compliance with NFPA 1710, which requires a minimum of four firefighters on an engine and pumper companies (NFPA, 2016).

- The greatest FES educational needs for the public were the following (NFPA, 2016):
  - A wildfire safety program based on a national model (84% of the U.S. population without a program)
  - Home fire sprinkler education (74%)
  - Car seat installation (70%)
  - An older-adult fire safety program based on a national model (67%)
  - Home safety visits (54%)
  - A youth firesetter program (48%)
  - Cardiopulmonary resuscitation instruction (42%)
  - Free installation of home smoke alarms (37%)
  - A school fire safety education program based on a national model curriculum (32%)
  - Fire prevention week activities (14%)

Challenges of the Modern EMS

A common challenge of modern EMS is that some of the country still cannot access EMS by calling 911, and people place many unnecessary 911 calls. Fees charged for transport are paid by Medicare, Medicaid, and private health insurance, or not paid at all. Just like anything that is free, emergency transport is demanded more often than it is actually needed. As stated by Pratt, "Prehospital 9-1-1 emergency patient medical care is a major part of the safety net for the American healthcare system. They might be the provider of last resort for the needy, yet they can be one

more mechanism for overloading the health care system" (Pratt et al., 2007, 14). The easy access of ambulance transport that is common in most densely populated areas results in high numbers of noncritical patients requesting medical care. Because emergency rooms cannot turn anyone away from medical care, they are overtaxed and may have to turn real emergencies away to an ER many miles away.

At the present time, fees can be charged only for transportation of a patient and not for emergency medical care provided by first responders. As a result, there are many private ambulance companies providing EMS care for free along with transport. The Patient Protection and Affordable Care Act of 2010, which is reforming the nation's healthcare system, does not have any specific requirements for EMS but contains regulatory authority for a federal agency to write rules affecting EMS providers. Many of the specifics of this legislation have not been implemented or are not fully understood. This Act is evolving and might be changed or repealed in the future. The Trump administration is supporting a legal constitutional challenge to this act and it has been found to be unconstitutional by a lower federal court. At the present time it is in an appeals court for review. More to follow.

Financial Challenges

The financial pain of the recession that started in December 2007 continued for many years. The U.S. economy is just now experiencing a substantial comeback. In the aftermath of this recession, a wave of political action with a distinct antitax and antigovernment sentiment swept the country. One leading example of this backlash occurred in Wisconsin in 2011. The state passed legislation that limited bargaining rights for public employee unions. The legislation also increased state employees' contributions to retirement and health insurance. Numerous other states are in economic peril and will have to take similar actions in the near future.

There are many persistent sources of concern for the U.S. economy, such as unprecedented deficits. Although the unemployment rate is falling to record lows and the gross national product is very healthy, the federal deficit keeps rising, with the total cost per taxpayer of more than $189,000 at this time (to access the latest numbers, go to https://usdebtclock.org/). The danger that inflation and increasing oil prices might cripple the economy is always present. And, finally, a worldwide pandemic is causing substantial damage to the United States and world economies. The overall extent of this pandemic, COVID-19, is not known at this time but has and will affect operations and funding of FES departments.

Unions will face several dilemmas in the future. Issues that are part of this discussion include:

- Union bargaining rights
- Supreme Court limits public sector unions from collecting fees from non-members
- Employees' salaries and benefits
- Revenue from EMS transport service
- Consolidations
- Reducing either staffing or stations (or both)

The only thing that is for certain is that change will happen.

Today's Fire and Emergency Services

What is the status of today's FES? Often, there is a large gap between the public's perception and the reality that many FES leaders encounter. Veteran chief fire officer Dr. Harry Carter states, "The fact is these leaders do not have the staffing, equipment, or most importantly, the financial support to back up their claims. Sadly, far too many of our folks in positions of leadership use their education and training to weave an intricate web of falsehoods and deceit. These folks are unwilling to face reality and provide a true assessment of their agency's capabilities to their citizens and their local governments" (Carter, 2005). Dr. Carter points to many departments that run fire companies with one, two, or three firefighters. This does not meet the minimum requirement of four firefighters as listed in several national standards, a requirement that was validated by a 2010 scientific-based study at the National Institute of Science and Technology (NIST) that showed that the size of firefighting companies largely affects the company's ability to protect lives and property (NIST, 2010). Dr. Carter also cites the example of Hurricane Katrina's aftermath, which exposed numerous issues involving mutual aid response, including failures of local, state, and national plans for catastrophic incidents.

As is true in any profession, there is always room for improvement; however, a number of positive trends should be recognized. For example, although careless smoking habits are a leading cause of fire deaths, these deaths are decreasing thanks to the increased effectiveness of smoke alarms and a decrease in the number of smokers (from 42.4% in 1965 to 15.5% in 2016). In addition, the Coalition for Fire-Safe Cigarettes has worked to save lives and prevent injuries and devastation from cigarette-ignited fires through state passage of fire-safe cigarette legislation. A report released by the NFPA on July 2013 stated that "the long-term trend in smoking-material fires has been down, by 73% from 1980 to 2011, helped by the decline in smoking, the effect of standards and regulations that have made mattresses and upholstered furniture more resistant to cigarette ignition, and more recently, the adoption of fire-safe cigarette requirements throughout the country" (Hall, 2013).

Furthermore, fire departments are now placing more emphasis on fire prevention, specifically through the use of modern building and fire codes. Automatic sprinklers, long used for property protection, are very effective tools for reducing or preventing loss of life or property (Evarts & Stein, 2020). In addition, fire safety education, which was started in the 1970s, has now become part of many schools and fire departments. Public safety education by fire departments includes information about vehicle accidents, swimming pool and water safety, and other common accidents that cause injuries and deaths (Evarts & Stein, 2020). There is also the trend of departments adding new services, such as rescue, hazardous material response, emergency preparedness, and EMS.

Still, the chief officer should never underestimate the difficulty of making changes. Preparation is the key to success. Remember to review the beginning of this chapter to provide background to contemporary issues. If we do not study history, we are doomed to repeat the same mistakes or discount successful solutions. Useful solutions to common FES issues, and techniques to implement change, can be found in the following chapters. The first step in preparation is being informed and educated in identifying a problem **FIGURE 1-5**.

FIGURE 1-5 Fire chief talking to firefighters.
© Jones & Bartlett Learning. Photographed by Glen E. Ellman.

Wrap-Up

CHAPTER SUMMARY

- There are many lessons to learn from the history of fire and emergency services. Consider the famous quotation by William Shakespeare that "the past is the prologue to the future" as really good advice.
- From many organizations and individuals, numerous progressive changes have been discovered and adopted to make up the present day FES.
- More changes are sure to be discovered by analyzing problems, including solutions in the FES.
- No FES organization is an island unto itself. Look to the FES community at-large to coordinate and identify problems and solutions for common issues.

KEY TERMS

Accidental Death and Disability: The Neglected Disease of Modern Society 1966 publication that serves as a blueprint for the national effort to improve emergency medical care.

America Burning Report A compilation of 90 recommendations issued in 1973 by the National Commission on Fire Prevention and Control that began the national effort to professionalize the fire service.

Commission on Fire Accreditation International A self-assessment process that measures fire department capabilities, response times, and compliance with nationally recognized standards.

Federal Emergency Management Agency (FEMA) The nation's incident manager for all hazards and major disasters, run by the Department of Homeland Security.

Fire Research and Safety Act 1968 legislation that established the National Commission on Fire Prevention and Control.

FirstNet The nationwide public safety communications system dedicated to America's first responders.

Insurance Service Office (ISO) Grading Schedule A document that grades municipalities from Class 1 to 10 based on 1 being the best and 10 indicating long response times and no water supply.

International Association of Fire Chiefs (IAFC) "The International Association of Fire Chiefs (IAFC) represents the leadership of firefighters and emergency responders worldwide; our members are the world's leading experts in firefighting, emergency medical services, terrorism response, hazardous materials spills, natural disasters, search and rescue, and public safety policy" (https://www.iafc.org/about-iafc).

National Board of Fire Underwriters "The National Board of Fire Underwriters was a group established in 1886 with the aim of preventing loss from fire and to protect property and lives from fire-related harms. It gathered statistics from fire incidents across the United States to provide a template for fire insurance" (https://www.insuranceopedia.com/definition/3060/national-board-of-fire-underwriters).

National Fire Protection Association Group that uses fire service experts and public input to create consensus standards for firefighter training and education, equipment, and emergency operations with an inclusive emphasis of safety and efficient emergency service.

National Highway Traffic Safety Administration The NHTSA is responsible for keeping people safe on America's roadways. Through enforcing vehicle performance standards and partnerships with state and local governments, NHTSA reduces deaths, injuries, and economic losses from motor vehicle crashes (https://www.nhtsa.gov/about-nhtsa).

National Incident Management System (NIMS) The default emergency management system in the United States.

REVIEW QUESTIONS

1. Describe how the fire in Baltimore in 1904 led to most fire department's interagency cooperation that we see today.

2. Look up the ISO Grade for your municipality. How do you think this affects individual homeowners, corporations, and fire departments in your area?

3. The chapter describes how the TV show *Emergency!* affected Americans' perception of EMS. Describe how a TV show or movie you've seen has influenced your perception of the FES.

4. List and briefly describe at least four government agencies that oversee FES as well as what kind of standards they control or influence.

5. What do you think is the greatest challenge facing the modern FES? Explain.

CHAPTER ACTIVITY #1: AN EXAMPLE OF PROGRESS—MAYBE?

When the DHS promulgated the NIMS (command and control at large scale emergencies), there were those in the FES community who opposed it. Some noted that it was not designed to address the multi-alarm fire command system that was typically used in parts of the country.

Another problem identified was that no common terminology for fire and EMS resources existed. One simple example is the definition of the basic unit of a fire department: the engine company. The name of this unit can vary by department, such as engine, pumper, wagon, or unit. One state plan has the following definition of a Type I Engine: 1,000 GPM (3785 L/min) pump, 750 gallons

(2838.75 L) of water, 1,200 feet (365.76 m) of supply hose, 200 feet (60.96) of handlines, and four firefighters. Each firefighter on a Type I Engine must be state-certified as a FireFighter II. Even in this state, many departments do not conform to this standard.

The NIMS currently does not specify minimum qualifications and certifications for personnel, and no consensus or consistency exists in the fire service for determining competency of fire personnel. Although a few states have mandatory firefighter certification systems, most fire departments operate independently, and each decides how to train its members.

Discussion Questions

1. Do all FES organizations have to comply with the federal emergency management plan?

2. Would your department's engine companies measure up to the Type I Engine requirements listed previously? If not, would your department change, or do you think that the requirements are not realistic and should be changed? Please provide a complete justification for your answer.

3. Are there any major differences between the plan outlined in the DHS document and your department's standard operating procedures for incident management? If so, please list the differences.

4. After reading the FEMA document, what challenges do you see in your fire or emergency services agency's ability to respond to a large catastrophic event? Please provide a detailed point-by-point analysis.

CHAPTER ACTIVITY #2: A SECOND EXAMPLE OF PROGRESS

In 1976, the U.S. General Services Administration identified minimum requirements for new EMS ambulances, and medical equipment companies began catering to the prehospital EMS market. As a result, patient care devices became smaller, lighter, and more efficient. Local governments standardized patient care protocols within their jurisdictions based on state and national requirements and recommendations. Original patient care protocols were based on what little research had been done at the time. Data collection from paper patient care reports was

difficult to understand, and the information acquired was incomplete and often inaccurate.

In the mid-1990s, electronic data collection systems, made possible by advances in medical equipment, came into use. With the ability to collect good data and the emergence of quality improvement programs to evaluate data, EMS providers learned that some accepted patient care practices were not only ineffective, but potentially harmful. Today, decisions concerning diagnostic and therapeutic equipment and patient care protocols are based on research showing accurate, definitive conclusions about the effectiveness of a product, drug, or treatment

Discussion Questions

1. Evidence-based medicine has changed long-held beliefs in patient care treatment. How has evidence-based medicine changed EMS protocols in your community?

protocol. This practice is known as "evidence-based medicine."

2. Are the EMS protocols the same in adjoining cities, the county, or the state? If no, why not?

REFERENCES

Carter, H. 2005. *Perception and reality*. Retrieved from http://www.firehouse.com/topic/leadership-and-command/perception-and-reality

Evarts, B., & Stein, G. (2020). *U.S. fire department profile through 2018*. Quincy, MA: National Fire Protection Association.

Hall, J. R. 2013. *Smoking materials fire problem*. Retrieved from http://www.nfpa.org/assets/files/PDF/OS.Smoking.pdf

ISO, Fire Suppression Rating Schedule. 2012. Jersey City, NJ: United States.

Jennings, C. 2008. *ISO system obsolete*. Retrieved from firechief.com/biog/iso-system-obsolete

National Academy of Sciences. 1966. *Accidental Death and Disability: The Neglected Disease of Modern Society*. Washington, DC: National Academies Press.

National Fire Protection Association. 2008. *NFPA Fire Protection Handbook* (20th ed.). Quincy, MA: National Fire Protection Association.

National Fire Protection Association. 2015. *Smoke alarms in U.S. home fires*. Retrieved from: http://www.nfpa.org

National Fire Protection Association. 2016. *A fourth needs assessment of the U.S. Fire Service*. Retrieved from http://www.nfpa.org

National Fire Protection Association. 2016. 1710, *Standard for the Organization and Deployment of Fire Suppression Operations, Emergency Medical Operations, and Special Operations to the Public by Career Fire Departments*. Quincy, MA: National Fire Protection Association.

National Fire Protection Association. 2018. 1500, *Standard on Fire Department Occupational Safety and Health Program*. Quincy, MA: National Fire Protection Association.

National Institute of Science and Technology (NIST). 2010. *Report on residential fireground field experiments*. Retrieved from http://www.nist.gov

Pratt, F. D., P.E. Pepe, S. Katz, and D. Persse. 2007. *Prehospital 9-1-1 emergency medical response: The role of the United States Fire Service in delivery and coordination*. Retrieved from http://fireserviceems.com/wp-content/uploads/2015/04/White-Paper-Pre-Hospital-9-1-1-The-Role-of-the-United-States-Fire-Service-in-Delivery-and-Coordination.pdf

Project 51. 1999. *History of the TV show* Emergency. Retrieved from http://www.squad51.org

United States Fire Administration. 1973. *America burning—The report of the National Commission on Fire Prevention and Control*. Retrieved from https://www.usfa.fema.gov/downloads/pdf/publications/fa-264.pdf

CHAPTER

2

Introduction to Administration

LEARNING OBJECTIVES

After studying this chapter, the student will be able to:

- Recognize how both management and leadership are integral to effective administration (pp 18–19).
- Outline professional qualifications for fire and emergency medical services (EMS) administrators (pp 19–20).
- Comprehend how problems caused by staff–line distinctions can be overcome (pp 20–21).
- Describe the pros and cons to selection of administrators through the process of election or appointment (pp 21–22).
- Examine how national consensus standards and standard operating procedures (SOPs) affect the consistency and effectiveness of emergency services (pp 23–24).
- Know the value of direct supervision and standardization in firefighting and EMS (pp 24–25).
- Explain how administrators can work to bring about change and what common challenges they face (p 25).
- Describe how to gather feedback from staff members and peers as part of the decision-making process (pp 25–27).
- Recognize how to engage influential sources of power, including tradition, political groups and individuals, unions, and the public (pp 27–32).
- Explain how to gain influence over the political process (pp 28–30).
- Discuss the process and challenges of negotiation (pp 33–34).

Case Study

Resolving Staff Divisions

In 1996, the Madison, Wisconsin, Fire Department made a decision to reduce its budget by decreasing the number of division chiefs and combining staff and line functions. Madison's Assistant Fire Chief Phillip Vorlander recognized that the distinction between line and staff functions had created divisiveness in the department, alienating shift commanders from other administrators. The department changed its organizational structure so that each battalion chief worked two 24-hour shifts and six 8-hour shifts per 2-week period. Between the 8- and 24-hour shifts, each battalion chief spent 4 days a week serving as a staff supervisor.

With this new organization, staff officers were more able to participate in the administrative decision-making. This increased the quality of the decisions and their acceptance, and the commitment to the management team. This structure also had a positive effect on the cohesiveness of the department because chief officers no longer had allegiance to a particular battalion, division, or shift (Vorlander, 1996).

Discussion Questions

1. In your department (or a department in your vicinity), are staff and line functions separate? If yes, would departmental decisions be better accomplished by obtaining input from all officers?

2. In the traditional organization of a fire department, would mandatory transfers of battalion chiefs have a positive effect on administrative decision-making?

3. Although not mentioned in the Madison case, would input from all personnel be helpful? If yes, give one example.

4. Identify three techniques to improve decision-making in a multi-station department.

Combining Management and Leadership

Effective administration requires two skills: management and leadership. Management ensures that the organization is prepared and able to accomplish its goals by establishing that sufficient personnel and equipment are available for the organization to perform its duties. Leadership can be observed when trained personnel safely and efficiently complete their mission using the resources provided by the fire and emergency services (FES) organization. At the organization level, leadership is required to "sell" new programs, the need for additional resources, or progressive changes. After the resources become available or the new change is implemented, effective management is needed to use them to support the mission. Administrators should possess management and leadership skills, even if they feel more comfortable in one of the roles than the other, which is common.

A person who excels in management skills will be very good at making the existing organization work efficiently. Many U.S. companies have trained great managers who propelled the country to the top of the industrialized world. For example, Disney—a great visionary company—had a period of time when the management was focused on increasing profits and dividends for stock owners in lieu of truly visionary media products. This caused a decrease in truly innovative new entertainment products. Although this sequence of events does not occur in every organization, it is common in many industries. For instance, management-oriented people generally represent most chief officers in FES organizations. Management skills are necessary for this role, because the chief officer often needs to be a good caretaker of public funds, staff, and resources.

However, managers may not be proficient at leadership. A person with strong leadership skills has a clear vision of where the organization needs to go and the courage to attempt the journey. There is no guarantee that the leader will arrive at the visionary goal, but because good leaders are inherently risk takers, they try anyway. Conversely, good leaders may not be good managers because they typically are not detail oriented and may miss those items that must be completed to achieve the overall goal because they are big picture visionaries. A leadership-oriented administrator may focus on a visionary goal 5–10 years down the road but forget to plan for tomorrow. In contrast, a management-oriented person generally is not comfortable with risk taking and would rather not get involved in visionary goals that have no guaranteed outcome. Managers may try to implement change slowly or make it voluntary; be aware that a voluntary plan can foster inconsistencies because members and supervisors set their own goals. When the voluntary phase results in a commitment of 10% to 15% or more members, this is a good time to phase in a mandatory policy.

A leadership-oriented person is more of a risk taker when it comes to new ideas and change. Many people drawn to the FES have a reduced sense of fear and are risk takers in physical actions. There is an old saying that highlights the heroic behavior of firefighters: when firefighters are entering a burning building, everyone else is leaving it. However, when it comes to changes in the organization, fear of the unknown may make many FES and EMS chief officers reluctant.

Although these descriptions of leadership and management may be oversimplifications, it is important for administrators to recognize the value of management and leadership skills and work to develop them through formal education, self-study, seminars, mentoring, and experience. The National Commission on Fire Prevention and Control's *America Burning Report* of 1973 further comments on the duality faced by the chief officer:

> Presiding over this tenuous alliance [fire department] is the fire chief, who wears two hats—one, the administrative hat required to run the organization; the other, the helmet he dons when the alarm is sounded to lead his firefighters in the suppression of a fire. Because the fire chief usually has come up through the ranks, the second hat probably fits comfortably (United States Fire Administration, 1973).

The *America Burning Report* uses the term *leadership* when discussing the ability to lead firefighters at the scene of an emergency incident. In this text, the term *leader* is used at the organizational or administrative level, which calls for a special set of skills and knowledge. Leaders are problem solvers; therefore, this text discusses the many aspects of problem solving. More precisely, this text is about the very special challenges to problem solving and/or policy choices that FES and EMS chief officers encounter in the public sector.

Professional Qualifications for Administrators

Career and volunteer organizations sometimes struggle with selecting the best person for the chief officer's position. Whether the person is elected or appointed, the result may be the same: the administrator (chief) may be ineffective and in some cases disruptive to the organization. FES is evolving as a profession, and the selection of chiefs should strive to be similar to other professions, such as doctors, engineers, and lawyers.

Professional qualifications for fire and EMS administration include three areas of competency: (1) training, (2) experience, and (3) education. In all three areas, fire and EMS administrators should meet minimum levels of proficiency; these minimal qualifications should be the same throughout the country, just as they are for medical, engineering, and legal professionals. For example, a medical doctor must complete college-level education and medical school, and then an internship/residency (training and experience). Only after completing these requirements and passing a comprehensive examination is the doctor granted a license to practice.

Many FES organizations can improve their departments' reputation and ability by adopting minimum professional requirements for officers. Although this is a complex and controversial subject, the following are some examples of recommended requirements for fire administrators:

- Minimum of 3 years of experience at each prerequisite level—firefighter and fire company officer
- Certification as a Fire Officer III or IV
- Training and certification as a Fire Fighter II and Fire Instructor I
- A degree from a regionally accredited college or university that is recognized by the U.S. Department of Education (an associate degree for smaller departments—those with less than six stations—or a bachelor's degree for larger departments)

HIGHER EDUCATION TIP: REGIONAL ACCREDITATION VS. NATIONAL ACCREDITATION

Generally, "regionally" accredited higher education institutions (colleges and universities) are preferred and are typically more academically rigorous. Nationally accredited schools are predominantly for-profit and offer vocational, career, or technical programs. When considering higher education institutions, it is usually better to attend a "regionally" accredited school.

EMS administrator qualifications are not as standardized as those of fire departments and vary across companies or positions. The three areas of competency—training, experience, and education—may be less defined in the EMS world. Fire departments have a hierarchical structure based on standardized training, but this is not the case with many private ambulance companies, because EMS supervisors have no national oversight organization to define those supervisory standards.

Ambulance companies may offer supervisor or management positions to paramedics with comprehensive field experience, but they may not require management experience or higher education. A paramedic who may be an excellent clinician does not necessarily have the skill set to be a competent administrator. The extent to which a new supervisor, manager, or administrator is offered training opportunities also varies greatly from company to company and may be considered the responsibility of the individual.

As EMS professionalism continues to increase, qualifications for management positions are becoming better defined. Common and preferred requirements for EMS administrative positions include:

- High school diploma or GED equivalent (college degrees may be preferred)
- Advanced Cardiac Life Support and professional-level cardiopulmonary resuscitation certification, with specialized training (e.g., hazardous materials, emergency vehicle operator, National Incident Management System, International Trauma Life Support, Pediatric Advanced Life Support) preferred
- EMT-P license and state driver's license
- Three to five years active prehospital experience, with supervisory experience preferred

Staff–Line Distinctions

Most FES organizations have a distinct dividing line between staff and line functions. This classification is especially apparent in larger departments where it is common to have specialized divisions, such as fire prevention, training, dispatch, logistics, or human resources.

There are several reasons why this traditional organization can cause conflict, especially when those on the line are asked to move into staff positions. First, firefighters and EMS providers often enjoy working at emergency incidents, so they may resist a transfer to staff positions if doing so takes away their ability to respond to emergency calls. Second, the crew can almost become a second family, and individuals may

be hesitant to leave that family to move into staff positions. In addition, the daytime work schedule often required in staff positions may not be appealing for those who have second jobs or enjoy having extended time off. In the common 24-hour shift, an employee works for 24 hours straight, and then is off for 2 or even 3 days. Firefighters and paramedics might become comfortable working this shift and look forward to having time for their second job, family, or social commitments.

To address these concerns, some organizations have combined the staff–line jobs at the mid-supervisory level. For example, a fire department may have a battalion chief in charge of each shift. This same battalion chief would also be assigned to supervise certain staff functions. During the weekday, the battalion chief is the administrative head of prevention but is available to respond to major emergencies when needed. Another alternative organizational strategy is to rotate command officers every few years to supervisory positions of both staff and line divisions. Serving in different commands ensures that the next chief officer has a well-rounded knowledge of the entire organization.

- Stay up-to-date by attending seminars and conferences, and by reading FES, government, and management periodicals.
- Read local newspapers, web sites, and publications from influential organizations. Be familiar with all controversial issues in the community, not just those that relate to FES.
- Visit other FES organizations and learn from their experiences and progressive ideas.
- Have high personal and professional standards for ethics and morality.
- Be very careful not to lie, including misinformation and omissions that would affect credibility. It is very difficult to recover from this damaging issue.
- If appointed to a top position, choose your initial proposed changes carefully. Choose only those changes that are certain to succeed. It is very important to start out as a winner.
- Last, but perhaps most important, lead by example.

A similar model may be found in EMS organizations. Paramedic or emergency medical technician supervisors are much like battalion chiefs in terms of job function and span of control. During a 24-hour shift, they might be responsible for administrative tasks, meetings, and field oversight and also might be expected to respond to certain types of calls or function as a provider when staffing levels are low.

Some companies have a controversial rule requiring supervisors to staff an ambulance while supervising some staff functions. Some EMS providers believe that, although supervisors should be required to maintain field skills and certifications, there should be a clear distinction between staff and line roles. In addition, union affiliation may create some conflict, because in many companies field providers are union members but supervisors are not members. This is a typical division found in many union organized companies where supervisors are separated from the workers. In companies where union field employees work as part-time supervisors, they may be restricted in their ability to discipline other field employees, handle sensitive company information, or investigate complaints.

Selection of Administrators

Most fire and EMS administrators are selected by election or appointment. Both methods present unique challenges.

By Election

In many volunteer FES organizations, administrators are elected. These organizations are often independent and self-regulated private corporations that are operated without formal scrutiny by the municipality or public (customers) they serve. Without public accountability or oversight other than the members named as its commissioners or board of directors, they may be more likely to make unintentional faulty decisions. To counteract this lack of accountability to the public, it is suggested that elected officials for the area served could appoint the chief officer from a list of qualified applicants, for example.

In addition to issues of accountability, these organizations may also have less stability and consistency than those of municipal FES organizations. This is because appointed officials typically have more latitude than elected officials in enforcing, interpreting, and changing the SOPs, training requirements, and equipment and apparatus needs and specifications.

In practice, volunteer chiefs may initiate change each time a new chief is elected. The resulting inconsistency is not caused by bad or incompetent people; it is the result of a system in which officers are elected by a democratic process.

Furthermore, because most volunteer chiefs are typically elected for a 1-year term, they must start working toward reelection as soon as they are elected, tempering their ability to be strong leaders. A similar situation exists in the U.S. House of Representatives, in which representatives are elected for 2-year terms. Almost as soon as representatives are elected and take office, they must start planning for their reelection. Thus, they spend a considerable amount of time and effort on reelection strategies, taking away from their key mission, which is to represent the people in their district and the interests of the United States. In most cases, these are honorable people who are placed in a system that has limitations encouraging policies and limits of the enforcement of rules and regulations that may be opposed by those supervised.

By Appointment

Just as there are limitations to volunteer organizations that elect their officers, there are problems when chief officers in municipal departments are appointed by a government's elected officials or administrators. While most are appointed from a list of qualified applicants based on formal education, experience, and a comprehensive interview process, the FES administrator might feel conflicting allegiances to those who appointed him or her, the labor group, and the public. It is not uncommon to find FES or EMS administrators who feel alienated from their organizations. The chief officer may believe that the labor organization, the elected officials, and other appointed administrators all want to convince or force their particular ideas on the organization. Government officials and labor organizations have powerful self-interests, but the public may only have a weak influence because of its lack of knowledge about emergency services. Therefore, the chief should counteract this by being an advocate for the public interest.

Rules and Regulations

One of the main duties of a fire or EMS administrator is dealing with rules and regulations often created as SOPs. This includes analyzing, reviewing, creating, determining, and enforcing rules and regulations for the entire organization. In some cases, such as local, state, or federal regulations, rules may be mandated by a higher authority in government. These rules may also be initiated by a task force of officers and members empowered by the administrator to create SOPs for safety and efficiency during emergency operations. It is important to recognize, despite what some critics proclaim, including lawyers, that SOPs do not leave an organization legally liable if they are enforced strictly. If SOPs are promulgated and enforced rigorously and consistently, there should be no concern for liability issues. Vigorous enforcement actually helps to protect the organization from lawsuits.

Other than requesting additional funding from elected officials, enforcing rules and regulations is one of the most difficult duties of the chief officer. There are always some members and supervisors in fire departments and EMS organizations who do not like rules and regulations or who disagree with the rules and regulations. However, rules are absolutely essential for accountability and consistency, especially during emergency operations. Too often, support for rules and regulations comes only after a major tragedy. For example, after a devastating hurricane in Florida, the construction of new buildings required the use of specific techniques to reduce or eliminate the damage from these powerful storms **FIGURE 2-1**. Building and fire codes are rules and regulations that chief officers can use to accomplish their mission of fire safety, and it is often easier to find justification and public support for strict enforcement and changes after a major disaster.

Fire departments should have a set of rules and regulations that outline operational actions and safety expectations for its members and disciplinary action that may be taken for failure to follow these policies, including SOPs. Rules and regulations and subsequent

orders from the chief should be in writing and distributed in such a method as to ensure all persons are made aware of them either through web sites or email distribution to members. Ongoing training sessions for new and existing SOPs should be scheduled throughout the year. There is no room for misunderstanding, especially of the safety regulations (NFPA, 2008).

As a general guideline, rules and regulations can fall into two major categories: management and emergency operations. For emergency operations, basic rules and regulations are absolutely necessary to provide consistent, safe, and efficient FES, especially in the first few minutes of emergency operations.

Additionally, employees who are unionized want to have a written contract that contains rules on personnel management. Such items, as when overtime is to be paid and many other conditions of employment, are typically included in a labor agreement. Most of these agreements are necessary to prevent inconsistent and unfair actions in personnel management. However, some agreements also contain items that can restrict effective administration. For example, a mid-west fire department's union contract severely limited the requirement for union members to train during inclement weather. This was specifically defined as below 40°F or above 85°F. When a new million and a half dollar ladder truck arrived in the middle of the summer, the firefighters were not allowed to train on this new truck because the temperatures would be above the high temperature mark. The very expensive ladder truck was stored for several weeks until the fall. One

FIGURE 2-1 Rules and regulations regarding building construction are created to prevent future tragedies.
© James Steidl/ShutterStock, Inc.

Administrator Dilemma

A supervisor receives a complaint from his superiors that a crew "abandoned" a patient. The supervisor is told that the crew was treating an abused child when the abuser returned and threatened them with bodily harm if they did not leave. The crew left, parked around the corner, and then called the police. The local news has a very negative spin on the story, and the public is outraged because the patient was an injured toddler. Therefore, upper administration is planning to suspend the crew over this incident to appease the media and the public.

Should the supervisor advocate for the crew based on the threat to their safety at the scene, or does he act on the need to mitigate the political firestorm before it has a negative impact on the organization's public reputation? Administrators have a responsibility to the organization to enforce the rules and the mission of the company, but they also need to be an advocate for the people doing the work that makes the organization viable and may adversely affect moral of the troops.

question that would challenge this requirement would be the expectation that firefighters typically respond and work at emergency incidents that occur at these adverse temperatures.

National Consensus Standards

National consensus standards, such as NFPA standards and Occupational Safety and Health Administration (OSHA) regulations, are relatively new to the fire service. Although the necessity of such standards is well accepted, FES organizations are still in the process of implementing changes to comply with many of these standards.

For example, an April 2001 article in the magazine *Fire Chief* asked four safety officers if their departments had completed a safety audit as described in the appendix to NFPA 1500 *Standard on Fire Department Occupational Safety and Health Program*. None of the safety officers interviewed reported that they were in full compliance with this safety standard, which was first adopted by NFPA in 1987, 14 years before the interviews. Instead, these safety officers estimated anywhere from 60% to 90% compliance (*Fire Chief*, 2001). Although compliance probably has improved since then, seldom does a department comply with

100% of the safety requirements. Because this type of change can be expensive and potentially disruptive to the organization and its members, slow improvement is very common.

Some NFPA standards are designed to reduce inconsistency in operations, staffing, training, and communications. These inconsistencies in the FES result in differences from agency to agency, town to town, and state to state. These differences add to the public's difficulty in judging the service provided and determining if the community is truly receiving professional-quality emergency service. At this time, a sign on the side of an engine that proclaims FIRE DEPARTMENT does not guarantee the same level of service for all similarly marked fire trucks throughout the country. Although the fire service has made a lot of advances, the pursuit of quality and consistency is still a work in progress. Administrators should work to implement these NFPA standards to ensure their organizations are delivering professional FES and be able to demonstrate that to their residents.

Some fire and emergency officials, along with elected and appointed municipal officials, outwardly resist complying with standards and regulations that they personally disagree with or that have fiscal impact. In some areas, local FES officials want the freedom to provide the level of public fire protection that is determined necessary by the local community. However, in many cases, the citizens of these communities have not done enough research to determine the needed local level of protection or quality of service (e.g., number of stations, training, companies, and staffing). Whatever resources and types of organizations exist at the local level are the result of a very complex and independent evolutionary process, not a systematic planning effort. That said, FES organizations are entering a new era of rules and regulations. In the EMS field, responders are now certified to national standards for emergency medical technicians and paramedics. In the fire services arena, national standards for training and safety are now available.

One example of a national standard for the fire service is NFPA 1021, *Standard for Fire Officer Professional Qualifications*. The chief may propose a new program to certify all officers to this standard. This type of program may require training, incentive pay, and overtime pay to cover classroom attendance (especially if education is a requirement for promotion). The selling point for this proposal is that, when completed, the city will have a more professional, competent fire department—a goal on which the municipal administration, the public, and elected officials can easily agree.

Standard Operating Procedures

Within an FES organization, the level of service can vary by shift or battalion. Freelancing or independent goal-setting can be deadly on the scene of a major fire; the public deserves an emergency services organization that can operate with a high degree of consistency and conformity. Standard operating procedures (SOPs) further this goal by achieving a high degree of conformity and consistency throughout the organization as well as ensure emergency scene safety and encourage situational awareness, which is sometimes a problem for a younger workforce that has little life experience and feels invulnerable. In particular, emergency response driving is a high-liability issue, and fatal accidents can occur. For that reason, many public agencies and private companies have developed SOPs regarding emergency response driving, such as those requiring units to stop at intersections with red traffic lights even in areas where emergency vehicles have exemptions from this traffic law. Infractions of this safety SOP are taken seriously for the safety of the crew, the patients, and the public.

Administrators must enforce the SOPs that guide the decisions made in emergency situations—when first responders have only a short period of time to consider all the alternatives—because it is critical for the safety and success of emergency operations. Although there should be some room for exceptions based on the judgment of the officer, an appropriate system to critique these emergency field decisions ensures that this latitude is not used without justification. This can be accomplished by using a standard written report submitted to a department safety committee. The committee would be directly authorized by the chief to analyze the incident along with the actions taken by the officer and/or crew. A written report would be completed and submitted to the chief for further action, including changes to the SOP or disciplinary actions. And finally, a formal report of the investigation should be circulated to all members.

For example, at a fire in a large vacant house, the first due engine stretched its hose line to the front door and began to enter. The department's SOPs required the first due engine to attack the fire from a position on Side 1 (the front), where the front door is found in most buildings. On entering, the firefighters noticed that the floor had burned through, leaving a large hole into the basement, which was initially blocked from view by heavy smoke and fire. The firefighters quickly retreated and radioed this information to the Incident Commander (IC). Based on the IC's judgment, a revised plan of attack was implemented, resulting in successful fire suppression.

Direct Supervision and Standardization

Many modern entrepreneurial businesses are organized around the principle of employee empowerment. Business administrators may argue that empowering employees is the best way for individuals to learn new skills and deal with changing environments. Employee empowerment can improve a business's service and competitiveness in the world where revenue generation is the goal. However, although the empowerment theory may work very well in a corporate or industrial organization, it does not lend itself to the FES organization at the company level.

FES organizations could not operate effectively if each company officer was empowered to determine what actions the company would take at an emergency incident. Although the overall incident strategy is the responsibility of the IC, the company officer is empowered at times to determine what tactics shall be used to complete a task assigned to them by the IC. For example, if the IC orders a company to ventilate the roof, it is usually incumbent upon the company officer to determine the most effective, efficient, and safest way to complete that task in a timely manner. Employees or companies acting independently (freelancing) at the scene of an emergency cannot be tolerated. Standardization and SOPs are needed to coordinate efforts, and direct supervision is required to ensure that each member works as part of the overall team **FIGURE 2-2**.

Direct supervision and standardization are especially critical for the FES organization during the first few minutes after arriving on the emergency scene. This is the time before the incident commander has arrived and has been able to evaluate and formulate a comprehensive strategy for the emergency. Consistent and reliable emergency operations at all times are

FIGURE 2-2 Safety committee meeting.
© Jones & Bartlett Learning. Photographed by Glen E. Ellman.

very important, if not critical, to the successful and safe handling of emergency incidents, which in most cases requires strict adherence to the SOPs. However, because a few incidents will not fit neatly into SOPs, officers are empowered with latitude for independent judgment.

Bringing About Change

Do not underestimate the chief administrator's power to make change. In most cases, the administrator's power goes beyond simply giving orders that may or may not be followed. When chief administrators speak and initiate a new project or change, the attention of the organization and its members is immediately focused on the proposed change.

The chief officer also has the power to set the agenda for change in the organization. Generally, changes come about when the chief officer initiates the process, with the exception being changes that are forced from the outside by a court order, a new state or federal regulation, union pressure, or elected officials. In most cases, initiating change takes real courage and detailed preparation.

The chief officer's ability to get the administrative job done, identify problems, and, when necessary, make changes in the organization makes up a dynamic process. To comprehend these dynamics it is critical to understand the chief's power, the interdependence on professional capabilities and agency reputation, and the chief's ability to influence others. The following questions help determine if a chief officer is ready to manage and lead the march to initiate change:

- Is the officer knowledgeable about all aspects of emergency operations?
- Does the officer have the appropriate formal education?
- Does the officer lead by example?
- Can the officer gather all the facts needed to make an informed decision?
- Does the officer really know what is going on in their department and community?
- Can the officer schedule and set an agenda for change that is reasonable and evenly paced?
- Will the officer's proposed change improve the service to the public (customer)?
- Will the officer take risks for the sake of making real progress in improving service to the public?

The dynamic process of bringing about change involves several complex elements, including getting feedback, understanding the various sources of power

at play, and the process of negotiation as discussed in Chapter 4, *Leadership*.

Getting Feedback

The chief officer must attempt to maintain a professional, independent position, which at times can make the job a very lonely one. Although this loneliness may make it difficult for an administrator to get unbiased advice and feedback on controversial administrative decisions, it is important to continue to solicit this feedback from staff members and peers.

In fact, effective fire chiefs will find they do more listening and reading than talking. Before seeking feedback, administrators should gather information about new policies or equipment by reading and studying the issue. Take care that your informational sources are nonbiased and based on expert opinion and facts. Reference sources such as NFPA standards are a good place to start. Doing so may be a challenge in the FES, where most periodicals are not peer-reviewed for accuracy, veracity, and legitimacy.

Feedback from Staff Members

To be a part of the decision-making process, it is important that staff members demonstrate loyalty to the agency. Knowing that staff is loyal allows administrators to communicate honestly, trusting that any sensitive or potentially controversial information remains private.

Administrators should solicit opinions in an open format. Preliminary discussions with staff may be held in meetings behind closed doors to facilitate an open, honest dialogue. Never equate a privately expressed negative or opposing view as a sign of a person's lack of loyalty. In fact, this person may be giving the chief officer the advice needed to stay away from a bad decision. However, after the administrator has chosen a particular proposed policy, staff should be expected to support the change.

The chief officer should realize that there are situations in which the decision-maker does not know all the facts or circumstances. This lack of knowledge is sometimes referred to as imperfect information. Imperfect information may occur, for example, when the top staff members are reluctant to make an argument that opposes what the chief officer would like to do and thus do not inform him or her of an important obstacle, or when the officer is facing pressure from elected or appointed high-level officials to act before knowing all the facts. Behind a desk, it is easy for administrators to forget the field perspective **FIGURE 2-3**. Making decisions without an accurate understanding of the current situation is not effective leadership. Decisions need to be balanced among administrative

FIGURE 2-3 Even in personnel counseling, jumping to conclusion can be very unfair to the employee. Do your homework.

© Jones & Bartlett Learning. Photographed by Glen E. Ellman.

goals, public needs, fiscal constraints, political realities, and the needs and concerns of staff.

When the subject is one of major importance to the organization, it is best to empower a task force to study the issue and make recommendations. The task force should include members that represent diverse opinions and all levels of the organization, including experienced firefighters, field providers, and officers. A diversified task force opens communication and prevents information obstructions that can occur between different levels of management.

Chief officers should encourage open and honest dialogue and explain that it helps them to make fair policy decisions. Staff members should feel safe bringing their concerns to the chief officer; angry reactions from the administrator can discourage staff from raising negative information in the future.

A management style called management by walking around is also useful in fire and EMS administration and is discussed in Chapter 3, *Management*. In some cases, although the chief officer has the authority to order a change, members may make this change only in name, without actually changing their actions. Departments may report to their peers and community that they are progressive, but chief officers can use management by walking around—literally visiting with members of the staff (along with appropriate accountability techniques)—to find out what is really going on in their organizations.

Administrators should visit stations often and float ideas for informal discussion. It is important to remember that there may be some members who are unable to differentiate a brainstorming session from a formal announcement of a new policy, so administrators need to develop a keen listening ability and effective communication style. During actual implementation of a decision, an administrator

should be their own intelligence officer, confirming that all the details to fully implement the decisions are complete.

Feedback from Peers

Often, other chief officers or professional acquaintances may be able to help in evaluating and discussing changes, especially if these discussions are better kept confidential in the preliminary discussions, research, and evaluations. While attending conferences and professional development courses, administrators can meet with professional peers who may be able to provide advice and insight into difficult decisions, issues, and contemporary problems.

Although some administrators may feel uncomfortable asking for help, it is important to work to overcome this reluctance. Administrators must be able to gather and analyze information to present a full picture to their peers. It is also important to be able to ask more than one person for advice to gain different perspectives on the situation. To illustrate this process, it is helpful to look at how a patient would deal with a life-threatening medical problem. The decision to undergo major medical surgery should not rest on the opinion of one surgeon. Just as in the field of medicine, consulting with two or more experts provides a second opinion and ensures that individual biases are weighed against one another. Furthermore, speaking to administrators in other parts of the state or country can provide a wider perspective and help prevent the conversation from being repeated locally.

Although feedback from peers is important, talking freely on a day-to-day basis about potentially controversial changes is discouraged. The casual nonstop conversation around the dining table in the firehouse kitchen is a great socializing time for firefighters and company officers, but the minute the officer puts on the fire bugles, everything they say takes on new meaning. As a firefighter, talk about controversial change may end with a good-natured argument; but, as the fire chief, this kind of talk may scare members to the point of damaging morale, resulting in strong opposition.

Real change always has opponents. It is not a good idea to give opponents a heads-up on the change, because this gives them more time to fight the change. One example of an effective way to combat this is the NFPA-sponsored annual meeting of Metro Fire Chiefs. It sends invitations to fire chiefs in some of the largest fire departments in the United States and several international departments. Only fire chiefs—not deputy assistants or other chief officers—can attend. This exclusion is intentional, designed to allow fire chiefs to feel safe that all conversations and discussions at the conference are confidential. The NFPA realizes that the fire chief position is special and that those underneath and those above the chief's position can become adversaries when the chief attempts to initiate change.

For those chiefs who cannot attend or who do not command a metropolitan department, networking with fire chiefs in other departments can be very helpful for honest discussion of controversial subjects. A good place to meet other chiefs is at the National Fire Academy or national conferences such as Fire Department Instructors Conference, International Association of Fire Chiefs, and the National Fire Protection Association. An exclusive environment of peers gives the fire chief a safe venue to discuss potential or controversial changes.

EMS administrators also have avenues to gather feedback from their peers. For example, the National EMS Advisory Council (NEMSAC) is the nationally recognized council of EMS representatives and consumers that provides advice and recommendations to NHTSA in the Department of Transportation and to the members of the Federal Interagency Committee on EMS. They meet three times each year and discuss contemporary issues and also provide a venue for EMS administrators to share ideas and experiences. In addition, various EMS conferences and seminars are presented throughout the United States with dates and times at https://www.ems.gov/meetings.html.

However, administrators from private EMS companies must recognize that competition for 911 contracts makes peer discussion of decision-making more challenging. EMS administrators may know each other well but share little proprietary information. EMS administrators may learn from others by examining errors other companies have made; doing research; being honest in determining if a choice is cost effective; and making a decision based on the best interests of the company, the community, and the work force they serve.

Sources of Power

One of the challenges that FES chief officers encounter in bringing about change in the public sector is related to other sources of power that affect decision-making. This includes the power of tradition, political entities, employee unions, and the community, although administration comes with its own power.

Supervisors Power

The chief administrator's direct supervisor may be a city/county mayor, deputy mayor, public safety director, elected, appointed, political, or merit and in some cases a board of directors. The names and how they achieve their position are many. What is important is that the FES administrator reach out to the individual(s) to achieve a friendly and professional relationship.

This is easier said than accomplished. Most new chief administrators have been working in a highly specialized technical group of likeminded acquaintances. In addition, it is common to be working with other members who are almost like family. And, there are those administrators who are hired in from outside and this process becomes even more complicated and difficult but vital.

Personal face-to-face meetings work best and should be varied in their structure. A once a week lunch with the boss can be very effective. One of the goals should be to see each other as a human being, not the enemy. The conversations ought not be all business but include sports teams, family, vacations, and any other personal topic, but not to too personal. Eventually, you may become close friends. This can help in two respects: it makes it easier to approach your supervisor to implement controversial changes, including when they have to be funded and, for your long-term tenure, it is more difficult to fire a friend.

Between face-to-face meetings, electronic correspondence can be very effective. Updates of important actions or accomplishments should be sent to the boss and, if critical, followed-up with a phone call immediately. This is especially critical for any controversial issue that may end up with media attention. Make sure your staff is aware of your need to communicate even bad news timely.

Blogs and news web sites are also a good place for the FES administrator to keep up with the latest trends and news about FES, including other city/county and departments even if they do not directly affect FES. In addition, obviously, the Internet is a good place for research. Be cautious of the source of information because it is not uncommon for web sites to have a biased viewpoint. For example, at the U.S. Cattlemen's Association web site, information on health benefits of red meat may be biased and should be questioned. All in all, this news and research information is always good material for conversations with the boss.

It is imperative that these relationships be achieved in a respectful manner. Keep in mind that your boss has many pressures on their ability to distribute taxpayer resources. The political, merit, or appointed supervisor also has a boss and may not be able to decide in your favor. Do not get belligerent when the decision does not go your way. Assume that there was a logical reason and save your push for another day, month, or year. Also, your boss does not owe you a detailed explanation of the reasons behind the decision, so do not press the issue. Be respectful of their judgment. Your reaction should communicate that you will try again in the future. Perseverance can be the key to long-term improvements in the FES department.

Power of Tradition

The power of existing traditions within fire and EMS services can provide a barrier to implementing change, often with tragic consequences. For example, a career fire department in Pittsburgh, Pennsylvania customarily staffed its engine companies with two or three members. At a fire, the officer was often physically operating the nozzle. This custom was caused by the lack of adequate staffing and the privilege of the position. After increasing the minimum staffing on each engine to four, on February 14, 1995, an engine company from this fire department responded to a house fire and found heavy smoke emanating from doors and windows. With a lieutenant on the nozzle and the two firefighters behind on the hose line, the officer was unable to supervise the operation and assess the situation on the floor where they were operating. When the hose line burned through, there was no longer any water to control the extensive fire on the floor below, and, tragically, the fire had also burned through their only known avenue of escape—the stairs leading up to the ground-level floor from the basement area where they were operating, and the three members on this hose line became the victims of the fire. This tragedy may have been prevented had the officer been supervising the operation, not physically operating the nozzle.

Too often, organizations are driven not by their missions, but by their rules, traditions, or budgets. Although rules are important, too many administrative rules and traditions can restrict an organization from being mission- or customer-focused. Therefore, administrators must understand the power of tradition in order to overcome it.

Political Power

To some, the term *politics* may bring up negative stereotypes of uncaring elected officials only taking care of special interest groups. In some communities it may seem that the big money contributors have more influence than ordinary citizens. Although some individual elected officials might abuse the power entrusted to them, this does not mean that the entire political system is not legitimate. Remember that politics is the foundation of the United States' representative form of government.

Although elected officials have many goals, one goal that is fairly common is to get reelected. In most cases, this means taking care of the voters' or the

---

CHIEF OFFICER TIP

Do's and Don'ts for Working with Political Power

In your relations with elected officials, DON'T:

- show animosity toward their opinions or decisions
- support their election efforts (the other person may win and you will assuredly have an opponent)

DO:

- invite them to departmental events
- conduct basic orientation presentations in the fire house
- demonstrate fairness, honesty, responsiveness, openness, and accountability
- provide accurate and timely information on a regular basis
- follow-up as soon as possible on any communications or requests
- meet with them privately on major issues
- get to know them better personally

© Smeby Jr, Charles L/Jones & Bartlett Learning

---

campaign contributors' needs. For this reason, many elected officials are heavily influenced by the local voters. Because many voters are concerned about the amount of taxes they pay, elected officials almost always question any request for additional funding for the department. These politicians are acting as the voice of the individual taxpayer who may believe that taxes are already too high. This is especially common in jurisdictions with large numbers of taxpayers who are on fixed incomes, such as Social Security.

Furthermore, the view that fire and EMS services are monopolies has given elected officials justification to impose regulations, just as they do with local law enforcement agencies and public utility companies. A monopoly occurs when only one provider or company has complete control of the market and can charge prices that are above the market value—that is, what they would be worth in a competitive market. If a house catches fire, for example, the occupants do not go to the yellow pages to choose a fire department to call. FES are provided to anybody who calls, but only one primary provider is responsible for the service provided. In the United States, monopolies are illegal unless they are a government-regulated industry or service. In this way, the government can meet the public's expectations that it will maintain high-quality emergency services. Public FES can be viewed as monopolies either provided by a government or allowed to operate without control by the local government (e.g., independent volunteer departments). Although municipal departments may have no trouble understanding the issue of government oversight, independent volunteer companies might strongly resist any oversight by elected representatives. This may be a mistake because administrators should respect and work with elected officials who represent residents in the community.

In addition to formal political authorities, such as elected or appointed officials, fire and EMS service administrators must be aware of informal sources of political power. This includes external groups of politically influential citizens such as:

- Employee or member associations (e.g., unions and volunteer fire companies)
- Media
- Special interest groups
- Professional organizations
- Businesses and industries
- Homeowners' groups
- Social clubs
- Religious organizations

Fire and EMS administrators should be in touch with other agency administrators or staff in these organizations and work to maintain positive relationships, because these groups can influence political decisions that affect the service's budget. In addition, the general public may affect approval of any proposed changes—yet another reason why public opinion is so important.

Some administrators may find it challenging to achieve the changes they desire in the face of the many opposing goals of these political groups. Before attempting a policy change, the fire and EMS administrator should consider the following:

- What political party is now in control?
- Are the elected officials short-term thinkers, long-term thinkers, or both?
- Are there any elected officials who are expected to lose or not run for reelection in the upcoming elections?
- Has a single-issue candidate recently been elected?
- Are elected officials elected at-large or from districts?
- Are there any signs of taxpayer disapproval?
- Is there any expectation of what is politically correct?

- Are there any elected officials who would have a reason to not support the FES organization?
- Are there any officials who would like to be a champion or strong supporter of the fire or EMS service?
- Do any of these elected representatives owe a special interest group that may oppose your changes?

An administrator must be willing to show courage, sacrifice, and determination to achieve their goals. The difference between caution and courage can be the difference between the status quo and successful progressive change. However, be cautious not to jump into a decision on change without fully analyzing all the potential opposition. It could defeat you or even in some cases cost you your job.

Always taking the safe path and insisting on having all the answers before making a decision prevents the chief officer from leading change in any organization. Seeking new frontiers for your vision and thriving on the unpredictability of the future helps achieve real change. Courage, planning functions, and vision reduce the unexpected changes that tend to push many organizations into crisis management mode. It is relatively easy to be a hero at the emergency incident; it takes a lot more determination to be a courageous leader who ventures out to make significant positive change in the organization.

Remember, there are always some people who are against change simply because it is change. As an administrator, one needs to be prepared to accept some inevitable criticism. Keep in mind that it may be unsubstantiated, self-interested, or fabricated. There are those that believe that the ends justify the means (i.e., lying, etc.). Be prepared for this type of opposition. The following abilities are helpful for administrators to overcome political opposition:

- Accepting the legitimacy of politics and elected officials
- Understanding the structure and process of politics and government
- Building political alliances, including personal friendships
- Obtaining the support of special interest groups
- Acquiring public support through effective marketing
- Practicing open and honest communications with employees, citizens, media, and appointed and elected officials
- Using conflict resolution, negotiation, and bargaining techniques

- Identifying the various stakeholders and any benefits they may receive or lose as a result of a change
- Developing trust between members in the organization, the public, the media, and government representatives

The Power of Unions

In FES organizations, unions work to protect the interests of their members and can have substantial influence over administration decisions. For example, the International Association of Firefighters (IAFF) is a labor union representing nearly 313,000 firefighters and paramedics who protect 85% of the nation's population. Since 1918, the IAFF has worked to unite firefighters and achieve better wages, improved safety, and greater service for the public **FIGURE 2-4**. Heading this organization is the general president, who is elected to a 4-year term by the membership. It is typical for the general president to serve for many years because there is no term limit. The lack of a term limit is advantageous because it emphasizes long-term goals over short-term success.

The IAFF actively lobbies in Washington through its political action committee FIREPAC, which has influenced OSHA regulations, NFPA Standards, and federal legislation. The IAFF contributes to congressional election campaigns for individuals who support its agenda and goals. In addition, the IAFF provides research and advisors to assist government officials in legislative decision-making. At the local municipal level, the IAFF's influence varies greatly throughout the United States, having less influence in the 27 "Right to Work" states **TABLE 2-1**. In a "Right to Work"

state, an employee cannot be required to join a union or pay union dues. This reduces the number of members and their influence on wages and benefits.

For many FES workers, the union becomes a social fraternal organization along with a platform for employee benefits and safety. However, in "Right to Work" states, the influence of the union varies depending on whether the administration formally bargains with the union. In any case, the real influence is primarily derived from the union president's ability to impact public policy. This is strongly affected by the same items that affect the chief administrator's power.

There are also national, state, and local efforts that aim to reduce the influence of public-sector unions. For example, in March 2011, Wisconsin's state government passed legislation restricting union bargaining rights for public workers (except for police and firefighters). These efforts are often driven by economic considerations and the desire to reduce expenses in the forms of worker wages and benefits, especially pensions, which the unions aim to protect. The union's job is to represent the workers in gaining better wages and benefits and safer working environments.

Although in difficult economic times unions have agreed to take a reduction in wages or benefits to save the jobs of union members, it is not the union's duty to save money for the community or government by declining wage or benefit offers. It is also important to understand that in other industries, employees who do not believe their wages, benefits, or working conditions are fair or safe may change jobs relatively easily. In FES, this is rare, because the system is designed to offer rewards and benefits based on seniority, discouraging employees from seeking other opportunities.

Many disagreements between the government and unions stem from each group not understanding the duties of the other. The job of chief officer, appointed official, and elected representative is to provide the best services to the public, given a specific tax revenue base. This is a very complex task. Offers of future retirement and health insurance benefits that once seemed realistic may now be problematic. For example, economic issues at the national level, such as Social Security and Medicare, more than likely will require changes to address their solvency.

Although there are many issues on which the government and unions may agree, there are also disagreements. These should be settled by principled negotiations, in which all parties should act in good faith toward their stated purpose. FES administrators should avoid being in a position where they feel forced to accede to union demands.

Unions like to use a process of binding arbitration to settle disagreements. In the process of binding

FIGURE 2-4 IAFF logo.
Courtesy of IAFF.

TABLE 2-1 Right to Work States	
Alabama	Nevada
Arizona	North Carolina
Arkansas	North Dakota
Florida	Oklahoma
Georgia	South Carolina
Idaho	South Dakota
Indiana	Tennessee
Iowa	Texas
Kansas	Utah
Kentucky	Virginia
Louisiana	West Virginia
Mississippi	Wisconsin
Missouri	Wyoming
Nebraska	

Courtesy of L. Charles Smeby, Jr.

arbitration, both parties choose an impartial arbitrator or third party to represent their position and reach a compromise, which is binding on both parties. Binding arbitration should be considered for all disagreements, especially safety issues; however, elected and appointed officials often do not want wages and benefits to be considered for binding arbitration. This practice is receiving more scrutiny and resistance from voters, senior municipal officials, and political leaders, especially when financial demands are sought by unions or have been gained in prior agreements that are not supported by tax revenues. One common example is underfunded pension plans, which is of great concern for local unions and taxpayers. Municipal negotiators may have made promises that were not founded in factual assumptions of future tax revenues. Many times, these negotiated decisions are driven by politicians who know that they will not be around when the actual bill comes due. Local politicians encourage these agreements to gain votes and donations to their campaigns.

Union negotiators should demand professional actuarial analysis to guarantee sufficient funding for any monetarily based benefits. Check any financial

assumptions very carefully, such as growth of investments to support trust funds that provide for future payments. For administrators and union officials involved in negotiations, the book *Getting to Yes, Negotiating Agreement Without Giving In* (Fisher et al., 2011) provides additional guidance and recommendations.

The Power of the Community

Administrators should identify and become acquainted with the community's power elite at the earliest opportunity. The power elite are a small number of highly influential citizens rarely seen at public meetings. These are local people who have easy access to elected officials as a result of longstanding friendships, wealth, or influence at the local, state, or federal level of government. The powerful elite can either help with the adoption of a new policy or, with active opposition, doom the policy to failure. In most cases, it is unusual for the power elite to get directly involved in FES policy decisions. However, for major policy proposals, such as the construction of a new fire station or a request for funding that requires an increase in taxes, the chief officer should consider approaching these individuals privately and explaining the request personally. Although the help of the powerful elite may not be needed in all cases, they may have the potential to influence policy changes.

The following are examples of potential conflicts with the power elite:

- A new fire station on a vacant piece of property is proposed. The information about the neighboring properties contains the name of one of the power elite. However, the chief fails to make a personal contact with this person before announcing publicly that a site has been selected.

- A new physical fitness program is proposed that contains reasonable goals along with an implementation plan for those existing members who cannot comply immediately. However, the firefighter's union was not consulted. The president of the union is a long-time resident of the jurisdiction and knows many of the power elite by first name.

- The chief has just finished a high-rise fire seminar at a national conference. On returning to the city, a reporter asks for any revelations on fire safety discussed at the conference. The chief announces to the reporter that he is now convinced that all existing high-rise buildings should be equipped with automatic sprinklers, and that a proposed local ordinance will soon be submitted to require it. What the chief is not aware of at that time is that the power elite own many of the existing high-rise buildings.

Private ambulance companies are not in the same situation as fire departments in terms of community power elite having direct impact on internal policy decisions. However, EMS management is wise to seek out and court the power elite who may have influence with those who have political clout. When contracts are up for bid, community stakeholders can have a big influence. Incumbent companies may lose contracts even if they have a great response time record or few patient care complaints. The winner is usually the company that comes in with the lowest bid. Recently, companies with a history of 40 or 50 years of good service have been ousted by others that offer a lower bottom line or are politically connected. This includes the pressure by firefighter unions to take over ambulance transport, encouraged by the International FireFighters Union.

When faced with a conflict with the powerful elite, the chief officer might not always be fighting a losing battle. Approach the potential change by slowly building a consensus among those people who will have an influence over the final approval. In general, arguments that are backed up by professional judgment and solid research have the best chance of approval. If national consensus standards support the request, it will be easier to gain the support of the power elite and others who shy away from public debates.

Remember, the power elite and special interest groups (e.g., firefighters' unions and volunteer fire associations) can have a big influence over elected officials. Approach these groups and their officers at the earliest opportunity to gain their support. Although it may not always be necessary to gain their support to be victorious, in many cases their support can make the process smoother, resulting in a better chance of success.

Administrative Power

Administrators should know that power is never guaranteed. Being promoted to chief officer does not ensure that all members of a company(s) will follow all of the chief's orders. A FES administrator, when managing and leading change, would do well to study how the most powerful administrators in the world use position and power to make changes.

The FES or EMS administrator is familiar with the power of the direct order or command. The following criteria are necessary for a command to be followed without question:

- The reason for involvement is instantly recognizable.
- The orders are widely publicized.

- The words are clearly understood, coherent, and unambiguous.
- The supervisors and members who receive the order have control of everything needed to comply.
- There is no doubt about the authority to issue the command.

It is rare that orders on the emergency scene are not carried out verbatim, although this does occur. Because the FES are quasi-military organizations, it is well understood by all members that they must obey the orders of a higher-ranking officer. Officers are trained and educated to follow orders even if the justification is not provided or understood. Individual officers and their company assignment fit into an overall plan that may not be clear to a specific unit, but fits into an overall strategy to mitigate the emergency. Therefore, orders must be obeyed without question, unless it is clearly an unsafe directive.

Private ambulance companies are less likely to have the same level of quasi-military supervision, but field crews know there is a chain of command they must follow (e.g., medical and trauma care is approved by online or offline medical control). Offline medical control is accomplished by protocols, policies, and standing orders approved by the local medical director. Online medical control occurs when a provider calls a base station and talks to a Mobile Intensive Care nurse or emergency department physician to ask a question, get an opinion, or seek approval for a situation that might not be covered under written protocols. Seeking approval from a higher-level provider provides legal protection for the paramedic.

However, in nonemergency situations, the ability to effectively administer the FES organization ultimately depends on whether the chief officer can persuade or convince others of what ought to be done for their own good or for the betterment of their service to the public—essentially, their level of administrative power. In addition, follow-up deadlines and delegation all play a very important role in everyday managing and changing projects.

The ability to recognize a problem and take action to make sure it does not happen again is a great attribute for an administrator. The chief officer should establish an informal information-gathering system made up of friends and confidantes throughout the organization. In addition, it is critical for fire and EMS administrators to maintain their professional reputation. If the chief has the reputation of not following up on assignments, poor communication skills, or low empathy, then when a request for members to comply with one of the department's orders is issued—either

Employee Noncompliance

A good example of how employees may have their own agendas that keep them from complying with an order occurred during President John F. Kennedy's administration. In 1962, the President became aware that the Soviet Union was constructing nuclear missile facilities in Cuba. Because Cuba was so close to the mainland of the United States, these missiles could be launched and hit targets in the United States in an alarmingly short time.

As it became clear that his negotiations with the Soviets were not making any progress, President Kennedy ordered a blockade around Cuba. A blockade can be construed as an act of war, and the crisis brought this country to the brink of nuclear war with the Soviet Union.

During the negotiations that followed, an offer was received from the Soviet Union—if the United States would remove nuclear missiles from Turkey, the Soviet Union would remove its missiles from Cuba. At the time, Kennedy remembered that he had previously given a presidential order for the U.S. military to remove its missiles from Turkey. Kennedy had received expert advice that the missiles were not necessary for the defense of the United States and could safely be removed.

The military officials—who were trained to obey all orders from superiors—had deliberately failed to carry out an order by the most powerful person in the United States, the commander-in-chief. The military officers believed that these missiles were critical to the defense of the country. Fortunately, the military's disobedience provided Kennedy with a bargaining token to make a deal with the Soviet Union. The United States still had the nuclear missiles in Turkey that Kennedy then used in negotiations to trade for the removal of the Soviet missiles from Cuba.

However, after this experience, Kennedy set up a system to follow-up and collect information about what was really going on in the many levels of the government. The President obviously could not manage such a large organization by walking around, but he established direct contacts with friends and trusted confidantes throughout the government.

Accurate feedback can be hampered either by subordinates following their own agendas or by high-ranking government officials not wanting to give the boss any bad news. This type of behavior by subordinates can be prevalent in departments where the administrator has a lot of influence over who is promoted. Many subordinates see their chances of being promoted as being based on their ability to do a great job and keep the boss happy. Colloquially, this is called surrounding yourself with "yes" people or "groupthink" of the close staff. Kennedy did not want to be blindsided again and therefore made extensive changes to provide impartial avenues by which information could reach him. (Neustadt, 1990)

verbally or in writing—many will not comply. Noncompliance is much more likely if members do not agree or are anxious about the consequences of the change or request.

The Process of Negotiation

In most cases, when a new vision or policy is proposed, the chief officer must negotiate or convince others that the change is good. Most administrators would think that they have the advantage in bargaining because they are the recognized leader. However, bargaining power only starts with the formal position; it also includes professional reputation and knowledge and skill in debating.

There may be opponents to the new vision because some members may lose privileges or simply because they resist any change. Member opposition does not mean that it is impossible to make policy changes, but the chief officer must be prepared, informed, and patient. Officers must have a plan to compensate those who stand to lose (if possible) and be able to prove that the change is based on solid research, logic, and professional judgment.

Chief officers who have a trustworthy and justifiable proposal may be able to negotiate an agreement with the members to implement the policy. For example, faced with irrefutable justification for the betterment of the public and the individual, many labor organizations cooperate with a negotiated agreement. If administrators have the facts and can prove them, they have the bargaining and influence power.

Administrators in government agencies should stay in contact with a number of individuals outside of their formal organization to help support major changes in policies, such as staff members of legislative committees, researchers from nearby think tanks, and representatives of lobbying firms and public interest advocacy groups. In addition, other government agency administrators and their staff need to be contacted on a regular basis to facilitate information flow and, when needed, solicit support for policy changes.

Informal organizations can be used to help facilitate the negotiations and data gathering that must go along with the decision-making and consensus building revolving around a proposed public policy objective. Informal organizations can help overcome barriers to change and facilitate communication between different groups; however, if not approached correctly, they can also sometimes initiate and facilitate resistance to change. An effective administrator must take full advantage of the informal organization to gather facts and acquire support for any policy changes.

Wrap-Up

CHAPTER SUMMARY

- Understanding that administration is made of two components, management and leadership, will help the reader understand how to be an effective and efficient administrator.
- Recognizing that professional standards are the key to successful administration and service to the public.
- Understanding that SOPs are key administrative tools to guide the members in consistent, professional, and safe service to the public.
- Comprehending the many aspects that improve the administrator's ability to recommend and successfully adopt progressive change despite the many challenges.
- Awareness in how to gain influence over the political process using feedback and support from members, including other agencies and outside influential people and organizations.

KEY TERMS

Administration The practice of supervising and leading an organization to achieve its goals.

Battalion chief A chief officer who supervises a number of fire companies or a subdivision of the fire department.

Chief officer The principal administrator of a fire department.

Emergency medical technician supervisors A supervisor who has the responsibly of several EMS units or a section of the organization.

International Association of Firefighters (IAFF) A labor union representing nearly 313,000 firefighters and paramedics.

Imperfect information The lack of knowledge required to make an informed decision.

Leadership Inspiring trained personnel to safely and professionally complete their mission using the resources provided by the FES organization.

Management Ensures that the organization is prepared and able to accomplish its goals by establishing that sufficient personnel and equipment are available for the organization to perform its duties.

Management by walking around Refers to administrators spending some part of each day listening to problems and ideas of their staff, while physically wandering around an organization.

Standard operating procedures (SOPs) Rules and regulations for safety and efficiency during emergency operations.

Task force A group of employees assembled to discuss an important decision before implementation and to make recommendations.

REVIEW QUESTIONS

1. Compare and contrast management and leadership. Do you feel your personal strengths lend themselves toward one or the other? Explain.

2. Do you think it's more effective when administrators are selected by appointment or elected into their positions? Defend your answer.

3. Describe how rules and regulations (SOPs) help reduce freelancing on the emergency scene. Research a specific example from a national consensus standard that is applicable to your experience.

4. Review the checklist on page 25 to help determine if a chief officer is ready to manage and lead change. Which of these questions do you feel is the most difficult for new administrators to check off? Why?

5. Provide an example of a past incident in which you solicited feedback in order to solve a problem. Whom did you ask for the advice? What strategies did you use to obtain effective feedback?

CHAPTER ACTIVITY #1: AN EXAMPLE OF POLICY ANALYSIS

A voluntary physical fitness program increases the average physical fitness of firefighters. It is well recognized that firefighting requires a high level of strength and endurance. Therefore, the assumption is that the public would be better served by firefighters who can perform at more advanced physical levels.

Firefighting is a team activity, so the weakest link limits the entire company or team. When a two-person team enters a structural fire and one individual is in good physical fitness and the other is not, the total team effort is limited by the less fit person. For example, a person in good aerobic condition can have a useful work time with a self-contained breathing apparatus of around 20–30 minutes, whereas the unfit person may not be able to stay more than 10 minutes before the low air alarm sounds. Therefore, the team of firefighters is limited to 10 minutes because when one firefighter leaves, the other must also leave for safety reasons.

Conclusion: A mandatory physical fitness program is needed.

Discussion Questions

1. Make a list of management and leadership goals that have to be accomplished to implement the conclusion from the above case study.

2. Do you have a management or leadership preference? How would this preference affect your actions necessary to complete the change process? Give several examples.

3. If you are stronger in either management or leadership, describe how you would select and incorporate people with the other preference into your implementation process.

CHAPTER ACTIVITY #2: CHAIN OF COMMAND

There is a shooting incident in your district and two victims need emergency medical treatment. After police secure the scene, your Basic Life Support (BLS) first responder engine arrives and Incident Command System (ICS) is initiated. As your company begins assessment, an EMT/paramedic unit from the private transport company arrives. With the help of firefighters, the paramedic assesses the patient with a bullet wound to the abdomen. In addition, EMTs assess the patient with an apparent extremity injury. The officer of the BLS engine determines that the first patient with the gunshot wound to the abdomen meets trauma center criteria.

The company officer calls for a medical helicopter per county policy because ground transport to the trauma center would exceed 30 minutes. He is given a 15 minute ETA. In 10 minutes, he calls the fire dispatch center to confirm the ETA of the helicopter and is advised that the request was canceled by the ambulance provider through their dispatch center.

Discussion Questions

1. As the FES administrator, how would you handle the paramedic canceling the helicopter without consulting IC, thereby failing to use the chain of command?

2. Who was ultimately responsible for the delay in patient care?

3. What can be done to assure this type of problem doesn't occur in the future?

REFERENCES

Fire Chief staff. 2001. What's new on the safety scene? *Fire Chief.* New York: Penton Publications, Volume 45, April 2001. Retrieved from http://firechief.com/mag/firefighting_whats _new_safety.

Fisher, R., B. Patton, and W. Ury. 2011. *Getting to Yes.* New York: Penguin Books.

National Fire Protection Association. 2008. *Fire Protection Handbook*, 20th ed. Quincy, MA: National Fire Protection Association.

Neustadt, R. E. 1990. *Presidential Power: The Politics of Leadership*. New York: John Wiley & Sons.

United States Fire Administration. 1973. *America burning: The report of the National Commission of Fire Prevention and Control*. Retrieved from http://www.usfa.fema.gov /downloads /pdf/publications/fa-264

Vorlander, P. 1996. An innovative approach to fire department command staffing. *Fire Engineering, 149*(8):131.

CHAPTER 3

Management

LEARNING OBJECTIVES

After studying this chapter, the student will be able to:

- Understand the role of management in fire services and emergency medical services (EMS) (pp 38–39).
- Know the four key managerial tasks of planning, organizing, directing, and controlling (pp 39–49).
- Discuss the value of experience and intuition in decision-making (pp 39–40).
- Describe techniques managers can use to increase staff compliance with rules and regulations (pp 45–46).
- Explain techniques for motivating staff members (p 45).
- Apply the three-step control process to accomplish organizational goals by measuring performance, comparing against a standard, and taking corrective action (pp 47–49).
- Discuss the management strategies of technology-based management, total quality management, leading by example, broad-based empowerment, and managing by walking around (MBWA) (pp 49–52).
- Comprehend managerial performance standards and methods of performance assessment (pp 52–53).

Case Study

The Devil's Advocate at Work

A fire prevention program manager would like to increase the number of home fire inspections performed by station personnel. The station personnel have reported that they do not have any time available as a result of emergency calls, in-service training, physical fitness, and station and apparatus maintenance. Firefighters work a 24-hour shift during which the required nonemergency duties are performed during daylight hours. The time after dinner until the morning, by tradition, is reserved for personal time or call response.

However, the prevention manager would like to suggest that the time after dinner (and during weekends) be used to do these inspections. The manager believes that this would be a good time because many residents who work would be at home during the evening hours and weekends. Because these inspections are voluntary, the requesting residents would be advised of the time of their appointment with the understanding that an emergency call would necessitate rescheduling.

Discussion Questions

1. How popular would this idea be for most members in your departments?

2. Would proposing this idea be controversial or resisted by the members and their labor unions?

3. What type of information or justification would hopefully be able to lead the group to a progressive change?

What Is Management?

Management is the glue that holds an organization together on a day-to-day basis. Management is what enables an organization to achieve its goals, efficiently and effectively, with the necessary staff, equipment, and resources. As explained in the chapter *Introduction to Administration*, management and leadership skills are two very different components of fire and EMS administration. Leaders provide the vision and are change agents for an organization, whereas managers supervise and facilitate the work of others within the organization.

Although many people may gravitate toward management for its status, being in a supervisory position comes with a great deal of responsibility.

There are four main responsibilities of management:

1. **Planning** includes creating and defining goals, and making important decisions and establishing strategic plans to achieve these goals. Without proper planning and implementation, the change required to meet new goals can be painful for the organization, its members, and its customers. Planning is discussed in greater detail in the chapter *Public Policy Analysis*.
2. **Organizing** is the process of determining what tasks need to be done and who is to perform these tasks. The manager determines the organizational structure (who reports to whom) and which parts of the organization respond to day-to-day problems. A good organizational chart indicates the major groups and who they report to for direction.
3. **Directing** the organization involves many tasks such as motivating and supervising staff, communicating with members, spotting noncompliance, administering appropriate disciplinary action, and resolving conflicts. These actions are what most people think of when they picture a manager.
4. **Controlling** refers to monitoring performance, measuring outcomes and progress toward goals, and making any adjustments necessary to keep the agency moving toward those goals. As part of this process, managers also provide feedback for future planning.

There is an old saying in management that to manage anything you must be able to measure it. Generally, effective management can be achieved if you set a clear goal and find something to measure that will indicate progress toward that goal. This may seem basic, but it is amazing how often it is not done and how hard it is to get right. A very simple example happens every day in many fire departments: At the beginning of each shift, the total number of firefighters present is reported to a manager. If a company is below minimum staffing, firefighters are detailed in or called back on overtime to meet the department's goal. This is the basis for all management control functions: valid measures.

These four responsibilities are discussed in greater detail in subsequent sections.

Planning

One of the key responsibilities that is critical to excellent service of a manager is to make day-to-day decisions to respond to situations that are not expected or planned for in the strategic plan. This section explores key aspects of the planning process—routine decision-making, including time sensitive decisions during emergency situations.

Decision-Making

In many cases of problems or identified issues needing change(s), the Fire and Emergency Services (FES) culture may default toward the use of intuitive decision-making by the chief officer. However, this approach generally is accompanied by a bias toward the tradition of the individual department. It also has been observed that the largest "big" city is imitated by the surrounding departments, which is another bias. In addition, a TV show such as *Chicago Fire* probably has an influence on many departments. Even the color scheme of the fire trucks is being copied by many departments (i.e., black top). This severely limits the potential to consider more innovative ideas.

There are many techniques or analytical tools that can be used to identify and organize research toward finding solution(s). Managers should be careful at selecting these analytical tools because many have been derived for complex issues in "for profit" businesses. Because these unknown issues are relatively rare, the use of an outside management consultant may be helpful.

1. As noted throughout this text, many operational issues can be answered by reviewing professional standards such as those promulgated by the National Fire Protection Association (NFPA). When needed for issues not resolved or addressed, a classical and very helpful process, used to organize decision-making, is listed below:
   a. Recognizing a critical need
   b. Defining the problem
   c. Collecting data and observations
   d. Analyzing the data
   e. Developing a hypothesis
   f. Testing the hypothesis multiple times to assure consistency and validity.

Probably the most key item is "defining the problem." It is very common to identity a symptom or not take into consideration that the problem may be the result of numerous causes. Also, in the FES it is not unusual to attempt to look for a quick and easy solution. This is the result of an action-oriented approach to everyday operations that requires immediate action to resolve an emergency incident.

This process is not intuitive and can be very complex. There are numerous educational sources to help understand this process as well as municipal government professionals or outside consultants.

Intuition in Decision-Making

Standard operating procedures (SOPs) cannot cover every situation at the scene of an emergency, so managers must be able to make quick, critical decisions. One of the qualities of an effective manager is the ability to make the right decision in the face of unfamiliar circumstances. These successful officers are often described as having good instincts, intuitively sensing when something is dangerous or wrong, or listening to a "gut feeling." Although intuition should not replace consideration of the facts, accepted knowledge, or prior experience, it can (along with common sense) contribute to decision-making.

Many experts believe that a person's subconscious mind holds all of the acquired knowledge, ethical values, past experiences, training, and skills that are used for complex problem solving. This has a large influence over what we call intuition. That is why it is always better when making an important decision to "sleep on it" if possible and allow the subconscious mind to process the information.

In part, intuition comes from experiences and acquired knowledge. In his bestselling book *Outliers: The Story of Success* (2008), author Malcolm Gladwell claims that the key to success in any field is, to a large extent, a matter of training, education, and about 10,000 hours of professional experience. Although a lot of experience is needed, it is not necessary to get all the experience oneself; the experience can be acquired from others through the review of case studies and writings of those with years of experience. Learning from others allows managers to gain knowledge of difficult situations without having to put themselves or their staff in danger, which is

Fire Prevention

An example of group decision-making can be found in in the following case. In the 1980s, the Prince Georges County, Maryland, Fire Department was experiencing numerous large-scale fires in garden-style apartment buildings and identified this situation as a critical problem because they were very difficult to extinguish, often resulting in firefighter injuries, and would culminate in whole rows of buildings burning down. These multi-dwelling structures were either three or four stories with four apartments per floor. They were accessed by a single central stair and the buildings were of combustible construction. Many of these buildings had two uninhabitable basement areas that were used as shared laundry and storage rooms for the tenants. These areas were below grade and had no windows; therefore, the only method of attacking a fire in these areas was to extend hose lines down a half story of stairs and into the door leading to these rooms—a tough and unsafe tactic. The storage areas were commonly used to store combustible items, such as renters' personal items, furniture, and even car tires. Most of these fires were deliberately set and would, without rapid extinguishment, spread to floors above via voids in the walls. The end result would find fire in the attic and lead to complete destruction of the entire building and in many cases adjoining buildings.

The fire department had gathered computerized fire data for many years, and one of the items captured was whether there was a delayed or immediate notification of the fire. After analyzing the data, a strong correlation appeared between the ultimate property loss and the detection stage. As one might suspect, delayed detection (caused by the lack of heat or smoke detectors) resulted in higher property losses.

By analyzing this data, the department was able to identify a problem along with a solution and implement special fire code requirements to address the problem. The owners were either required to install a limited water supply sprinkler system using the domestic water supply or vacate and lock the storage rooms and install a smoke detector tied to the buildings' fire alarm. This change resulted in the significant reduction in these large-loss fires. In this case, the data identified a measurable problem that was analyzed and used to ultimately improve fire safety for both the occupants and firefighters extinguishing these large fires. Mission accomplished.

especially advantageous in emergency operations. To gain knowledge, managers can learn from textbooks, reports of actions at emergency incidents, visits to other FES agencies, investigations into emergency responder deaths and injuries, articles about incidents, interviews with other emergency responders, professional websites, and professional qualifications and safety standards.

Managers may wish to make decisions on their own as individuals, or they may make decisions based on group input. Group decision-making occurs when individuals, typically those that are affected by the decision, collectively study and make a choice from the potential realistic options. In most cases, the chief manager provides a definition of the problem, some background information, and any limitations on the group's decisions before the decision is handed over to the assigned group.

When to Use Group Decision-Making

When the final decision directly impacts organizational activities, members' benefits, or performance benchmarks, group decision-making may make acceptance easier, particularly when sacrifice is required. Groups may find unique solutions to problems; for example, unions have voted to accept pay decreases instead of layoffs. With group decision-making, managers are less likely to feel blindsided by a lack of information about possible (or preferred) solutions. One of the major reasons noted for bad decisions made by management is lack of factual information. Group decision-making ensures that many diverse opinions are discussed. This process also results in the collection, verification, and analysis of a lot of background information regarding the problem, dramatically reducing the chance of a poor decision being made.

The collection of verified background information from reliable sources is especially important in today's world, when major media communications organizations have been found to misrepresent facts and in some cases promote outright propaganda. For example, social media platforms are struggling with blocking "hate speech" or falsehoods that can inadvertently or deliberately exclude real truths as a result of their new policies. In order to make effective decisions, managers must be aware of information from biased organizations or advocacy groups. Rigorously check any facts or data you use for veracity. There is no easy method to this process; it just requires decision researchers to be very conscientious and diligent.

Emergency Incident Research

When conducting emergency incident research, consider using the following Internet resources:

- *Firehouse Magazine,* https://www.firehouse.com/
- *Fire Engineering Magazine,* https://www.fireengineering.com/
- *The Fire Chief,* https://www.firechief.com/
- *JEMS: Journal of Emergency Medical Services,* https://www.jems.com/
- *EMS World Magazine,* https://www.emsworld.com
- Fire Fighter Close Calls, http://www.firefighterclosecalls.com/
- U.S. National Institute of Science and Technology, Fire Research Division
- *NFPA, https://www.nfpa.org/News-and-Research*
- U.S. Department of Homeland Security, https://www.ready.gov/preparedness-research
- Disaster Information Management Research Center https://disasterinfo.nlm.nih.gov/
- U.S. Fire Administration, https://www.usfa.fema.gov/data/Disaster Information Management Research Center
- *CSB Chemical Safety and Hazard Investigation Board, https://www.csb.gov/*

How to Select Group Members

Effective group decision-making should include a group of individuals from all levels of the organization. Outside experts such as elected officials or other agency representatives can also contribute to the process. Group members need to be selected carefully based on their openness to discussion, absence of bias, expertise, and ability to come up with rational solutions. Managers must be cautious not to be predisposed toward selecting group members who are in favor of the administration's preferred solution. The advantage of increased acceptance is lost if members believe the decision-making process was manipulated. Group members should have a minimum amount of experience, because there are some idiosyncrasies of the profession that simply take time to learn; however, they should also not be near retirement. Members nearing retirement may not have a stake or interest in solving the problem. One common policy is to allow groups, such as the union, to appoint one or more members to the decision-making group. These selections should be accepted by the administration even if these members might be biased and would not otherwise have been selected.

Optimal Group Size

Although there is no unanimously accepted number, groups of 5–15 members seem to be the most effective size for timely group decision-making. Whatever number is chosen, it should be an odd number to enable the group to reach a final decision using a simple majority of members. Some organizations that use group decision-making to determine consensus standards (e.g., the NFPA) may have committees with up to 30 members. Larger groups may have greater difficulty arranging in-person meetings, and in most cases all members will not be able to attend. Although electronic audio-video conferencing can help, ideally meetings should be in person. With the COVID-19 virus, temporary video conferencing meetings may be the only option.

Techniques to Aid Group Discussions

Several different techniques may be useful in group decision-making.

Understanding Limitations

Managers are responsible for communicating any limitations of resources or other relevant areas (e.g., specific laws or regulations) to the group members. This may include mandates from elected officials, legal restraints, and state or federal regulations. In addition, the manager may wish to outline possible solutions and describe any that will not be accepted by the administration.

Gathering Information

Managers may wish to encourage group members to consider ideas for solutions that have demonstrated success in other departments. Group members may wish to consult research articles or visit departments that have had similar problems. This allows group members to gain insight into what has been tried elsewhere and what has succeeded or failed. Managers should encourage members to have an open mind; just because a solution failed somewhere else does not mean it will not be successful in their department.

Newer technologies and communication sources are almost infinite. Research using https://duckduckgo.com/ and https://www.google.com/ will serve up a large list of Internet resources. First, it is a good idea to use more than one search engine because some edit online content. Be cautious and review as many as you can to find truthful nonbiased information.

For larger very dispersed departments, video conferencing combined with face-to-face meetings can be very effective and travel time advantageous. This takes a little planning that should include background research information. Also, any limitations on any suggestions that would not be acceptable need to be revealed to the participants.

And finally, the old fashioned phone can be very helpful. Using the Internet, find phone numbers of members in other departments that may have experienced the same issue. For example, in a suburban highly populated county, a new indoor sports arena was being planned in the early 1970s. Because this department had no experience with the fire protection features or fire prevention rules for this type of venue, they sent fire prevention officers to visit the fire departments in major cities that had these buildings. They were welcomed by fire officials in these jurisdictions and were provided tours and conversations on effective fire prevention procedures for the safety of inhabitants. This turned out to be a critical help.

Reviewing National Consensus Standards

Fire services and EMS may be required to meet standards defined by national organizations, state governments, or local contracts to ensure quality and accountability. The guidelines provided by federal agencies, state authorities, professional organizations (e.g., NFPA, National Highway Traffic Safety Administration [NHTSA], Occupational Health and Safety Administration [OSHA]), and union agreements should be considered when determining company policies. For example, it can be argued that if a department complies with NFPA 1500, *Standard for Fire Department Occupational Safety and Health Program*, it will be empowered to deliver superior emergency service to the customer.

Brainstorming

In the process of brainstorming, members may suggest any solutions without initial judgments; this is essentially a technique to generate ideas **FIGURE 3-1**. Anonymous input should be arranged, if possible, to encourage those members who may not want to publicly advocate their ideas for fear of criticism. However, this may not be necessary because only ideas—not criticism—should be allowed during brainstorming sessions. After all of the suggestions have been documented, the group should be given time to consider the available options before discussing them, ideally until the next meeting.

FIGURE 3-1 Brainstorming allows any and all solutions to be recommended without any initial judgments.
© ASDF_MEDIA/ShutterStock.

Discussion

At the next meeting, members may discuss the options and offer criticisms of the proposals. Members or staff should be selected to research the ideas, including all previous successful or unsuccessful experiences. Pros and cons should have solid data or reasoning behind them and should not be dismissed just because they have not been done before. Going through all the available options may take several meetings, but it is time well spent to ensure that group members have examined all possible solutions.

Devil's Advocate

Only through questioning the status quo (e.g., the culture or tradition) does real change occur in organizations. For this reason, one or more group members may be assigned or voluntarily take on the role of "devil's advocate" during the decision-making process. The devil's advocate is a person who pretends, in an argument or discussion, to be against an idea or plan that a lot of people support, and is an effective way to make other group members justify their support of a preferred idea. This role is not for everyone; the desire to be liked by others and not take an unpopular stance is one of the most challenging issues a person may face in his or her private and professional life. To express one's concerns without angering others, it is best to keep the focus on how the customer, member, or department will benefit from the new policy or decision and how alternate solutions may be harmful for the

Examples of Group Decision-Making Techniques

The following case studies provide examples of the previously described group decision-making techniques.

Gathering Information

One major metropolitan fire department appointed a committee to design a new fire station. It had been many years since they had the funding for a station, and all of the existing stations were built by volunteer companies before the county consolidated. Because there was no uniformity in these designs and they were outdated, the committee decided to visit newly built stations in adjoining departments. Eventually, they found a station that was almost perfect for their needs. They obtained permission from the other department to use the same architect, which saved a great deal of time and money by eliminating the need for new engineering and architectural plans.

Delayed Implementation

A department adopted a physical fitness program and testing system; they used delayed implementation to allow incumbents (who may not have exercised for years) extended time to comply. For example, in the first year, physical fitness equipment would be purchased for each station, and medical evaluations and physical fitness testing would be completed. A plan for improvement would be given to each member. During the second year, members would start exercising at work using their individual goals as a guide. Between years 2 and 10, the existing members would be expected to achieve a percentage of the stated goal starting at 20% and increasing at 10% per year. At the end of 10 years, all members would meet the same fitness requirements, achieving the desired goal.

Forward Thinking

A large ambulance company purchased hydraulic gurneys for all its ambulances across the country. It was a monumental expense on the front end, but the company realized the gurneys would all but eliminate career-ending back injuries caused by incorrect lifting. The future goal was that hugely expensive worker's compensation, and possibly patient lawsuits, would be significantly reduced. Field personnel were happy with the decision, which would help protect their backs and extend their careers.

safety of the customer or efficiency of the department. The devil's advocate is someone who thinks outside the box, is a risk taker, and is not afraid of change. In many respects, the devil's advocate is a true leader who may lead the department through substantial, controversial, or progressive changes. He or she should aim to present input without angering the group, because this could lead to ostracism and being excluded from future group decision-making processes.

Consensus

At the end of the discussion, members may determine if there is sufficient support for any of the ideas. It is possible that a combination of proposals may be the best solution. Managers should encourage group members to compromise; sometimes the best compromise includes some form of future implementation (e.g., allowing several months or years to achieve the required change). Group members may determine how they will reach a consensus at their first meeting, or this process may be designated by the administration. Consensus may be reached by unanimous decision-making, a simple majority, or a specified majority (e.g., the NFPA requires a two-thirds majority and the U.S. Senate requires a three-fifths majority for most important laws). Although not necessary, having a majority consensus is preferable. It is easier to obtain as opposed to unanimous agreement, and it indicates that most members agree with the group's decision.

Delayed Implementation

A delay in initiating the new decision may be necessary because of funding or some other practical limitation. Delayed implementation is sometimes combined with a "phased-in implementation" to smooth out the adverse consequences of some projects that create substantial change.

Forward Thinking

This technique is used to ensure that the decision takes into consideration what may happen in the future. Many decisions do not consider the future, and, therefore, it is common to have unintended consequences.

Disadvantages of Group Decision-Making

Group decision-making offers many benefits; however, it does have some disadvantages. For example, groups almost always take substantially longer than an individual to reach a decision. Also, as demonstrated by the saying "a camel is a racehorse put together by a committee," groups do not always reach ideal decisions. Poor results can be caused by failure to focus on the customer or department's goals, or by a lack of leadership, organization, or preparation.

Even if the results of the group process are valid, they may be unnecessarily complex and have internal inconsistency. For example, if a department's procedure for the first due engine on a structural fire contains more instructions and decision points than an officer could realistically complete in the typical 7 to 10 seconds before action must be taken, this may be the result of poor group decision-making. Because meetings occur in a nonemergency setting, the group has plenty of time to discuss and analyze all possible actions, often resulting in overthinking and complications.

There is also a danger of minority domination in group processes. A charismatic group member or several members representing a particular group may dominate decision-making.

Other members may feel a need to go along with the crowd to avoid conflict. This is called "group think" and can be very influential. This desire to go along with the group can be very strong in emergency response agencies where strong personal bonds often exist between members. Managers can help avoid this situation by providing clear direction and limitations, as well as by providing every member of the group an opportunity to express themselves. Each group member should be called upon by the chairperson to make a comment.

Organizing

In addition to planning and decision-making, another key task of most managers is determining the organizational structure. Departmental organization in FES is usually very similar in all departments, although there may be some minor variation in job titles based on tradition or department size, simplifying this task **FIGURE 3-2**. Budgetary considerations have led a few organizations to create alternative departmental structures, but this is rare.

The typical department is organized based on a principle called unity of command—the idea that each individual reports to only one manager. This clear hierarchical structure is especially necessary to ensure order at the emergency scene. Unity of command becomes even more critical at large incidents where there are several different departments providing mutual aid. In national emergency incidents, the National Incident Management System (NIMS), discussed more thoroughly in the chapter *Introduction to Administration*, is used to ensure unity of command at catastrophic emergency incidents. FES organizations that use this system at everyday incidents will be comfortable with it and better prepared when large-scale incidents occur. Making anything a routine is more likely to result in it being done and done correctly.

At the top of the department, managers may report to individuals outside of the organization. In municipal departments, managers report to a mayor, another elected official, or the appointed public safety director. In special taxing districts, a committee of elected officials is formed to oversee the FES department and represent the public interest. However, these elected officials may not have sufficient knowledge or expertise to judge the service, which can present a challenge for department managers. Unlike municipal departments, many volunteer departments do not report to anyone. This lack of accountability is usually offset by the high degree of commitment and motivation volunteers have to provide high-quality emergency service. However, managers must keep in mind the following quote from historian and politician John Emerich Edward Dalberg Acton: "Power tends to corrupt, and absolute power corrupts absolutely."

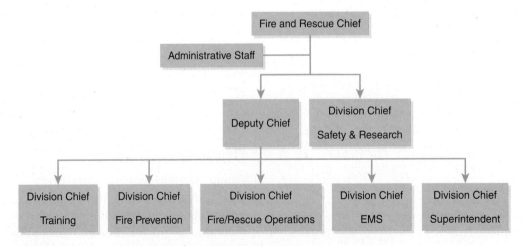

FIGURE 3-2 Example of a basic fire and rescue department.

Directing

The third major responsibility of managers—directing—can take many forms, such as orders, controls, and instructions. Directing is the process of a manager instructing the members under his or her authority to accomplish an organizational objective. Instruction may occur from person to person or take the form of written procedure or policy. At its simplest level, it is the verbal order of an officer to their crew on an emergency scene.

Directing can occur through direct supervision or voluntary compliance and varies with the objective to be accomplished. For example, fire inspectors normally operate under voluntary compliance (self-direction) as a result of the nature of the role. Generally, they are specially trained and educated members who operate individually when performing a fire inspection. When members are motivated and informed, enforcement methods are seldom needed. However, to ensure total compliance, mandatory methods of direction, including follow-up and discipline, may be necessary.

In addition, managers can achieve compliance through the use of rewards, positive role modeling, supervision, and a general culture of compliance. Ideally, managers would not have to closely supervise every member, because this opens the very real possibility that members are noncompliant when they are not being watched. Self-discipline is preferable for functional and safety issues. Close supervision may be needed when a new change is implemented, but this level of supervision is not needed after the culture and individual behaviors are changed.

Managers can use several of the following techniques together to achieve compliance.

Creating Rules and Regulations

Rules and regulations are the backbone of directing. They achieve consistency by ensuring that all managers are following the same guidelines and policies. Important organizational rules and regulations should be in line with the goals of the organization, measurable, and written down. In creating an agency's SOPs, a manager should strive to limit the number of SOPs required and to ensure that they are simple and easy to understand. Creating SOPs is often a complex process that identifies relevant state or federal regulations, local laws, traditions, and emergency practices affecting the department. Other departments' SOPs can be a great resource to help check the accuracy and completeness of new or existing SOPs for your department.

Motivating Staff

One of the goals of management is to motivate members to achieve certain tasks or goals. Ideally, members should be motivated to the point that they no longer require direct supervision. Motivated members are more inclined to cooperate, complete tasks willingly, and follow departmental rules than dissatisfied or unmotivated members. Members who are self-motivated can also add to the motivation of those around them. These members might be good candidates for management positions because they demonstrate management skills and leadership traits.

Fire and EMS providers can be motivated by many considerations, including a desire to gain security in their membership status, obtain the respect and trust of supervisors, or safety. They may also be motivated by the administration's commitment to their well-being, increased salary and benefits, or pride they have in the organization. Members may be self-motivated by their job duties, a sense of achievement, or a desire for recognition.

For example, external motivation can take the form of public recognition or ceremonies that recognize heroic actions **FIGURE 3-3**. However, managers should be careful when defining heroic acts; unsafe actions should not be rewarded because receiving a reward or recognition elevates the act to an aspirational goal. In addition, appreciating employees through recognition of non-emergency actions has great motivational

---

CHIEF OFFICER TIP

The chief officer may find some of the characteristics of younger generations challenging to manage. Page (1999) explains this generation gap in the article *Your 21st Century Firefighter*:

> "Our future firefighter will have precious little experience with teamwork, self-sacrifice, personal organization and respect for authority. It's not necessarily his fault. Young people have grown up in the soft, affluent society that was created for them. We sold our farms to corporations and won the Cold War so our kids didn't have to bale hay or get drafted."

Although a disparity does exist, blame should not be placed on the younger generation. As Page goes on to acknowledge, the ultimate question is, should the fire service change to accommodate the differences between generations, or should the new generation change to meet the traditional fire service standards?

FIGURE 3-3 Awards are opportunities for excellent public relations, but be careful that the acts rewarded are not unsafe actions.
© Jones & Bartlett Learning. Photographed by Glen E. Ellman.

benefits that can last the entire year or more. These rewards should be for actions that are measurable. Some corporations have used goals exhibiting success by up to 85% of employees for recognition. Listed below are several possible categories:

- Number of smoke detectors installed
- Number of home fire safety inspections
- Citizens instructed in CPR
- Members meeting minimum physical fitness criteria
- Fire officers becoming nationally credentialed
- Fire companies evaluated for emergency operations proficiency such as those complying with stated time goals in NFPA 1410 *Standard on Training for Emergency Scene Operations*

Additional vacation time or other incentives can also be a great tool to encourage members, especially when sacrifice may be needed. For example, it is common to offer incentive pay for advanced certifications that require many hours of studying to achieve, for example, becoming a certified paramedic. Furthermore, if a potential policy would be met with a great deal of opposition, incentives may help smooth its acceptance.

However, according to David Javitch from Entrepreneur.com, managers often make the mistake of assuming that workers are motivated by money alone. Studies show that recognition and status have a more lasting impact on output than raises or bonuses. Furthermore, although employers may not worry about motivating their smartest employees, it is often these employees who are the most vulnerable to boredom

and frustration (Javitch, 2009). Money may not be an effective motivator for these employees or professionals, who typically already have good pay and benefits; instead, they may be looking for increased responsibilities, challenging work, or professional growth opportunities.

Motivating Diverse Generations

This can be extremely complex because there are essentially five different generations that need to be supervised as follows:

Traditionalists: 2% (1925–1945)
Baby Boomers: 25% (1946–1964)
Generation X: 33% (1965–1980)
Generation Y: 35% (1981–2000)
Generation Z: 5% (2001–2020)

Each has the same non-inclination to money as a motivator that Javitch discusses; however, they have many differences in motivating items and preferred communication. There is not a simple answer to be found. These differences can be very important to take into consideration, especially when they involve major changes. In addition, when having one-on-one conversations with employees, supervisors, or elected or appointed officials, including personal relationships, these differences can make your communications either understood or not make any sense, almost like you are speaking a foreign language to the other person.

For a detailed and comprehensive explanation of these idiosyncrasies, the Purdue University web page at https://www.purdueglobal.edu/education-partnerships/generational-workforce-differences-infographic/ is a great place to start. This complex subject is presented in a bulleted format as a result of reviewing extensively many scholarly studies on these generations.

One system that encourages professional growth and responsibility at the firefighter level is a tiered promotion ladder (e.g., recruit firefighter, firefighter, and senior firefighter). This system is based on professional standards for training, education, and experience, and includes salary increases. Because managing is about getting the most and best out of each member regardless of rank, not every member should be encouraged to seek promotion. Promotions should be evaluated based on clearly defined requirements and responsibilities of the new position. Promoting the wrong person can easily lower morale and upset other members of the organization. Furthermore, not everyone can be an officer, and many members are more comfortable not being one. However, every department needs well-trained professional firefighters, and there should be a system that recognizes and rewards these vital members. This system can lead to improvement in the quality of the service provided to the public.

Controlling

Controlling is the management function that ensures completion of all steps to accomplish the organization's goals and provide feedback when unintended consequences occur. If managers do not control, there would be no way to know if policies, plans, and procedures are being achieved correctly and on time. The control process is made up of three steps: (1) measuring performance, (2) comparing the performance results to a standard, and (3) taking corrective action to ensure compliance to the standard.

For example, an EMS company requires that all patient care reports be completed within 24 hours of the call or before the end of the member's shift. To control compliance with this requirement, the supervisor researches the time that patient care reports are completed and the time of the corresponding incidents. These results are compared to the standard of 100% compliance, although the company might allow time variances for certain complications such as technical difficulties, multiple late calls, vehicle breakdowns, computer aided dispatch issues, and so forth. Corrective action occurs when the manager identifies a crew member not in compliance with the policy; the manager determines the reason why the patient care

reports were not completed and provides an opportunity for the provider to correct the problem. Disciplinary action is taken for multiple infractions or those that were clearly misfeasance, nonfeasance, and malfeasance.

Measuring Performance

The first step in the control process—measuring actual performance—seems relatively straightforward. However, managers must determine the correct outcome to measure to avoid unintended consequences. For example, if a manager monitors and measures the time it takes for turnout, members will likely show improvement in this area but may resort to unsafe methods to shorten turnout times or falsely report shorter times. Some outcomes can be measured quantitatively (e.g., a specific number); others may be measured categorically (e.g., yes or no); and others may be measured qualitatively (i.e., no absolute answer but a general measure of several different criteria). However, most "outcomes" should be measurable. What managers measure can determine what members will focus on to improve their performance. See the chapter on *Public Policy Analysis* for more examples.

Comparing Results to a Standard

If comparing the results of the case study presented in this section to a standard, data would need to be collected. These data would need to note the total number

---

Facts and Figures

Incident Command System

One method of controlling can be found in the Incident Command System, now a subcomponent of NIMS, which is the original model of command used at local emergency incidents. It contains all four main responsibilities of management discussed previously. With the exception of large disasters, all emergencies start as local emergencies. When the situation gets more complex and starts to involve multiple local, state, or federal jurisdictions, then an expanded, more contemporary model of management is necessary. NIMS compliance is required of a wide variety of first responders, including EMS, and is the official emergency management system of the Federal government.

NIMS establishes a uniform set of processes, protocols, and procedures for all responders, and works with the National Response Framework. NIMS provides the template for incident management, whereas the National Response Framework provides

structure and mechanisms for national-level policies. Key incident command functions include:

- Incident command
- Operations
- Planning
- Logistics
- Finance and administration

A NIMS-compliant Incident Command System should be used on every incident involving more than one resource. Initial command is assumed by the first arriving unit. This could be a police vehicle, private ambulance, or fire engine. As other agencies appear on scene, command should be transferred to the most appropriate and qualified person and agency. Which agency takes that role depends on the type of incident and also on local structure and resources.

of fire incidents and the property lost per incident over a calendar year for storage room fires in garden-style apartments, and would have to be collected for several years before and after full implementation. The results would need to be compared to the historical data before improvement was confirmed.

Furthermore, in that particular example, there was also anecdotal evidence of success because many firefighters openly lamented the lack of multi-alarm fires. The data supported these observations.

Taking Corrective Action

When a manager finds it is necessary to take corrective action, the first step may be a feedback session with the employee. Feedback is an important part of a manager's responsibilities because it can help correct the performance of an individual or identify problems with a policy. Regular feedback meetings can be used for managers to get complete and honest information. Feedback meetings are like a form of counseling and should be scheduled in advance to discuss specific concerns in private at a time with minimal interruptions. In these meetings, the manager notifies the employee member of the issue and cites an example of when the issue occurred. The goal of these meetings is to discuss the concern and possible solutions. The manager must determine if the employee's noncompliance is the member's responsibility or indicates a problem with the rule, policy, or SOP. Although disciplinary action may be necessary later, the manager should aim to put the member at ease during these sessions and establish a supportive atmosphere in which positive feedback is given to help correct noncompliance. During feedback sessions, a manager should focus on the facts of the issue and not generalize or use subjective comments **FIGURE 3-4**. The purpose of the meeting is to correct job-related behavior; personal

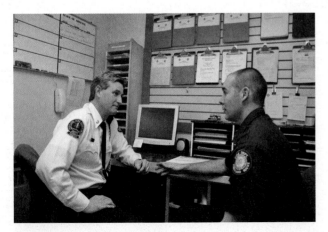

FIGURE 3-4 Private counseling meeting.
© Jones & Bartlett Learning.

judgments or criticisms have no place in these sessions. Any negative feedback should be directed toward the job-related behavior and noncompliance backed by examples, facts, and verifiable observations of the inappropriate actions. The member should be encouraged to present his or her side of the story and explain any extenuating circumstances. The manager should listen to what the member has to say even if it is not factual or accurate. Excuses should be acknowledged, because they may point to issues that the member needs to correct to ensure that the infraction does not happen again; however, the facts should carry more weight than excuses. After presenting the issue, the manager should explain in clear terms what the member can do to correct the negative job-related behavior. If the member cannot correct the problem, the manager may need to discuss disciplinary action. Although rare, there are members who are not able to meet job performance requirements; these members should be quickly separated from the department because this is a safety issue for the other members and the team. Job performance requirements (JPRs) are those detailed in NFPA Professional Qualification Standards or EMS certification standards and may be based on failure to perform certain skills, medical issues, or physical fitness. The feedback session should end with a clear understanding of the issue and any solutions necessary. Some discipline policies require a written report of the conclusion.

When feedback sessions do not provide the desired result, disciplinary action may be necessary. For example, consider the use of the three-step control process to address the concern of firefighter safety. The manager would first collect data on firefighter death and injury. (This information is well documented by the U.S. Fire Administration and the NFPA.) Next, the manager would work to determine a standard of comparison for safety goals (NFPA 1500 *Standard on Fire Department Occupational Safety and Health Program* and the National Fallen Firefighters Foundation sponsored "Everyone Goes Home" 16 Firefighter Life Safety Initiatives may provide useful guidelines to develop these goals). After the data are compared to the goal, the manager would take corrective action to work toward the goal. The manager could start by creating a policy or SOP in consultation with a group of members. After the policy was created and approved by the administration, then compliance with this policy could be measured. For example, if the policy required an annual physical examination, members might be required to submit verification that they had completed the examination or face disciplinary action. Generally, this is something the management schedules and arranges, then notifies the member.

Unfortunately, many FES and EMS disciplinary systems are ineffective because managers do not take appropriate disciplinary actions or employees are not aware of what constitutes rule infractions. For a disciplinary system to be effective, a manager should be appointed as the point person, or for larger organizations, a small staff on all disciplinary issues; selection should be based on an exceptional reputation of ethical behavior in the manager's personal and professional life. Having one person ultimately responsible for all discipline negates any unfair actions by subordinate officers and ensures consistency. This manager must be trained on issues regarding discrimination, sexual harassment, and employee rights.

A list of common rule violations and appropriate consequences should be created and given to all members. For example, the following would be a good start:

- Not arriving at the shift on time
- Use of or under the influence of illegal drugs or alcohol
- Stealing from homes of victims or the department
- Failure to accomplish JPRs
- Failure to meet minimum physical fitness standards
- Insubordination
- Aggressive behavior on duty
- Any serious criminal act off duty
- Numerous or serious failures to comply with SOPs

The consequences should be developed in consultation with legal and labor representatives. It may also be useful for the manager to compare the policy to those of neighboring departments. Most disciplinary policies involve progressive consequences, requiring more severe disciplinary actions for repeat violations or serious offenses. These steps should be followed precisely without deviation. A formal training session on the disciplinary system should be presented to all new employees and, when first implemented, to all existing members. Chief officers should watch for violations of the rules and regulations by walking around to visit stations and emergency incidents. Any employee who violates a rule should be reported to the manager for disciplinary action. The manager should aim to be fair and consistent in applying disciplinary action. If it is obvious that a company officer is deliberately not paying attention to violations by members of their crew, the officer should be disciplined by the supervisor. This situation can be a special problem in the fire service, where officers may be close friends with members they supervise.

CASE STUDY

The Control Process

In Florida, an applicant for a firefighter position must sign an affidavit that states that he or she has not used tobacco for 1 year before the date of application. In the state, a diagnosis of heart or lung disease can be considered a job-related illness. Requiring applicants to sign this affidavit reduces the possibility that heart or lung disease was caused by smoking tobacco rather than job-related factors. The goal behind this initiative was to reduce the number of firefighters who were smokers with the hope that firefighters would remain smoke-free for the length of their careers.

However, in more than one fire department, the managers observed several new firefighters smoking. The manager wondered if they had started smoking after joining the department or had not been honest in signing the affidavit. The process of control (measuring outcomes, comparing to standards, and taking corrective action) could be used to solve this issue. The manager would measure new firefighters' compliance with this condition of employment by an initial nicotine test, followed by subsequent random tests (possibly combined with illegal drugs). The results of the testing would be compared to the tobacco-free standard. Any new firefighters found in violation of the non-tobacco use policy would be counseled about the policy on the first occurrence. If still noncompliant when tested again, they could face job separation.

Management Tactics

Several different strategies may be used by managers to help their organization achieve its goals.

Technology-based Programs

There are many technology-specific programs that are helpful for the manager. The use of technology is one of the most effective approaches to providing safety. Although not exhaustive, the following are some examples:

- Some of the best examples of the use of technology in FES have been implemented in the area of fire prevention. For example, residential occupancies are attributed with the greatest number of fire deaths. Although humans are the most sensitive detectors of smoke, they can be unreliable at times, such as when sleeping. That is why the simple and inexpensive smoke alarm (a measuring device) has been so effective. It is always ready to perform its life-saving function. Therefore, if the manager's goal is fire safety for residential occupants, this electronic device significantly and reliably makes progress toward this goal.

- Human resource systems software cover a vast amount of ground these days, including not only employee management, but policy and compliance management, shift scheduling, performance management, and even payroll.

- It is not uncommon for FES departments to have rules and SOPs that need to be enforced to be effective, because this is the only method to guarantee 100% compliance. For example, one popular mandate is to require all firefighters to be seated and belted before responding to a call. This is not easy to measure because the officer is always looking forward with firefighters riding in a back compartment. Recently, the NFPA standard for fire apparatus was updated to require a Seat Belt Monitoring System. This system provides an audible and visual warning when a seat is occupied without a fastened seat belt.

- Driving is the most dangerous part of an emergency response. Because some driving laws are suspended or altered for emergency response driving, the process of "due regard" must be honored. An intersection accident where the agency is held liable can cost them not only in money, but in lives, reputation, apparatus loss, and rising insurance rates. However, it is impossible for managers to monitor emergency vehicles every time they are running "hot." Therefore, some agencies have outfitted vehicles with "black boxes"; the information the devices provide has saved companies from losing lawsuits and provides a measure of the driver's adherence to driving regulations.

Total Quality Management

Total quality management is a quality improvement approach that was first extensively used in Japan's auto industry. It was created under the assumption that customers are more likely to buy products that are high in quality. Therefore, this approach strives for 100% perfection. It has been defined as consisting of efforts to have a situation where employees continuously improve their ability and ideas to provide products and services that customers will find of particular value. Generally, it is not in use in FES because it is business oriented.

Jeff Dewar, quality consultant of QCI International in Red Bluff, California, further argues for the need to eliminate defects altogether. To make his point, Dewar came up with various cases of what life would be like if things were done right 99.9% of the time. For example, being just 0.1% less than 100% would result in 500 incorrect surgical operations per week in the United States (Dewar, 1989)! This approach emphasizes the importance of striving for zero defects and is a good safety goal.

Leading by Example

Leading, or rather managing, by example is one of the most powerful methods of encouragement a manager can offer. For example, a manager who always uses a seat belt when responding to calls encourages others to do the same. If a manager does not lead by example, management efforts can be sabotaged and, in many cases, change will not be institutionalized, creating the possibility for efforts to be overturned in the future.

As mentioned in the chapter *Introduction to Administration*, a manager is not necessarily a leader. A leader embodies special behavioral traits that help achieve new and innovative changes. However, managers can strive to be leaders. Successful leaders are courageous risk takers who do not fear the criticism that often accompanies significant change. Leadership means speaking up before there is consensus on an issue or problem. Also, a leader has the courage to admit mistakes and adjust plans accordingly.

For management, job-relevant knowledge—one of the key traits of leadership—can be gained through higher education. Continued learning increases a person's confidence and intuition. In selling new ideas to others, confidence is extremely important. Another trait that could be added to this list is the ability to listen. An important skill for all leaders, listening helps managers gain vital knowledge about the support or opposition to their proposals. Listening also helps define any problems, which is the first step of solving any management difficulties.

Managers can be leaders by setting a positive example for others to follow. By serving as a role model, managers can gain voluntary compliance of members. Leading by example can also help managers create a positive reputation and gain the trust of their staff. Managers should always strive to:

- Be honest
- Display competency in professional knowledge and skills

Traits Associated with Leadership

1. **Drive.** Leaders are self-starters with high energy. They are persistent and strive for success.
2. **Desire to lead.** Leaders enthusiastically take on new responsibilities and enjoy influencing others to accomplish enlightened goals.
3. **Honesty and integrity.** The personal and professional ethics of leaders are above reproach.
4. **Self-confidence.** Leaders believe in themselves and have faith in their own beliefs.
5. **Intelligence.** Leaders must be intelligent enough to analyze large quantities of data and ideas. Perseverance and hard work make up for any lack of knowledge.
6. **Job-relevant knowledge.** Leaders have extensive knowledge about the agency and the profession from practical experience, training, and formal education.
7. **Extraversion.** Leaders are generally happy, optimistic, and energetic. When needed, they can be socially outgoing; but they also take time to be alone for deep thought.

Data from Judge, T. A., J.E. Bono, R. Ilies, and M.W. Gerhardt. 2002. Personality and leadership: a qualitative and quantitative review. *Journal of Applied Psychology, 87*(4), 767.

- Be loyal to the organization and to subordinates
- Be fair and consistent
- Show a commitment to sharing ideas and information
- Keep promises
- Keep confidences of superiors and subordinates

Broad-Based Empowerment

In some industries, empowering low-level supervisors can be helpful for the organization to run smoothly and achieve its goals. This management technique may work best in dynamic fields such as technology firms. However, broad-based empowerment is not appropriate for FES organizations, which require direct supervision and adherence to rules and regulations for efficiency, safety, and consistency of the overall emergency response.

The only time when broad-based empowerment can be used effectively in FES or EMS is when employees are given a voice in decisions that might benefit the department, such as purchasing equipment, organizing training sessions, developing SOPs, or scheduling shifts. In these situations, acceptance of change increases with employee empowerment. As discussed in the sections on group decision-making, people who believe they are a part of the decision-making process tend to be more productive and committed to the change.

However, besides these types of decisions, direct hierarchical decision-making is necessary, especially in emergency operations. Independent decisions at the company level can reduce the quality of response provided and create chaos at large-scale incidents where there are specific expectations of the capabilities and operations of all companies involved. The individuals within a fire or emergency service often do not have a sufficient understanding of the larger picture to make appropriate decisions. For example, in one major metropolitan fire department, a manager empowered company officers to develop their own in-service training program after making a local assessment of the hazards in their first due areas. One officer in a rural area devised a yearly schedule of drills that did not include the use of fire department standpipes used in high-rise firefighting. In the officer's mind, this decision made sense because the first due area did not have any high-rise buildings. However, what the officer overlooked was that members of the company could be asked to fill in at another station that did respond to calls in high-rise buildings. By empowering company officers to make the decisions, this manager reduced the quality of the department's service and put firefighters into a situation where they were not adequately trained.

Broad-based empowerment can be equally dangerous in EMS situations. Processes and protocols are required for ambulance maintenance, patient care, and operational safety; there can be no flexibility. This is especially true when EMS, fire, law enforcement, and other agencies are working together at an incident.

Managing by Walking Around

Managing by walking around (MBWA) is a management technique in which managers physically visit all areas of the organization. This can be extremely helpful for the typical FES organization whose stations are physically separate from headquarters. The chain of command can still be used to send and receive information from the different units, but MBWA allows for direct supervision and increased communication, because managers could spend up to 60% of their time out in the field. Information can become distorted or purposely suppressed as it travels from bottom to top and top to bottom through the communications lines of a large and dispersed organization. This opens up the possibility of honest feedback from frontline employees that is very difficult to obtain for the boss. To be an effective manager, helpful criticism and suggestions for improvement are necessary and should be encouraged. Using MBWA, FES and EMS managers can visit their stations to speak with personnel about issues they are having, make firsthand observations at emergency incidents, and create a formal critique procedure. Managers can more accurately assess emergency operations from these firsthand observations. However, for the few very large departments or companies, this can be very time consuming, but still should be the goal of the top manager.

Many MBWA managers consider themselves facilitators who strive to listen carefully during their out-of-office visits for any signs of things that may not be going well. Managers can use their senior position to remove any roadblocks to enable members to accomplish their goals. In this way, MBWA sends a powerful message to members that the manager believes that the work being done is important to the mission of the organization.

MBWA can also be effective in dealing with supervisors, local elected officials, neighboring officials, and other department heads in government. Using MBWA, the manager becomes personally and professionally acquainted with these influential officials and can increase support and cooperation from outside agencies. This is a very effective management technique both inside and outside the department.

Assessing Managerial Performance

A manager's performance should be measured by the two key criteria of efficiency and effectiveness. This control process is used by senior administrators to judge the proficiency of an individual manager. Efficiency is measured by the ability to perform well using limited funds or resources, whereas effectiveness is measured by the manager's success in attaining goals.

Management effectiveness is directly related to the power, expertise, knowledge, ethical reputation, and personnel and professional relationships of the manager. Formal rank is the most obvious sign of power; however, power and relationships are codependent and

rely on the manager's ability to create an effective power base. Creating this support is one of the most important goals for a successful career in management. To do so, individuals can undertake the steps listed below:

- Show respect to everyone you meet and treat others like you would want to be treated. Sometimes just saying "I need help" is enough to get cooperation.
- Give people the benefit of the doubt; assume they are really trying their best.
- Get to know other influential managers or associations at local, regional, state, and national levels of your profession.
- Create good relationships with your neighboring departments and unions.
- Develop skills and information sources that make you a unique member, such as the go-to person to research issues, generate policy papers to support departmental goals, or know the right person that can help solve a problem. Reading and studying this text are information-gathering tasks that add to expert knowledge. One word of caution, however. Tone down your level of pride and enthusiasm for your own self-worth; others may be resentful.
- Seek assignments in all segments of the department to gain a broad perspective of the entire organization.
- Never think you know everything and never stop learning and listening.
- Seek out a mentor; mentors are invaluable for helping you prepare for career advancements, new challenges, and other opportunities.

Building a power base is a long-term project. Managers should seek out opportunities as they become available and always keep their eyes open for possibilities to learn something new, meet other managers, or embark on a new challenge.

For managers to increase their efficiency, they can analyze basic tasks, seek to eliminate unnecessary motions or tasks, and aim to hire the best-qualified applicants. One of the most important management functions is selecting new members who meet or exceed minimum job performance levels. One of the most valuable assets a manager possesses is his or her staff, because a manager cannot be expected to perform every task that makes an organization run. Delegation is key to increasing efficiency. To delegate effectively, you must clearly define the task for your subordinates, state the time frame in which the task should be completed, and follow up to ensure that the task was completed in a satisfactory manner.

For example, a chief officer sends the supervisors of each operations area a memo instructing them to check their snow chains. In the memo, the officer states a time for completion and requests a written reply that the task has been completed. The officer must record this deadline and set up a reminder system so that completion of the task can be verified within the appropriate time frame. Follow-up is absolutely necessary in all cases, and it should be done as inconspicuously as possible so that the members do not think they are viewed as untrustworthy. If the delegated task is a long-term or ongoing project, other management techniques, such as MBWA or leading by example, can be used to supplement a written report.

Many managers believe that multitasking allows them to increase their efficiency. It would seem to make sense that the ability to perform multiple management functions at the same time would result in increased productivity. However, this is not necessarily true. A Stanford study compared two groups: one that engaged in extensive electronic multitasking and another that did not. Several experiments were conducted to test the capability of both groups to pay attention, remember information, and switch from task to task. The results showed that multitaskers do much worse in all three areas than those who focus on one thing at a time. In fact, the more the participants multitasked, the worse they did compared with the focus group (Gorlick, 2009). Multitaskers may be easily distracted and unable to perform well.

Further, new studies conducted at Stanford University indicate that multitasking may damage your brain and career. Frequent multitaskers performed worse than those who were not frequent multitaskers because they had more trouble organizing their thoughts and filtering out irrelevant information (Gorlick, 2009).

Researchers at the University of London found that, in addition to slowing you down, multitasking lowers your IQ. Managers should consider limiting their multitasking to improve their concentration and focus on the present problems, including personal communication (USC, 2020).

Multitasking can also negatively affect communication. If you want to make someone feel important, give him or her your complete attention. When a manager is texting, emailing, or answering the telephone during a face-to-face conversation, it may be insulting to the other person in that conversation. The only interruption that should be allowed is an emergency call.

To be effective, managers must recognize that there is always room for improvement and should seek out opportunities for improving their managerial skills, such as self-study through books, courses, or seminars. They should also be open to receiving helpful criticism and suggestions for improvement. Nobody is perfect; everyone has room for improvement. A 360-degree performance review may be a useful tool to obtain this feedback from employees. This type of evaluation uses surveys sent to employees in all directions: those below the manager (who report directly to him or her), other managers on the same level, and the supervisor directly above the manager. Where this is not possible, feedback can be provided by other staff members or supervisors who have direct knowledge of the manager's work. Through this review process, managers may gain information they can use for self-improvement and uncover criticisms that they need to address. The results of this evaluation may be part of annual performance reviews, but should also be substantiated by the supervisor's own observations.

Wrap-Up

CHAPTER SUMMARY

- FES management and its application are explained by their need and use in the contemporary department.
- The first and a key function of management is Planning.
- The second is the Organizing function that requires a structured organization for FES.
- Thirdly, Directing has many of the same components as business and government.
- Fourth and last function is Controlling that is critical to successful management.
- Special organization tactics may be necessary to implement efficient and effective FES management.

KEY TERMS

360-degree performance review A type of performance evaluation that uses surveys sent to employees at all levels to obtain a more complete picture of a manager's performance.

Brainstorming A technique to generate ideas in which group members may suggest any solutions without initial judgments.

Broad-based empowerment Empowering low-level supervisors to alter or change policies or procedures to achieve a better outcome such as better quality or less time to complete the task. Once this is determined to be successful, it is assumed that the new or improved policies and/or procedures will be distributed to the entire organization.

Controlling Monitoring performance, measuring outcomes and progress toward goals, and making any adjustments necessary to keep the agency moving toward those goals.

Compromise To compromise is to make a deal through negotiations between different parties where each party gives up part of their demand.

Consensus Agreement reached by unanimous decision-making, a simple majority, or a specified majority.

Delegation Delegation is the process of handing decision-making authority to lower-level employees and to be successful, an employee must be able to obtain the resources and cooperation needed for successful completion of the delegated task.

Directing The process of a manager instructing the members under his or her authority to accomplish an organizational objective.

Effectiveness The manager's success in attaining goals.

Efficiency A manager's ability to perform well using limited funds or resources.

Group decision-making Occurs when individuals, typically those that are affected by the decision, collectively study and make a choice from the potential realistic options.

Group think When some members of a group go along with majority opinion to avoid conflict.

Intuition Intuition is commonly thought of as instinctual or even magical phenomenon, but is formed in our subconscious from the influence of past experience and cumulative knowledge.

Malfeasance Wrongdoing, including illegal actions or misconduct, especially by a public official.

Managing by walking around (MBWA) A management technique in which managers physically visit all areas of the organization.

Minority domination When a charismatic group member or several members representing a particular group disproportionately influence decision-making.

Misfeasance The performance of a lawful action in an illegal or improper manner.

Multitasking Performing multiple management functions at the same time.

Nonfeasance Failure to do what ought to be done.

Organizing The process of determining what tasks need to be done and who is to perform these tasks.

Planning Creating and defining goals and making important decisions as well as establishing strategic plans to achieve these goals.

Progressive consequences A disciplinary system requiring more severe repercussions for repeat violations or serious offenses.

Total quality management Constant process of detecting and reducing or eliminating errors in manufacturing, streamlining supply chain management, improving the customer experience, and ensuring that employees are up to speed with training.

Unity of command An organizing principle based on the idea that each individual reports to only one manager.

REVIEW QUESTIONS

1. Compare and contrast individual versus group decision-making. Which do you believe is more effective in the long-term? Provide a specific example from your own experience.

2. Detail at least three ways a manager can ensure compliance with his or her directions. Which of these methods would be most likely to work for you as an employee as a form of self-motivation?

3. The control process is made up of three steps: (1) _____, (2) _____, and (3) _____.

4. Select from the following list of emerging technologies and describe in about 500 words how your chosen technology could help the FES managers of the future:

A. Biometrics (e.g., facial recognition)

B. Virtual reality applications

C. Robotics and automation (e.g., drones)

D. Three-dimensional printing

5. List the pros and cons of a 360-degree review format.

CHAPTER ACTIVITY #1: MBWA

A safety chief for a large metropolitan fire department was in a staff vehicle traveling to a meeting when he heard a dispatch for a townhouse fire over the radio. At this time, the department had just started a comprehensive safety program. General orders covering safety policies and SOPs had recently been signed and issued by the fire chief. According to the SOP, the safety chief was dispatched only to second-alarm and higher emergency incidents, but he was in the neighborhood and decided to proceed in a nonemergency mode to observe the companies arrive and operate at this fire. About two blocks from the scene, the safety chief was passed by the first due engine. After it passed, he noticed a firefighter standing on the back step—a violation of the department's safety general order for emergency vehicle operations.

After arriving on the scene, the safety chief continued to watch the firefighters work at the townhouse, which had heavy smoke pouring from the front door and second floor windows. He observed one firefighter place a ladder to the front of the structure, climb the ladder, and start breaking out the windows on the second floor from the outside. This is a common ventilation procedure to remove heat and smoke from the structure. However, this firefighter, who was the driver for the first due engine, had on a helmet and no other protective clothing—another violation of the safety rules.

The fire was quickly extinguished by the first alarm assignment. After the fire was extinguished, the safety chief continued his trip to the meeting without speaking to anyone on the scene. The battalion chief in charge of this incident was a personal friend, so when the battalion chief returned to his office, the safety chief called to advise him of the observations. Formal disciplinary actions could have been initiated, but that would have breached the normal chain of command. As the two spoke, the battalion chief immediately volunteered that the firefighters had already admitted to the violations of the safety rules. The firefighters had recognized the safety chief and knew that they had been observed. As a result of this exchange, these individuals followed the safety rules more closely in the future. This commitment is a preferred result of MBWA—to gain compliance with rules and regulations while demonstrating support for the worker's safety.

Discussion Questions

1. If this behavior reoccurred, what actions should be outlined in the department's disciplinary procedures? Be specific.

2. Would a comprehensive education safety program help to prevent these types of SOP violations, or is strong discipline the only effective method? Or a combination?

CHAPTER ACTIVITY #2: GROUP EFFECTIVENESS

In most FES agencies, groups are commonly assigned to make decisions regarding controversial or change-based issues. However, some management experts doubt the effectiveness of groups to make good management decisions. For example, the management expert Robbins states, "Research comparing participatively set goals with assigned goals has not shown any strong or consistent relationship to performance…Groups almost always stack up as a poor second in efficiency to the individual decision maker" (Robbins, 2011). Still, others continue to defend the value of group decision-making.

Review the following resources, keeping in mind the comparative decision-making process of groups with the unilateral decision-making of the fire chief or other agency head. In 400–800 words, discuss the best way to determine the goals of your organization. A minimum of two reference sources is required.

- "Group Decision-Making" by Andrew E. Schwartz, https://www.questia.com/magazine /1P3-11347/group-decision-making

- "How to Help Groups Make Meaningful Decisions," https://managementhelp.org/groups /group-decision-making.htm
- "Group Decision-Making," https://en.wikipedia .org/wiki/Group_decision-making

CHAPTER ACTIVITY #3: CONTROL SYSTEM

One goal of the FES is to provide a professional level of fire suppression services to the public. As a management function, this service can typically be measured by looking at outcomes. However, as explained in this chapter and the chapter *Public Policy Analysis*, outcomes are hard to measure. Many of the measurements that are kept by the traditional FES organization are measurements only of workload, not of outcomes. For example, FES organizations keep accurate statistics on the number of emergency incidents to which they respond, but that does not show how many lives and how much property was saved, which are outcomes.

Assessing the number of lives and properties saved is very difficult, if not impossible. For example, if the department responds to a small trash can fire in a high-rise building and extinguishes the fire quickly, could the department claim to have saved the entire dollar value of the high-rise building?

Outcomes for a FES organization must be measured using other items that are easily and verifiably measured. For example, if it can be demonstrated that response times can be reduced by either changing existing procedures (e.g., streamlining the call-taking process) or adding new resources (e.g., increasing the budget for new staff or new facilities), the argument can be made that improved service to the public has been accomplished. Even in volunteer organizations, implementing a system to have members on duty in the station substantially reduces total response time, resulting in a better outcome.

Use the following resources to research measurement control systems:

- https://www.inc.com/bruce-eckfeldt/you-cant -manage-what-you-dont-measure-here-are-11 -ways-to-measure-any-aspect-of-your-business .html
- http://www.nfpa.org

Then, devise a measurement control program that ensures that EMS providers are proficient in the skills and knowledge needed to provide outstanding EMS.

CHAPTER ACTIVITY #4: POWER BASE CONTACTS

Management effectiveness can be directly related to power, expert knowledge, and relationships. Identify a "power base" that could increase your influence and effectiveness as a manager: use references of your choice to create a list of at least three people and/or organizations that a chief officer should network with for management effectiveness at the local, regional, state, and national levels.

CHAPTER ACTIVITY #5: TRUST

Trust plays a large part in the ability to be an effective manager and leader. One of the best ways to understand this concept is to look at examples. Review the following articles for examples and descriptions of trust:

- "The Supervising EMS Officer" by Tray Hagen, https://www.emsworld.com/article/10284162/leadership-best-practices-supervising-ems-officer
- "Mentoring Tomorrow's EMS Leaders" by Jay Fitch, https://www.jems.com/articles/2010/08/mentoring-tomorrow-s-ems-leader.html

After reviewing the articles, write a short story exemplifying four of the following eight items:

1. Practice openness.
2. Be fair.
3. Speak your feelings.
4. Tell the truth.
5. Be consistent.
6. Fulfill your promises.
7. Maintain confidences.
8. Demonstrate confidence.

These stories may be from your personal or work life and can be either positive or negative examples.

REFERENCES

Dewar, J. 1989. *Almost perfect isn't good enough*. Retrieved from http://www.shsu.edu/~mgt_ves/mgt481/lesson9/tsld029.htm

Gladwell, M. 2008. *Outliers: The Story of Success*. New York, NY: Little, Brown.

Gorlick, A. 2009. *Media multitaskers pay mental price, Stanford study shows*. Retrieved from http://news.stanford.edu/news/2009/august24/multitask-research-study-082409.html

Javitch, D. 2009. 5 employee motivation myths debunked. *Entrepreneur*. Retrieved from http://www.entrepreneur.com/article/202352

Judge, T. A., J.E. Bono, R. Ilies, and M.W. Gerhardt. 2002. Personality and leadership: a qualitative and quantitative review. *Journal of Applied Psychology*, 87(4), 765–780.

Page, J. 1999. Your 21st-century firefighter. *Fire Rescue*, 17(11), 10.

Robbins, S. P. 2011. *Fundamentals of Management: Essential Concepts and Applications* (9th ed.). Upper Saddle River, NJ: Prentice Hall.

University of Southern California (USC). 2020. *To multitask or not to multitask*. Retrieved from https://appliedpsychologydegree.usc.edu/blog/to-multitask-or-not-to-multitask/

CHAPTER

Leadership

LEARNING OBJECTIVES

After studying this chapter, the student will be able to:

- Understand the importance of progressive change in fire and emergency service organizations (pp 60–61).
- Recognize the influence the chief administrator has on the likelihood of change success (p 61).
- Describe the steps to create change (pp 61–69).
- Explain the need for a sense of urgency in the change process (pp 61–63).
- Describe the ideal characteristics of a guiding coalition (pp 65–66).
- Describe techniques that can be used to communicate a vision for change (pp 64–65).
- Know the roadblocks that can defeat efforts to accomplish change (pp 66–68).
- Explain the importance of short-term goals within a long-term project (p 68).
- Summarize the steps to institutionalize change within an organization (pp 68–69).

Case Study

Institutionalizing Change

The chief officer of an emergency medical services (EMS) agency decided to transition to a staffing model that requires an ambulance staffing configuration of one paramedic and one emergency medical technician (EMT), which was a change from the normal of two paramedics. This proposal would result in reduced costs for the EMS agency because an EMT's salary is less than a paramedic's as a result of the less demanding certificate process. This staffing model made sense because the local fire department staffed their apparatus with firefighters/paramedics and could be used for calls when more than one paramedic was required for treatment or transport.

However, a few ambulance paramedics expressed discomfort with having an EMT for a partner, believing that EMTs would not have the skills required for advanced life support partner collaboration. The EMTs were also hesitant, because they only had experience with non-emergency transfer work, and expressed reluctance to accept additional responsibilities and increased expectations. Local politics and attitudes complicated matters, because the relationship between the public (fire) service and private (EMS) agencies was strained. The leader recognized that the change would be difficult for fire and EMS agencies, but knew that the economic situation required a more efficient and effective system of emergency response. The EMS chief was confident that with proper education and training, all involved would be comfortable in their new roles.

Discussion Questions

1. In preparation of an explanation for the need, including expressing a sense of urgency, what information should be included?

2. Who should be contacted to explain the justification?

3. What should be included in a plan to overcome any resistance by the paramedics?

4. What should be included in training and educating employees?

Embracing Change

Change is inevitable; it is also a necessary condition for a business to survive and prosper. To stay competitive, businesses must constantly change to meet customers' expectations. Businesses change to reduce costs, improve the quality of products and services, locate new opportunities for growth, and increase productivity. Change for an organization has many forms, including restructuring, reengineering, transforming, acquiring, merging, innovating, modernizing, downsizing, rightsizing, quality improvement, and cultural renewal.

For public service agencies, change is driven by the need to improve service. In this sense, progressive change may take the form of additional staff, expanded services, improved equipment, higher standards of training, and education for personnel. A good example of this in the fire and emergency services is the growing number of fire departments providing EMS.

Some members may be resistant to change, so leaders must educate members and convey the importance of change. Leaders understand that not to change is to remain stagnant. Public agencies can continue the status quo for many years without any outward indication that change is needed; in many cases, the public, elected officials, labor organizations, and fellow managers may not be aware that changes are needed to improve the quality or efficiency of service. However, the quality of service, equipment, training, strategy, tactics, and management practices suffer if a fire or emergency service does not continually seek and embrace change. Effective leaders must be willing to propose changes to correct any shortcomings in services. Leaders must recognize the ability of their organization to adopt change without forcing too much change too quickly, which drives people out of their comfort zones and promotes resistance.

Leaders must be patient and persistent. They should communicate the vision that justifies the change to the organization's members, the public, and government officials through words and actions. Permanent change does not occur until most of an organization's members believe in the new vision. To secure change, a leader must allow sufficient time to ensure that the next generation of officers embodies the new vision. When appropriate, one adoption technique is to make the change a requirement for promotion (e.g., minimum education standards). If implemented slowly, the change can be met by all members interested in being promoted, thus increasing acceptance and perceived fairness throughout the organization. In this manner, change initiatives encounter less resistance and have an increased chance of survival.

Traits of Successful Leaders

Truly **successful** leaders will have the upper hand over the doubtful and the unsure because they inspire others to follow them **FIGURE 4-1**. The following list represents those traits that result in the confidence a leader needs to be successful:

- They get their happiness from within.
- They don't pass judgment.
- They are effective communicators with both verbal and written messaging.
- They don't say yes unless they really want to.
- They listen more than they speak.
- They speak with certainty (this requires expert knowledge in the subject).
- They seek out small victories.
- They don't seek attention.
- They aren't afraid to be wrong [and freely admit mistakes].
- They celebrate other people.
- They are not afraid to ask for help (Bradbury, 2015).

And finally, they lead by example, recognizing how their actions influence others.

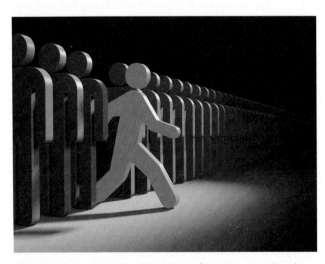

FIGURE 4-1 A leader will stand out from the crowd and can be identified by those they lead.
© Palto/Shutterstock.

Creating Change

Leaders at any level can help pave the way for progressive change. A newly appointed leader can create an enlightened climate by setting the tone from the start, helping members to think of change as normal and expected. More experienced leaders can use a national or local incident to take the organization on a more progressive course, especially a safety issue. The first time a leader institutes a new change, it may help to start with an issue that is meaningful but will not solicit strong opposition. Tradition in fire and EMS services is strong, and transformation may happen slowly. A leader cannot force change and expect that it will be accepted in all cases. The leader must follow a set of processes to create progressive organizations and allow them to adapt to changing circumstances and environments. Although changes can be implemented by issuing a new standard operating procedure (SOP) or order, to be completely successful at making the change last in the future, change needs to be applied using a comprehensive, inclusive plan.

After change is identified from the vision, leadership must take over to produce permanent change that can be successful in transforming an organization to provide better service. The following steps help a leader create change:

1. Identify the need for change and create a sense of urgency.
2. Develop a vision.
3. Communicate the vision.
4. Create a guiding coalition.
5. Overcome barriers and resistance.
6. Create short-term goals.
7. Institutionalize change.

Step 1: Identify the Need for Change and Create a Sense of Urgency

To identify the need for progressive change, a leader should perform or make arrangements for a group to perform a policy analysis as discussed in the chapter *Public Policy Analysis*. The results of the policy analysis would identify the problem and the need for change, assess the urgency, and recommend potential solutions.

With this identified change, the leader must be able to convince approximately 15% of the members that this is a good idea before formally starting the change process. This core of believers is necessary because some changes require adaptations of tradition and may require sacrifices. Further, to create a sense of urgency, the cooperation and willingness of a small group of core members is needed to support the change initiative. In a public service agency, the support of elected and appointed officials is also critical; the same approximately 15% support rule also applies to them. The change leader needs supporters above and below them in the organization to support the change.

However, mandating change is always possible when implemented with governmental regulatory power. In some organizations it may be possible to gather the support of a core group simply because of the members' loyalty to either the organization or a charismatic leader. Avoid falling into this trap of **blind loyalty**; make sure the change justification is based on facts and the core members' support is real.

A chief officer (leader) is easily identified on an emergency incident (see **FIGURE 4-2**). However, a leader for progressive organizational changes takes another set of aptitudes and knowledge along with some risk taking.

No matter how hard the change leader pushes, if others do not feel the same sense of urgency, the momentum for change will probably die far short of the finish line. In a paramilitary organization it is always possible to force change by issuing a new SOP, but if the SOP is not based on some credible justification supported by a guiding coalition, then the change may only be temporary until the next administrator takes charge.

Unlike for-profit businesses that must respond to economic pressures and stockholder demands, fire and EMS services often receive fairly consistent funding from tax revenues. It is common for the following subliminal message to be present: we are winners and must be doing something right, so relax and enjoy the status quo. As such, the natural economic forces that create a sense of urgency in the business industry are not present in government agencies. Still, factors that may

FIGURE 4-2 FDNY chief confers with other fire officials on scene of residential fire in Brooklyn neighborhood of Gowanus in daylight, May 2014.
© Andrew Katz/Alamy Stock Photo.

CASE STUDY

Identifying Change: A Safety Example

In Montgomery County, Maryland, the next time you call 911, you may not see red lights or hear sirens when fire and rescue personnel respond to your call for medical assistance. In addition, after assessing your medical issue, you might not be taken to a hospital but to an urgent care facility by non-emergency transport.

This new policy (2020) is called "cold response" and can be safer for EMS personnel, community residents on the roads, and patients. It means emergency vehicles won't force other drivers out of the way to respond to your location or when transporting to the medical facility. This change will only be used for non-critical patients and is more proficient by providing a less expensive medical treatment.

Finally, this new program also keeps people with less serious injuries or ailments out of busy hospital emergency rooms, leaving more room and beds for those who need them most.

© Smeby Jr, Charles L/Jones & Bartlett Learning

---

influence a fire service leader to instigate change include cuts in the budget that may cause salary or benefit reductions, layoffs, or the closing of fire stations or companies; lack of funding to replace old or unreliable fire or emergency medical apparatus; or some other threat to the membership. Recent shocks to the U.S. economy are pushing some fire and EMS organizations to reduce expenses, but they still lack the competitive influences of a market economy driving them to change. In the business world, where there is competition and free entry into the market, competitors are very quick to point out their better service, price, or product.

Because of this, many of the situations that a business leader can use to create a sense of urgency are not available to the fire and emergency services organization. This makes the leader's job more difficult, but not impossible. The chief officer may have to look very carefully and have the patience to wait for the opportunity to make a change. This opportunity may unfortunately come in the form of a crisis (e.g., a national catastrophe or death of a local emergency member) that ultimately catches people's attention and increases the level of urgency.

In the public arena, it pays to be prepared for these tragedies. Have a plan and justification prepared. These are opportunities to pump up the urgency for change. For example, procuring needed equipment and personnel, funding training sessions, or creating mutual aid agreements may be easier after an earthquake or hurricane. In one case, the State of Florida's fire service identified a major problem in the ability of mutual aid fire companies (also, communications with other agencies, i.e., police, emergency preparedness staff, air support, Department of Transportation [DOT], utilities, public works, National Guard, etc.) to communicate by radio during several hurricanes and major wildfire incidents. As a result of these catastrophic tragedies, the state funded portable radio equipment that could provide compatible radio communications using portable handheld radios (100 given out to local responder companies) and a portable base station with a 100-foot crank-up tower FIGURE 4-3.

Another example of a tragic event that resulted in major change was the unprecedented terrorist attacks on the United States on September 11, 2001. Among the many resultant changes within the federal government, federal grants for the fire service that were slated to be eliminated from the federal budget are now being funded again. It was agreed on by many fire service administrators that more resources were needed for first responders at everyday and

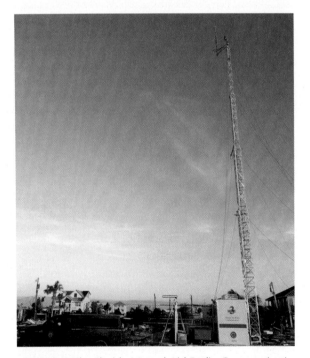

FIGURE 4-3 The Florida Mutual Aid Radio Communications (MARC) units are emergency deployable radio towers and radio cache that can be deployed into an impacted area to enhance or replace existing communications systems, providing interoperable communications systems for emergency responders. In this photo, MARC 3, hosted by Alachua County Fire Rescue, is deployed to Mexico Beach, Florida in response to Hurricane Michael, a category 5 storm that caused extreme damage. MARC 3 provided radio support for multiple USAR teams as well as engine strike teams and rescue strike teams operating in the area.
Courtesy of Kevin Rulapaugh.

catastrophic emergencies. These administrators then pulled together their visions and collectively applied pressure on their federal legislators, ultimately receiving the help they knew was necessary. Assistance to firefighter grants for fire prevention and safety and staffing for adequate fire and emergency response are available from the U.S. Department of Homeland Security, and in FY2019 totaled $700 million, which has not substantially increased over the last 10 years.

However, waiting for a tragic incident to initiate change is an unpredictable leadership strategy. Raising the public's interest during good times may not be easy, but it is certainly possible. Strong leaders can create awareness, helping people understand the rationale for the change without experiencing real losses. Leaders should help others understand the possible consequences of having inadequate staffing, insufficient training, or out-of-date equipment and the effect these shortcomings could have on the quality of service provided. In this way, the leader can create a sense of urgency through a hypothetical crisis rather than a real one. Members, elected officials, and the public need to be convinced that change is necessary to solve a problem or improve service.

Step 2: Develop a Vision

A comprehensive vision is one of the more important elements of a successful progressive change. Vision plays a key role in change by helping to direct, align, and inspire the actions of large numbers of people. A vision is like a personal goal derived by highly experienced and educated professionals.

A good leader has a clear vision of where they want to take the organization in the future. This vision is a result of the leader's formal education, intuition, common sense, and professional training, coupled with years of experience, which allows him or her to envision progressive changes that they would like to implement. In most cases, the leader is the source of the change proposal. It may not be an original idea, but it is their vision of the idea that is first proposed and communicated to the guiding coalition or to members and elected officials.

When a change vision becomes clear, the leader may feel uncomfortable because it may cause some pain to other members of the organization. In the typical fire and emergency services organization, new members start at the bottom and are promoted slowly to the top, making many friends along the road. Although the leader may not want to jeopardize these friendships, they must be confident in and have the courage to stand by their vision, even if it goes against tradition or has the potential to disrupt personal or professional relationships.

Step 3: Communicate the Vision

To achieve organizational goals, a leader must be able to communicate their vision to the members of the organization. The vision should be clear and concise to make this possible. Overly complex visions are not easily understood. Therefore, before attempting to communicate your vision to others, you may wish to write it down and have a person outside the field of fire or EMS review it. If someone with very limited knowledge or experience in the field can understand it, then members of your organization should be able to understand it as well. Remember that your vision will also be shared with government officials, the media, and the public, so keep it simple and straightforward. In most cases of major change, the approval of elected and appointed government officials is required and they must be convinced of the validity of the change.

Explaining your vision should take no more than 5–10 minutes. An analogy may be a useful shortcut to help others understand. Effective leaders are almost always great written and verbal simplifiers. For example, to illustrate the need for adequate staffing, it may be useful to compare the organization to a sports team (e.g., "Imagine trying to win a basketball game with only four players—rather than the customary five—on the court."). As another example, to relate the need to purchase a new ambulance, one may wish to make an analogy to the familiar need to replace a car that has too many miles and years of use. By using analogies, you have a simple request and clear justification for your vision that others can understand. It is also important for a leader to communicate the vision by example, serving as a role model for the steps necessary to achieve a set goal.

Written and verbal communication can be helpful to communicate one's vision. Most people do not internalize anything the first time they are exposed to a new idea or message. To counter this limitation, common techniques used in the advertising field are to use **repetition** and multimedia to communicate the desired message. A common rule of thumb in marketing is to expose the audience to the message at least three times, but preferably from six to twenty times.

For example, a new SOP on emergency driver responsibilities lists several mandatory rules for safety. After researching state motor vehicles law and actual crash experience, and consulting with department lawyers, your vision to reduce department crashes could be to educate and enforce rules that require drivers to use due caution in their speed and dangers of proceeding through intersections against a red light. Your presentation to the department members

FIGURE 4-4 Crash at intersection going through a red light.
© Gordon M. Grant/Alamy Stock Photo.

may include a Power Point presentation with several actual pictures of crashes (see **FIGURE 4-4**). This is also the appropriate time to specifically list the new driving rules using a Power Point slide and handouts.

Creating SOPs is an effective method to communicate a vision. SOPs are often a response to real and

CASE STUDY

Leading Change: A Safety Example

A fire rescue department issued driving regulations that contained a requirement that all emergency response vehicles must come to a complete stop at all red traffic signals and stop signs (as per National Fire Protection Association [NFPA] 1500 standard) during emergency responses. Each paramedic unit was equipped with an automatic speed recorder. A crash occurred one day when a paramedic unit failed to stop at a red traffic signal, striking a car that was proceeding through a green light.

The "stop at all traffic control signs and signals" regulations had been issued about 3 months prior, but had never really been enforced. It was common for many apparatus drivers to proceed through the stop signs and signals without coming to a complete stop. When these circumstances were relayed to the jurisdiction's lawyer, he recommended eliminating the driving regulation, stating that if the department was not going to enforce the regulation, it would make the jurisdiction more liable.

However, the chief was able to persuade the attorneys that the safety regulation was imperative and a plan was implemented to enforce the rule. By identifying a flaw in the department's management and deciding to enforce the safety rule, the chief communicated strong support for this safety change. This is genuine leadership.

perceived safety issues and the need for consistent, higher quality service. These are items that can cause liability exposure if not addressed by the department. For example, if there is a failure to provide for the safety of the members or to give adequate service to the public as measured by nationally accepted professional standards, there is an increased liability exposure.

Once committed to a change vision, the leader must monitor every detail of execution for signs or behaviors that are contrary to the new way of doing things. Watch for inconsistencies and identify them immediately. Again, the details can kill the proposed change. For a change program to work, it must have a high level of credibility and support from a substantial number of members, the public, and elected officials.

Finally, the level of urgency needs to be high enough to motivate the leader to find the time needed to effectively communicate the vision and for the organization to complete the transformation. Relevant government rules and standards should be referenced in written communication to relate the urgency for change. For example, if a fire department needs to purchase replacement protective clothing, the leader should reference the requirement to comply with NFPA 1971, *Standard on Protective Ensembles for Structural Fire Fighting and Proximity Fire Fighting* for the safety of firefighters.

At times, the chief officer may need to tackle more controversial changes, such as those relating to standards of training, incident command, staffing levels, and medical and physical fitness standards. These are admittedly aspects of change that can be costly or draw resistance. However, most of these changes can be phased in over time (as noted in the NFPA 1500 standard), which drastically reduces the potential impact and resistance, including any new financial requirements. Fostering change can be a time-consuming process and leaders need to keep a close eye on the progress to ensure the effort is on schedule.

Step 4: Create a Guiding Coalition

Why can't the chief officer just order a change and have everyone comply? The answer to this question is found in the nature of organizations and the people affected.

A strong guiding coalition, sometimes called an "advisory board," of which the "core of believers" should be prominently represented, with the right composition, level of trust, and shared vision is one key to successful change. Depending on a specific change issue, other members may be recruited from other agencies or even the public. The coalition has the responsibility to recommend implementation of a particular change and monitor progress toward that goal. Coalition members should understand the need for change and agree with the sense of urgency **FIGURE 4-5**. For example, one contemporary issue involving change in many fire and emergency services organizations is the implementation of a physical fitness program. The facts that support this type of program are very compelling and based on solid research. Reviewing the Firefighter Life Safety Initiatives from "Everyone Goes Home," sponsored by the National Fallen Firefighters Foundation, reveals a major goal to "… implement national medical and physical fitness standards" (https://www.everyonegoeshome.com, 2020). In addition, from a common sense perspective, having members in good physical fitness is good for their health and provides an emergency services worker who can perform at greater levels of strength and endurance. To implement this change, the leader would want to identify those people who could be appointed to the physical fitness committee (guiding coalition). One technique that can be successful in identifying committee members is to implement a pilot program and ask for volunteers. Any volunteers are good candidates for the guiding coalition.

The committee members, who are selected by the chief officer, should have position, expertise, credibility, and diverse perspectives. Members should be optimistic, confident, competent, and results-oriented. They should have the courage to say what is on their minds even if it may be unique or could hurt somebody's feelings. They should be dedicated to providing quality service to the public.

All members of a guiding coalition must be committed to achieving the same goal for real teamwork to be possible. Team-building events and exercises such as off-site retreats may be useful, especially in

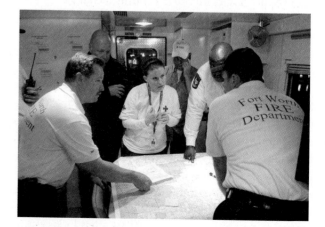

FIGURE 4-5 Guiding coalitions should embody expertise, credibility, and leadership.
Courtesy of Glen E. Ellman.

FIGURE 4-6 Team-building events and exercises may be useful for creating a unified sense of purpose.

© James Scott Stewart/ShutterStock.

large organizations, for creating a unified sense of purpose and priorities among coalition members **FIGURE 4-6**.

Some opponents may claim that guiding coalitions are stacked with "yes members." However, the purpose of this committee is to help implement the change, and without committed supporters it would not function properly and may fail. As an overall strategy, the leader could make a potential change list, prioritize the items on it, and then choose one item that is major, clearly justified, and easier than others to achieve. Creating a new tradition in the organization that expects and looks forward to change helps facilitate future changes.

Step 5: Overcome Barriers

A leader must work to overcome barriers to progressive change. Common barriers include tradition, lack of resources, insufficient staffing, and general resistance to change from members, government officials, or even the public. Before instituting a change, a leader should develop a list of potential barriers and devise possible solutions for overcoming them. For example, as discussed in the opening case study, many fire departments have added EMS to their scope of services. This change requires most of the members of these fire and emergency services organizations to achieve some new level of training, and certification such as emergency medical responder, EMT, advanced EMT, or paramedic certification. The barrier in this example is the resistance of members to this substantial change. Although each situation is unique, fire and emergency services organizations have offered benefits, such as incentive pay and paid time off to attend classes, to offset the sacrifices necessary to attain the new levels of job performance.

In the case of many volunteer organizations, these sacrifices are not offset by any monetary incentives. Because volunteers are very dedicated to their organization and the public they serve, they try to fulfill the requests of their administrators despite the sacrifice. However, it seems that the new EMS demands on volunteer fire and emergency services organizations may be resulting in a reduction of members. Increased call volume, services, and training requirements are the main cause of this adverse impact, and is an example of an unintended consequence of change.

The most important step of overcoming barriers is to identify them. At times, the barrier may be invisible to the organization and its administrators because it has been institutionalized into the organization's tradition. Therefore, the guiding coalition may prove to be an excellent resource for researching, identifying barriers, and developing possible solutions. After the barrier is identified, convincing members, elected officials, and the public that it is really an impediment to the new progressive vision becomes the paramount task. If the barrier is clearly defined, generally its effect on the change becomes apparent. Identifying the problem is always the hardest part of any change process. Looking at the experiences of other fire or EMS services also can be a useful approach. Great ideas can be found in small and large organizations, including those staffed by volunteers, career, and combinations of the two. Remember, what one department considers a barrier may already have been overcome by adopted practices or procedures in another department. Citing the experience of another department can be very helpful in selling a proposed solution.

One type of barrier that the fire and emergency services organization may encounter is that of existing tradition. Local structural barriers vary from department to department. Generally, some of these barriers are legacies of how things are done in a close major central city; the city's procedures become normal throughout the surrounding suburbs, even though times change. For example, when many central cities first created their paid fire departments, labor was relatively inexpensive. They were able to hire and staff their companies with large crews and many stations. Today, however, this is not normal, given the high cost of career firefighters and EMS personnel.

Another barrier that occurs in fire and emergency service organizations relates to supervisors who oppose change. Supervisors to a large degree are a product of their experiences. Some do not like change and may see the change as a personal attack on their safe, predictable, and comfortable environment. These supervisors may be very good at taking command and

CASE STUDY

The Barrier of Tradition

Two fire departments share a common border. One department is completely paid and the other is a combination department. When dispatching fire companies, the paid department expects that each company will respond without delay, staffed by a minimum of four firefighters per company. This is the typical organization of many completely paid fire departments.

On the other side of the border, the combination department was created by consolidating several volunteer departments into one county fire department and hiring additional paid members. Generally, the paid firefighters are assigned to stations to offset periods when an adequate numbers of volunteers are not available. When alerted to a call, this department has no guarantee of the number of volunteer members available, the time needed to respond, or even if they will respond at all (although the latter would be rare in most cases). Although a growing number of volunteer organizations have recently been requiring on-duty staffing by their members, this department does not require their volunteer members to spend on-duty time at the station or to be on call in the neighborhood, because this was never an expectation (tradition) of the original volunteer departments. Although a need for paid personnel has been acknowledged, the department continues to maintain as it always has with no guarantee of standard dispatch of companies and staffing.

Once this barrier to adequate response to emergency calls is identified, the solutions can be proposed and initiated. One such solution is to combine the paid personnel into independent companies and use the volunteer companies for extra resources. So the dispatch for a house fire might be three paid companies and two volunteer companies, for example. If the volunteer companies do not report by radio that they are responding within a reasonable period of time, additional paid companies would be dispatched. This would meet the two goals of immediate response and an adequate number of firefighters for effective and safe operations to extinguish the fire.

managing emergency situations; however, they can present an enormous obstacle when major changes are pursued. Easy solutions to this sort of problem are difficult to find. See the next section on Resistance for more information.

One should also look for interdependent systems that require additional changes. Whenever a new item is found that needs changing to support the original change, this insight helps solidify the original change. This process is called "facilitating," and the leader should be the main person to help make possible the change using their influence. When new obstacles are discovered, the leader must help find new solutions to overcome the obstacle.

Resistance

Change can result in fear of losing some existing benefit or privilege. It also threatens investment in the status quo (tradition). Leaders frequently encounter resistance to change from members inside the organization.

Members may be comfortable with the existing policy and have difficulty veering away from tradition, or they may fear that the change is not good for them personally or the organization as a whole. Because of this reality, one must be careful when celebrating or rewarding short-term wins. Although the opposition may acquiesce to the new change, there is the possibility that they might reassert their resistance at any time. Until the changes have truly been institutionalized into the organization's culture, which may take many years, permanent change can be very fragile (see Step 7 for more information about culture).

The longer tenure a person has, the stronger their opposition may be. For example, change may conjure up fears of losing status, money, authority, friendships, personal convenience, or other economic benefits. Change expert Lane Wallace (2009, 52) explains why some people cling to an opinion even when faced with overwhelming opposing substantiated proof:

> Social science researchers have found that people employ "motivated reasoning" to fend off evidence that their strongly held beliefs are wrong. Many people feel that they are their opinions, and hate to lose arguments. So, when confronted with new, troubling information, ideologues selectively interpret the facts or use "contorted logic" to make the conflicting evidence just go away.

In this sense, resistance can be associated with a personality trait and, therefore, may be near impossible to change. It should be expected that there will always be some members that oppose the change.

Even when the change is not in the members' short-term best interest, the leader should be able to explain the benefits and personal fulfillment from making the shift and the negative consequences that would emerge without the change. Leaders can help reduce members' resistance by:

- Starting a comprehensive education and communications program.
- Incorporating members into the decision-making process.
- Facilitating meetings where members can voice their opinions and concerns.
- Encouraging members to deal appropriately with apprehension and stress associated with the change.
- Rewarding members' efforts to overcome resistance and embrace change.
- Consistently and vigorously enforcing the rules and regulations that define the change.

When these efforts, combined with inspirational leadership and creative incentives, are not enough to persuade a member, discipline may be necessary. The final stage of the disciplinary process is separation from the organization. Faced with this consequence, most members comply given a reasonable amount of time to conform. Be sure that you do not rely on voluntary compliance or institute rules that are not enforced; this invites failure. Even in the largest organization, disciplinary action may only be necessary for a few members; word spreads quickly that the leader means business. Although the leader should primarily aim for voluntary compliance, disciplinary enforcement may be necessary to ensure full compliance.

This also may hold true with managers. A leader should approach resistant managers and try to convince them to buy in to the new vision. If the supervisor does not provide voluntary compliance, the leader should issue a written warning against interference, insubordination, or noncompliance. After that, disciplinary action may be necessary. Although disciplinary action may be possible, separation from the organization may not be an option because of union agreements. Think through your plan and consult with the organization's attorney before implementing disciplinary action.

Sometimes special interest groups, such as labor unions and volunteer associations, also resist change. If influential supervisors or groups are not involved early in a change process, they can undermine the entire effort. For powerful organizations, placing representatives on the guiding coalition may take care of some concerns or at least temper resistance. By including representatives as part of the process, they are exposed to the justification for the change, making it more difficult for them to strongly oppose it. Consensus-building techniques can also be very helpful in these situations.(see Chapter 12, *Policy Analysis, Consensus Building*).

Step 6: Create Short-Term Goals

While working toward long-term goals, a leader should have intermediate checkpoints to periodically monitor progress. For example, a person who is working toward significant weight loss by dieting and exercising can weigh in and celebrate each new pound that is lost. These short-term goals are necessary to carry the momentum for a long-term project. This is especially true with any behavioral change. For example, if an EMS service has adopted a goal of being more customer-oriented, then positive behaviors exemplifying good customer service should be encouraged, recognized, and reinforced.

Short-term wins are also very effective at keeping the urgency at a high level. They can be incentives, such as recognition, money, or both. If these positive incentives do not work, then disciplinary actions, following the standard local procedures, should be taken at these short-term benchmarks for those who fail to comply.

Short-term goals should have the following three characteristics: (1) visibility to the entire organization, (2) clear and unambiguous outcomes, and (3) outcomes that are a direct consequence of the change. It has been observed by change experts that "Most people won't go on the long march unless they see compelling evidence within six to eighteen months that the journey is producing expected results. Without the short-term wins, too many employees give up or actively join the resistance" (Kotter, 2012). It is important for leaders to have patience and perseverance on the path toward completing their long march.

Step 7: Institutionalize Change

For a change to become permanent, the underlying culture of the organization must adapt. The culture of an organization is very difficult to address directly because it becomes an invisible part of the normal day-to-day operations and has a powerful influence over the behavior of members who are expected to conform. In most organizations, individuals are selected and trained to meet the expectations of the existing culture. Shaping of a new member's behavior, attitudes, and values is done routinely by the incumbents without a conscious effort, reinforcing the existing culture. However, there are steps a leader can take to overcome this barrier.

First, the leader and guiding coalition should discuss the rationalization, justification, and evidence

AIDS and the Change to EMS Culture

In the early years of EMS, there was little credence given to disease transmission and cross contamination. Field responders were often covered in the blood and bodily fluids of their patients, and they often wore little if any protective gear. AIDS changed the landscape and practices of health care. When the first case of AIDS was reported in 1981, there was little information about how this virus was transferred, but there was a fear that contact with an infected person's blood would lead to certain death. The fact that the disease had a long incubation period left many responders wondering if they had already been exposed.

There was a lot of uneasiness as providers continued to care for patients during the massive research effort to understand this disease. In 1992, Occupational Safety and Health Administration (OSHA) standard 29 CFR 1910.1030 (Bloodborne Pathogens) became effective for all employers, with employees "reasonably anticipated" to have occupational exposure to blood and other potentially infectious materials. This OSHA standard required employers to determine exposure, create an exposure control plan, perform annual awareness training, and implement engineering and work practice controls. As a result, EMS developed a new standard of patient care and saw a dedicated effort by employers to meet the new mandates.

Initially, there was reluctance to change the culture of EMS, even in the face of potential death. Some providers thought the gloves were annoying and refused to use them. Some agencies were slow to adopt work practice controls until they were mandated. As new providers joined EMS, they were trained to a new, safer standard, but some of the seasoned EMS providers believed the new requirements were unnecessary. More recently, this focus on contamination from patients will be amplified by the coronavirus pandemic. It should be noted that there is no vaccine for HIV as is also true for COVID-19 at this time. Eventually, providers began using latex gloves, cleaning contaminated equipment, wearing HEPA respirators, and using safer sharps equipment on a regular basis.

FIGURE 4-7 One important cultural value is physical fitness.
© Glynnis Jones/ShutterStock.

bring on new hires that represent the values of the new culture and to promote members who are open to the new changes **FIGURE 4-7**. Throughout the process, the leader should carefully monitor progress across the organization to determine the implementation's success and the need for any corrections. By listening to members and practicing management by walking around (discussed in the chapter *Management*), a leader can identify problems and make appropriate corrections.

After a problem has been identified with the change initiative, correct it as soon as possible. Admit the mistake, unintentional consequence, or misunderstanding before others force their own modifications. Do not wait to address cultural barriers until forced to by the opposition. They may use these errors to bring down the entire change project. If the opposition identifies problems and forces modifications to the change program, the status quo forces will have won a battle, and they will be harder to defeat the next time.

Finally, the leader should lead by example, exemplifying the new cultural norms. Cultural change is the final step in institutionalizing a change. The true test of a leader may occur only after he or she leaves the organization; if the new practices and vision remain and the organization does not regress to past practices, then the full transformation has been completed.

Leading Change at All Levels

Even if you are not the administrator of the organization, you may be able to lead an effort for change. However, if you are presently in a role of middle- or

that support the change, recognizing how the culture has served the organization and acknowledging that its traditions are no longer efficient, safe, or useful in the context of this change. The leader also should aim to lower-level management and having a vision that requires change to accomplish, you may wish to keep these ideas to yourself or limited to those associates and friends who will not be threatened by the thought of change. In some organizations, being known for new ideas that threaten fellow members or superior officers may cause you to be overlooked for promotion.

Still, this does not mean compromising personal professional standards. A future leader must be careful to seek out those who are open to change for support. You may wish to start by finding a senior supervisor who would be open to the change and meeting with them privately to explain the need for the change. Depending on the topic and size of the organization, the leader may form a guiding coalition of supporters to propose a pilot program and test the change. Recognize that approval may have to come from the senior officers, lawyers, elected officials, and member groups.

In many cases, the most effective place for leading change is at the top. That said, even the chief officer may be limited by elected officials, city managers, or, for volunteer organizations, those who elected the officers. Further, it is not unusual for employee organizations to have influence. By following the change step process, leaders will have their best chance of leading an organization through significant change.

Leadership is truly a lifelong learning project. John Kotter, the author of *Leading Change*, points to five actions any leader can take to successfully lead change: (1) take risks, (2) self-reflect, (3) solicit the opinions of others, (4) listen carefully to others, and (5) be open to new ideas (Kotter, 2012). The more knowledge you have, the easier it is to find suggestions and ideas to overcome opposition. Ask questions and reach out for the opinions of others.

Organizational change is never easy, and you need plenty of patience to follow the seven-step change process outlined in this chapter, but progressive change is the only avenue to creating a truly competent professional fire and emergency services organization. Keep in mind that the process begins only after the problem has been identified. Be very careful to take the time and effort needed to ensure that your analysis and its conclusions are supportable when challenged. This is the hardest and most critical: Step 1: "Identify the need for Change and Create a Sense of Urgency." Again, patience and perseverance are the keys to success.

Wrap-Up

CHAPTER SUMMARY

- Change is expected and unavoidable for a progressive organization.
- Leadership is the key to any changes needed to achieve being a progressive organization.
- The leader must be able to have a clear vision of the progressive change and a valid justification for it.
- A methodical seven-step process is explained that will give the leader the best possibility to implement change successfully.
- Step 7: Institutionalize Change is challenging but critical to completing the change.

KEY TERMS

Analogy A comparison between something very familiar and a new concept, idea, situation, etc. This helps the reader quickly and accurately understand the new idea.

Barriers Obstacles to change.

Blind loyalty This behavior is seen in close acquaintances' of a leader, manager, or administrator. They are often seen supporting all actions even when strong evidence indicates the decision or information is wrong. Sometimes these individuals are known as "yes men"

and can be recognized as those that agree with everything that is said by the supervisor.

Change The process of altering, modifying, or making different an idea, device, or policy.

Checkpoints Intermediate, short-term goals implemented to periodically monitor progress toward a long-term goal.

Confidence A sense or consciousness of reliance on circumstances, faith, or belief that one will act in a right, proper, or effective way.

Core of believers 10% to 15% of the members of a given group who believe that a proposed change is a good idea.

Culture A set of shared attitudes, values, goals, and practices that are found in an institution, community, or organization generally unconsciously accepted by the members. Many times identified as "tradition."

Discipline The action taken by a supervisor, government official, or parent for conduct that does not follow a rule or system of rules governing generally accepted appropriate behavior, conduct, or activity. This may take the form of a verbal or written reprimand and may include or some type of punishment like a fine, time off, even termination. Generally it is designed to change the behavior so the individual conforms to the rules and regulations, including generally accepted conduct of the culture.

Facilitating Changing items or processes in interdependent systems to support the original change.

Guiding coalition A group of individuals who represent the total organization and are appointed to define and devise a plan for implementation of a major change effort.

Incentive Positive (e.g., a rule or system of rules governing conduct or activity, money) and negative (e.g., discipline) reinforcement provided to make change more attractive.

Leader A person generally in authority that guides and champions an organization or group through significant and often controversial changes.

Multimedia A technique using a combination of sound, video, and written text for conveying ideas or information.

Pilot program A small-scale preliminary study conducted to evaluate the feasibility, time, cost, and adverse consequences, many times unintended, prior to the implementation of a change in policy or equipment.

Policy analysis A process that identifies the problem and the need for change, assesses the urgency, and recommends potential solutions.

Sense of urgency This is the process to persuade others that the change proposed needs to be at the top of this list to start and complete in an expeditious fashion.

Special interest group An organization related to the fire and emergency services, such as a labor union and/or volunteer association. If these groups are not represented during change efforts, they might represent a barrier to change.

Tradition An inherited, established, or customary pattern of thought, action, or behavior handed down in an organization or society consisting of information, beliefs, and customs. It is passed down without written instruction or formal instructions of a culture of social attitudes, beliefs, customs, and institutions.

Vision A thought or concept formed by the perception of a need for change from the study of data, experiences of others or events, common sense, or intuition. In FES, it is frequently the result of analyzing and understanding the requirements of national recognized professional standards.

REVIEW QUESTIONS

1. Think of a problem you've encountered at school or work that could benefit from a change effort. Discuss (in three to five paragraphs) the need for change and the urgency of the problem.

2. Develop your vision for solving the problem you outlined in your previous response. Write out your 3- to 5-minute "elevator pitch" describing the problem and your chosen solution, keeping in mind the need for clear, concise communication.

3. What barriers do you anticipate to implementing your proposed change? List at least three potential barriers in a bulleted list, and identify the stakeholders involved in each barrier.

4. Identify one or two potential solutions for each barrier you identified in the previous question, including the positive or negative incentives that might work to overcome the problem.

5. Create a brief timeline for your vision and identify at least three short-term wins you could implement to maintain accountability and momentum toward your long-term goal.

CHAPTER ACTIVITY #1: LEADING BY EXAMPLE

A metropolitan fire department adopted a physical fitness program that required mandatory participation and a yearly test. The program, which covered new employees only, required participants to sign a condition of employment document stating that they would not be able to stay employed as a firefighter if they did not pass the annual test. Because these firefighters had to pass this test to graduate from their recruit school, they were all capable of passing the test the first day in the fire station.

Once each year, all battalion chiefs (BCs) had to oversee a physical fitness test for the firefighters in their battalions. One particular BC was a strong believer in keeping fit for both personal and professional reasons and was also able to successfully pass the yearly test.

During each shift the BC would participate in the department's physical fitness program and on many occasions would join one or more companies during their training. It was well known by all firefighters how the BC felt about physical fitness. For the yearly 1.5-mile run test, the BC would advise the firefighters that they would have to beat him to the finish line. The BC would then run at a pace to finish in 12 minutes, which was the pass-fail mark. In the 3 years that the BC was assigned this field command, he never won a race. In this period of time, all the newly hired firefighters passed the test.

Discussion Questions

1. Is this BC helping to lead change? If yes, explain in detail.

2. Is it fair that the new employees have to pass this test and not the existing firefighters? Discuss fairness in terms of the new recruits, fellow employees, the department, and the public.

3. Find (using an Internet search) and analyze three separate ideas or arguments that a leader could use to support or oppose a physical fitness program for firefighters. Provide complete reference sources.

CHAPTER ACTIVITY #2: COMMUNICATING THE VISION

Your ambulance company is in contract negotiations with the county for your 911 transport contract. Even though your company has been the exclusive provider in this area for a long time and has a record of exemplary service, it may not be enough. The final decision will depend heavily on who turns in the lowest bid. Competing companies will be very competitive to in order to win this large county contract.

Discussion Questions

1. In this example, how would you support and justify being the successful bidder? What data and/or logical arguments would you use to support your viewpoint?

2. Would the use of social media aid as a means of communicating with the public?

3. What other methods can your company use to convince the county supervisors and the community that your company is the best choice?

4. Are there any changes that would be helpful to support your proposal?

REFERENCES

Bradberry, Travis. 2015. *12 Things Truly Confident People Do Differently*, New York: Forbes Media LLC.

Kotter, J. P. 2012. *Leading Change*. Boston: Harvard Business School Press.

Wallace, L. 2009. *Why fact can't compete with belief*. Retrieved from http://www.theweek.com/article/index /101112/Why _fact_cant_compete_with_belief# Volume 9 (2009-10-09)

CHAPTER 5

Financial Management

LEARNING OBJECTIVES

After studying this chapter, the student will be able to:

- Understand the importance of a budget in fire and emergency services (FES) administration (p 74).
- Detail the differences between operational costs and capital expenditures budgets and funding. (pp 86–87)
- Explain the phases of a budgetary cycle: planning, submission, review, and administration (pp 75–78).
- Identify the roles of the financial manager and the public in the budgetary process (pp 78–79, 80).
- List common sources of agency revenue (pp 80–81).
- Outline different types of taxes and fees and how they can generate revenue for a FES agency (pp 81–85).
- State alternative revenue sources, such as government bonds, investments, and borrowing (pp 86–88).
- Discuss the financial considerations of providing FES as free public services (p 89).
- Explain techniques for increasing funding, such as providing evidence from national standards, performing a cost-benefit analysis, and defining potential sources of funding (pp 88–90).
- Know how to track expenditures (pp 91–93).
- Describe purchasing policies, total cost purchasing, cooperative purchasing, and legal purchasing considerations (pp 93–95).
- Understand the importance of financial accountability and the process of financial audits (pp 77–78).
- Discuss the difficulties involved with making fair and ethical budget cuts (pp 95–97).
- List ways administrators can try to protect their agency's budget (pp 95–96).
- Explain relevant tax considerations that affect FES administration (pp 82–83).
- Understand how the global economy can influence the budgets of local agencies (pp 84–85).

Case Study

Budget Cut for Training Programs

An example of a negative outcome caused by cutting training programs occurred on Memorial Day weekend, 2011, for the City of Alameda Fire Department in Northern California. A 52-year-old man, allegedly wishing to commit suicide, walked into the cold water of the San Francisco Bay. Although 911 was called and the Alameda City Fire Department responded, firefighters and police refused to go into the water and attempt to save the man. An hour later the man succumbed to hypothermia.

In previous years, the fire department for this island community had a comprehensive water rescue program that included shore-based and surface-based rescue techniques. After several years of struggling to balance the budget, however, the water rescue program was abandoned and decertified. That day in 2011, the fire department firefighters and paramedics were following a department policy created in 2009 that prohibited personnel from performing water rescues because they were not currently trained or certified to do so.

In the wake of this disaster, the chief changed the policy to give incident commanders more discretion on using personnel and equipment based on specific circumstances. The department's training division implemented water rescue training, and two shallow water rescue boats were put into service. Although not considered in this situation, mutual aid with a close outside jurisdiction may also help in these specialty resource situations. Unfortunately, however, the damage done to the department's public image can never be completely mitigated.

Discussion Questions

1. How would you endeavor to repair your public image?

2. Is this one of the special services that would better be provided by sharing with other local departments staffing and financial support needs? Please explain.

3. Please provide a list of three other special services that should be considered for sharing responsibility and costs between FES departments (KTVU, 2001).

Introduction

Financial management is the process of translating financial resources into services while providing fiscal accountability. To make the best financial management decisions for their organization, FES administrators must have a basic understanding of the budgetary process, knowledge of their agency's sources of **revenue** and **expenditures**, and an awareness of relevant tax considerations. Understanding these concepts is extremely important because one of the main criteria government officials and the public use to measure an organization's success is its ability to achieve its goals within its allotted budget. Although chief officers may not be experts on taxes, they must have a good understanding of all tax issues before proposing a tax solution to help fund a new program or enhance an existing one.

This chapter presents an overview of the process of fiscal management along with tools and techniques that FES administrators can use to identify the real effects of budgetary changes and ensure adequate funding.

Budgets

An administrator is responsible for planning and tracking the agency's budget—its revenues and expenditures. The administrator needs to be very cognizant about every penny in the budget to achieve a moral, ethical, and legal duty to the public, elected officials, and members of the department.

Types of Budgets

There are several different budget procedures. Each type of budget has its strengths and weaknesses. The most commonly used in the field of FES is the **line-item budget**. Line-item budgets are useful for monitoring expenditures and tracking financial accounts, although they are generally not as effective as other methods for measuring outcomes. Performance budgets, zero-based budgets, and program budgets are more useful for this type of goal. Administrators who wish to explore these techniques should review the chapter *Public Policy Analysis*, which discusses measuring outcomes in detail.

Line-item budgets are essentially a statement of used and available funds broken down into categories. And they are typically used in the approval process where one year is compared to other years. The categories start with the major agencies, such as police, education, and fire and rescue. Next, the budget is broken down in each department to represent major program areas, such as prevention, operation, and emergency medical services (EMS). In each area, costs are listed by major categories, such as salaries, retirement, and maintenance.

This budget type is used in the request for funding on a yearly basis. Some agencies use the fiscal year, which is a full year period starting on July 1 for local and state governments and October 1 for the federal government.

Budgetary Cycle

What is described here is the general budget process; budgeting for certain items, such as a new fire station or increased staffing, might start years in advance. After determining the most appropriate type of budget, an administrator can begin the budgetary cycle, which consists of the following general phases:

- Planning
- Submission
- Review and approval
- Management
- Accountability and auditing

Planning

The process of budget planning, which is discussed in greater detail in the chapter *Public Policy Analysis*, may begin with the selection of an advisory committee made up of representatives from the agency, the public, other agency heads, and elected and appointed officials. More commonly, this planning starts at least a half year prior and by an in-house review in the subdivisions of the department. Decision-makers may consider the agency's sources of revenue, operating costs, and expenditures, and any relevant guidelines or standards to determine the most appropriate budget

for the organization. Administrators may receive guidance from the city manager, county manager, or budget director regarding expected increases or decreases in the anticipated revenue or expenditures for the coming year, particularly in relation to salaries and cost of living increments. This is a good venue for the administrator and their subdivision managers to discuss a new vision and get valuable feedback.

Submission

After the proposed budget has been created by the advisory committee or a compilation of plans from subdivisions, the administrator needs to prepare and submit a formal budget to the budget director, legislature, or mayor or other elected official. This proposal is the administrator's responsibility and need not be identical to the recommendations. Although the recommendations are to be strongly taken into consideration, they are not mandatory and the administrator has the final say.

As part of the submission, the administrator needs to provide justifications for each major section of the budget. Any new items or enhancements call for detailed explanations and possibly their own separate budget forms. Consider reviewing successful budget increase requests from previous years including those of other agency heads; this can help the administrator submit a request that has the best chance of being approved.

Review and Approval

In most cases, the proposed budget submission receives a preliminary confidential review by the designated government officials, although this review process varies depending on state and local laws. Some areas may require that this preliminary proposed budget and justifications be made public, but in either case the administrator can anticipate receiving private input from the city manager, public safety director, or budget director, including employee groups. The administrator should carefully consider this input and make every effort to resolve any differences before moving on to the next phase or making budget wishes known publicly.

After this preliminary review, the agency's budget is consolidated into one municipal budget, which is analyzed for arithmetic accuracy and compliance with municipal or agency policies and objectives. If the total requests are above the estimated revenues, it is typical for the personnel in the budget office to make recommendations on budget cuts for individual agencies. The chief officer should stay closely involved with this

CASE STUDY

During a budget working committee meeting for a county, one of the council members identified a job position in the building department for review. The job was for a fire protection engineer to review construction drawings for compliance with the fire safety requirements of the building and fire codes. Not having any knowledge of code enforcement, the council person called the fire chief directly to ask for advice. They were friends and both had a professional respect for the other.

Code enforcement of fire safety items in the building and fire codes for permits had not been clearly defined in local ordinance, but the responsibility was mentioned in both ordinances: building and fire. The chief knew this could solve the longstanding issue of both fire and building plans examiners and inspectors enforcing the same requirements by completely transferring this plans reviewing engineer to the fire department. That is what happened as a result of a quick phone call.

© Smeby Jr, Charles L/Jones & Bartlett Learning

process and use every opportunity to express support for the FES budget. It is ideal to have good relationships with officials in the budget agency to help the process of negotiation and compromise go more smoothly. These relationships are not built up overnight but are generally the result of long-term friendships derived from mutual respect and professionalism.

After the budget has been reviewed and approved by the municipal executive officer and the budget agency, the budget goes to the legislative body for approval. Budget hearings are scheduled and made available to the public. At this time, the chief officer and department staff members should be prepared to answer and defend any inquiries from special interest groups, the media, citizens groups, and labor organizations.

Remember, at this point the budget is now a proposal of the senior elected or appointed official, and any attacks on this budget need to be addressed. The chief officer will be asked to appear before the legislative body to officially present the department's budget. The officer should be very familiar with the budget and its justifications. To prepare for this appearance, it is useful to role-play with senior staff and try to answer the most difficult questions that staff members can think to ask regarding the budget. The administrator will also want senior staff to attend the budget presentation to address specific questions in their areas of expertise.

TABLE 5-1 summarizes stages of budget review.

Management

After the budget passes legislative review, it is officially adopted. The next step, management, is the final element of the budget cycle. In this phase, the manager monitors revenues and expenses over time and revises the budget if necessary. He or she may also be asked to prepare periodic budget reports as required by the executive officer, elected official, or budget department head to monitor the expenditures and detect any deviation. These reports contain lines of funds for each category and read from left to right with such categories as Approved, Current Month, Year-to-Date, and Balance Available.

Each month, the administrator should examine the budget reports carefully. If any items are unbalanced

TABLE 5-1 Stages of Budget Review		
Stage	**Reviewing Party**	**Notes**
Preliminary review	Designated government official (city manager, public safety director, budget director)	■ Confidential, or may be made public depending on state or local laws
Municipal review	Municipal executive officer, budget agency	■ Budget is analyzed for accuracy, as well as compliance with municipal policies and objectives ■ Recommendations are given for individual agencies in the municipality at this stage
Legislative review	The legislature has public meetings to discuss the budget and receive input from all groups listed	■ Budget hearings are made available to the public ■ Chief officer presents the budget and answers questions from special interest groups, media, citizens, and labor organizations

(e.g., if more than half of allocated overtime funds for the year was spent in just 3 months), the administrator should examine the reason for this imbalance and make necessary changes to the budget or agency overtime policy, such as by requesting supplemental appropriation, transferring funds from other areas of the budget, or cutting back on overtime for the rest of the fiscal year.

When an agency's budget changes, the administrator can easily add or detract an incremental amount of money to a line-item budget to reflect changes in revenue or costs. For example, if the tax revenues increase or decrease for the next fiscal year, it is easy for the government accountants to simply cut all budgets by a specific percentage. This relieves them from cutting a specific program that may be viewed as a popular public service.

However, politically expedient cuts in budgets can be a lot more complicated than they initially seem. To illustrate the complexity of a reduction in funding, consider the following example. The accountants in the finance department of a municipal government have proposed to cut the budget, and the chief is given the latitude to propose a reduction plan. The department has 100 employees and the new authorization allows only 90 employees. In implementing this budget cut, the chief may choose to cut all of the fire prevention and training staff positions, which total 10 employees. This proposal would keep all existing fire stations open and seems to be the least controversial plan.

However, as the months and years go by after the cut, the real impact of eliminating fire prevention and training will become evident as firefighters lose their proficiency and injury rates increase. There will also be more severe fires as a result of neglecting fire protection features in buildings and lack of quality proficiency in firefighting.

An example of unintended consequences in financial decision-making can occur when a building owner considers the installation of an automatic sprinkler system. In many cases the developer opts to forgo the extra expense of a fire sprinkler system. However, if a sprinkler system is installed in a building, a potentially large and catastrophic fire in the community may be prevented and lives and property may be saved.

Furthermore, a fire of large magnitude may overwhelm the local fire department's resources and present a hazardous situation for firefighters. If the fire destroys the owner's building, the municipality loses a substantial amount of real estate tax, not to mention the potential for a number of local people to be without jobs if the building is a business. One of the major justifications for automatic sprinkler ordinances is the elimination of target hazards that would have this type

of adverse economic impact on the local community. These are benefits to the entire jurisdiction in addition to the property or business owner. Again, when market forces fail to bring a demand for the public good, this becomes a justification for government regulation. This subject is discussed further in Chapter 10, *Government Regulation, Laws, and the Courts.*

Accountability and Auditing

Financial accountability is confirmed through the auditing process. Financial audits are commonly performed on an annual basis to ensure that financial records are accurate and complete. As part of this process, an auditor examines assets to confirm that they are in their assigned place and checks the approved budget for congruence with actual expenditures.

Periodically, the department will be audited for its use of public funds. Auditors confirm that public funds have not been misappropriated, embezzled, accidentally doubled, or stolen and that there was no failure to follow procurement policies. The consequences of any improprieties or negligence can include prosecution, loss of funding, a negative reputation, and additional regulations.

The fire or EMS administrator should be alert for any random or scheduled audits and check with predecessors or government financial officials to obtain a briefing on the last audit. This is one of those areas where it pays to be very detail oriented. If there is an occasion when the purchasing policy was not followed and there was a good justification, make sure to thoroughly document the situation and all the details in writing. This documentation can be extremely helpful when auditors question emergency spending or other unapproved purchases.

The administrator would be wise to have an internal audit of the budget done to identify any items that could be deemed unnecessary or unjustified. These items may not undergo scrutiny during good financial times, but when budget reductions are necessary, the presence of any unjustified items rightfully decreases the cost-conscious reputation of the department, leading to additional oversight and cuts. If administrators can eliminate these items ahead of any budget crisis, their reputation of being a frugal steward of public funds may help the department hold onto more of its budget during hard times.

In addition to financial audits, a fire or EMS agency may also be asked to participate in performance audits, such as compliance or procedural audits. A compliance audit checks for adherence to a legal or policy mandate. This type of audit can include elements of fiscal management: for example, if a department charges a fee for fire prevention plan review, the audit may review fee calculation, collection, and storage. The compliance audit may also include a review of the agency's petty cash account for small purchases to ensure compliance with relevant spending guidelines. **Procedural audits**, designed to evaluate management policies and procedures, also can include financial policies and procedures.

Government officials may hire outside consultants to complete these audits. Administrators should always comply with the auditing process; their cooperation ensures a fair review. Remember that no department is perfect, and the audit may uncover areas for improvement. Auditors' comments and suggestions can provide a valuable opportunity for additional growth, new projects, or new funding possibilities. Look for ways to turn these suggestions into new requests for additional funding, if necessary. The administrator would be wise to create a management team to work with auditors to answer their questions and explain the rationale for certain decisions, policies, or procedures.

There are occasions when audit reports are used to justify increases in the budget. If an administrator encounters strong resistance to budget requests, an outside consultant can provide an independent professional opinion. It is a good idea to review the request carefully before making this proposal; if the consultant recommendations do not support the budget requests, the request may backfire and the administrator will lose credibility with superiors. Again, this strategy should be used only when one is absolutely sure that the request is unquestionably justified.

Key Players in the Budgetary Process

In addition to the many people involved in this ongoing budgetary cycle discussed in the previous section, the administrator must be in contact with several other key decision-makers in the budgetary process, including the agency's financial manager and other agency managers.

Financial Manager

The **financial manager**'s role in an FES organization is a busy one. After the budget has been adopted and funded, one budget cycle ends and the next cycle begins. In many cases this ends up being a multiyear activity. While administrators are busy with the never-ending budgetary cycles, they must rely on a financial manager to alert them to any potential problems that may interfere with the agency's budget plan. In larger organizations, this may be a separate individual, whereas in smaller departments it may be the duty of the senior administrator.

Problems that may need to be addressed include unexpected overtime, increased costs in capital items, tax revenues below predictions, insufficient fee collection, or **unfunded mandates**. For example, government regulations and professional standards may be changed to require updating of safety equipment. This was the case in the early 1970s when self-contained breathing apparatus was required to be updated to "positive pressure." This regulation, which protected the wearer from any air leaks inside the face piece, was not combined with any funding from the state or federal government. Therefore, the local governments had to pay the bill.

If the budget was approved on the basis of a fee collection, the financial manager should keep a very close eye on this collection process to make sure it does not fall below expectations. A revenue shortfall in fee collection is likely to directly impact the agency. A revenue shortfall in a municipal tax can be remedied, however, by a reduction in any or all of the municipal agencies' budgets.

Financial managers should also follow any proposed legislation or regulations from state and federal governments that might affect the agency's funding. To stay up-to-date with legislation and regulations, the financial manager may rely on an automatic referral system or a network of other financial managers, FES organizations, and professional organizations (e.g., the International Association of Fire Chiefs [IAFF] or the National Highway Traffic and Safety Association [NHTSA]) that publish this information.

The NHTSA's "Emergency Medical Services Agenda for the Future" includes information on EMS legislation and regulation and system finance. The National Registry of EMTs (www.nremt.org) published a document called "The National EMS Scope of Practice," which highlights scope of practice concerns. The National Registry of EMTs also establishes standards for EMS certification and licensure. State websites have documents specific to state legislation and regulations. For example, California has an organization called California Fire Chiefs (www.calchiefs.org), which is a statewide organization that follows legislation specific to fire service and fire EMS operations. The financial manager should review these types of publications and report in writing the existence of any proposed legislation or regulations that may have a financial impact on the department. Further, there are some opportunities to influence these regulations during the public input phases.

Monitoring the community for demographic changes that may affect the services requested and tax revenues available (e.g., aging communities experiencing a decrease in population and average taxpayer income) can be critical to avoiding financial surprises. These changes may not affect tax revenues for many years so it is easy to spot a future trend. This information is readily available from the U.S. Census Bureau. Year-to-year changes are monitored at the local, county, and state level by government agencies.

Finally, the financial manager may have to recommend budget cuts dictated by the local government. The administrator along with the financial manager should make a concerted effort to protect the FES budget from any cuts. This effort involves building political alliances and promoting the department's programs to key political officials and influential citizens. In some respects, the FES administrator is a salesperson who must constantly sell the department's products and services. Part of this sales pitch involves educating the public. The administrator should make regular public presentations using speakers to convey the important message of the department's services to public interest groups and organizations. When the government calls for budget cuts, the administration will need as much support as possible.

Bureaucrats and Other Agency Managers

The administration of an FES organization is one part of the government bureaucracy. Administrators from other agencies such as schools, police, public works, building inspections, library, solid waste, etc. are also in competition for tax revenues. These bureaucrats are also expected to implement the programs approved by elected officials. It becomes almost impossible to create an ordinance, etc. that is precise enough to guide the bureaucrat in implementing every aspect of programs. Therefore, bureaucrats end up with a lot of latitude to determine how to implement their programs.

In some cases, bureaucrats from several agencies can cooperate for a common goal such as increasing tax revenue. Raising taxes to fund ongoing and new programs does not occur routinely, but did occur in New York City when the real estate tax rate was raised by over 18% in 2002. The additional revenues reduced the proposed cuts to the Fire Department of New York for 2003, even though the department had to close six fire companies. Since 2000, nine states have enacted so called Millionaires Taxes. "One way to raise the necessary funds is to raise personal income tax rates on the highest incomes, a policy choice sometimes referred to as a 'millionaires' tax.' This approach makes sense: evidence indicates it can generate substantial revenue for public investments that boost a state's productivity in the long run, without harming economic growth in the short term." Proponents of these taxes argue that "Even modest rate increases on high incomes can generate significant revenue from a relatively small number of taxpayers." This was the result when "…policymakers began looking at targeted rate increases for affluent households to help close stubborn budget shortfalls stemming from the 2001 recession, as well as to respond to the increasingly skewed concentration of wealth and income at the top" (Tharpe, 2019). This is a highly controversial public policy and statistics used to justify these tax increases can be debatable. However, rarely do additional taxes benefit the FES community.

Most government agencies have very few standards for the services provided to the public. For example, if the police chief (a bureaucrat) proposes a budget increase to cover the cost of new police officers, the chief typically justifies this by reporting that the department would be able to reduce some type of crime with more personnel. The police chief may select a crime that has recently received some notoriety. Because there is no standard on the number of police officers needed, the chief's request will probably be judged by the elected officials on the justification provided, the influence of the police chief, and the potential for gaining votes in the next election. Public fire protection and EMS are fortunate to have national standards for their service levels, which are not held by police and education. In general, the following three areas are covered quite comprehensively in the national standards:

1. Training for firefighting and EMS personnel: National Fire Protection Association (NFPA) professional qualifications standards for firefighters and Department of Transportation (DOT) standards for emergency medical technicians and paramedics
2. Safety of personnel: NFPA 1500, *Standard on Fire Department Occupational Safety and Health Program,* and Occupational Safety and Health Administration (OSHA) Fire Brigade and Respiratory Protection standard
3. Deployment and staffing of FES units: NFPA 1710, *Standard for the Organization and Deployment of Fire Suppression Operations, Emergency Medical Operations, and Special Operations to the Public by Career Fire Departments*

The more chief officers rely on national standards, a formal planning process, or cost-benefit analysis, the better chance there is to convince elected officials to budget for new and existing programs.

The Public and Their Representatives

Recognizing that the government must choose among many conflicting interests and striving to present strong arguments to support the fire or EMS department's budget requests are important roles of FES administration. Without valid guidance, governments do not always do what is right, efficient, and in the best interests of citizens. If successful, FES administrators will convince the elected and appointed government officials to spend taxpayers' money on the department's needs.

In addition to convincing government officials responsible for this decision-making, administrators must also aim to convince the public of the justification for the department's funding needs. Although most communities use elected representatives to make budgetary approval decisions, some towns use a process of direct voting in which citizens can directly approve or oppose budgets, ordinances, or bond issues. It is assumed that when all else is equal, voters vote according to their pocketbooks. Most elected representatives behave in a like fashion. If they watch the financial needs of the voters very closely, they have a good chance of being reelected.

When requesting additional funding, administrators must convince voters or their representatives that increased spending is of value to the public. The justifications should contain a very clear explanation and evidence that the department will be able to increase the quality of FES, including safety and efficiency issues with the new funding. In general, public polls show that most citizens believe that FES departments are doing a good job. Therefore, it may be difficult to convince the public that there is a need for additional resources without a convincing argument that the requests are justified. This is commonly called a "paradoxical situation." It will take extra effort to provide the justification of the need because the public and elected/appointed officials believe you are doing a good job now. And, for many administrators, it is not comfortable to bring up issues needing improvement because this can be characterized as a weakness. One way to do this without any criticism is to frame the improvement as being the result of a new progressive safety standard, for example.

Just as when speaking to government representatives, the administrator must clearly explain to the public how the extra funds benefit the community by enabling the agency to increase the quality of its

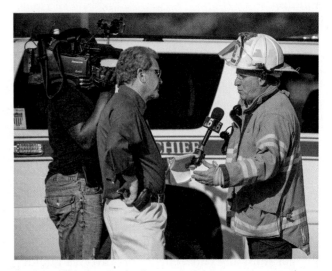

FIGURE 5-1 Fire chief talking to a TV news crew to update the public on an emergency scene.
Courtesy of Glen E. Ellman.

services. For example, if the administrator is requesting funds for a new station and increased staffing, he or she has to demonstrate that the increased spending decreases response times as a valid justification. Several FES organizations may join forces to appeal to the public. A cooperative, regional approach may be best to gain public approval for services, such as hazardous materials, terrorism planning, response to local catastrophic events, and technical rescue teams. This is especially true for smaller FES departments.

Administrators should seek out influential private citizens in the community and work to cultivate an ongoing friendly business relationship with as many of these important people as possible. In many cases, these private citizens have as much, if not greater, persuasive power over the decision-making process of local government than the elected officials **FIGURE 5-1**. Remember, statistics may be the best way to demonstrate the benefits of existing service levels and the impact of reductions, so it is a good idea to identify and document all benefits of current services to the community and know this information well.

It's important for the fire chief to keep the community informed.

Revenues

To protect an agency's budget and/or to obtain funding for new programs, administrators should have a solid understanding of common sources of revenue. The main source of revenue for most FES agencies are government taxes, such as **property tax** (sometimes referred to as millage rate), **income tax**, and **sales tax**. Increases or decreases in these taxes may be caused by economic variations or just represent normal fluctuations of these sources. For government officials to approve a tax increase for a specific new or expanded program, the administrator must be able to convince them that there is an overwhelming need for the new program and to present a solid justification that would be supported by the voters.

Taxes

For example, property tax (or real estate tax) is the most common funding source for local government programs, particularly in states that do not have a state income tax. This tax is assessed as a percentage of the total value of property and improvements, or in some cases the rate is based on income for a commercial business. One big advantage of the property tax is that it is relatively easy to collect. Nobody can hide a house or a store from the tax collector.

For a property that has structures on it, there is a direct connection with fire services because there may be a need to extinguish a fire in the building and the need to provide emergency medical services for the occupants. Residential property taxes are assessed at a flat rate based on the value of the property, as opposed to a progressive tax like the federal income tax, which increases with the income of the individual. All property owners pay the same percentage based on the value of their property. If a homeowner lives in a million dollar house on a horse farm, that owner pays a lot more tax than a family living in a mobile home on a small parcel of land. The greater the value, the greater the tax.

Commercial property taxes may have different rates and can be based on the income of the business and are therefore subject to wide fluctuations. When property values fell in 2008 during the downturn in the housing market, so did tax revenues. This kind of reduction usually does not occur quickly because the tax changes only after the property is reassessed by the tax collector's office. Some states assess properties on a multiyear basis; therefore, the impact is not realized for several years. There is a lag between the cause and the effect until property values can be reassessed. This is also true on the other end of this trend, when the property values start increasing. This type of variation in tax revenues can be predicted ahead of time by monitoring the sale price of properties in the community.

Income tax is the federal government's leading source of funds, followed very closely by the payroll tax (Social Security and Medicare). Income tax is a progressive tax, which means it is taxed at a higher percentage for higher wage earners. The Internal Revenue Service reported that in 2014 the top half of U.S. taxpayers paid 97.3% of all federal income tax. The bottom 50% paid only 2.7%, which points to a progressive tax. In addition, the top 1% paid over 39% of the federal income tax (Internal Revenue Service, 2015).

Many states also have income taxes. Several of these states link their state income tax to the federal calculations for taxable income. Some cities also have income taxes. Because income tax revenues fluctuate with the economy, this source of funding is affected very quickly by significant economic changes. Furthermore, this type of tax has one major problem: tax evasion. One study reported that 18% to 19% of income is not reported to the Internal Revenue Service (Cebula and Feige, 2008).

Sales tax is applied to consumer purchases, such as electronics and vehicles. It is affected very quickly and is the most widely fluctuating revenue source, because personal spending levels vary greatly based on season and individual considerations. The sales tax is a common source of revenue at the state and local level. Most states have a sales tax, although there are five states that do not: Alaska, Delaware, Montana, New Hampshire, and Oregon. For the other states, there are differences in the percentage rate and the items covered from state to state. For example, some items that may be exempt are food, clothing, and medicine. In several states, either a county or city government can add a local sales tax on top of the state's sales tax.

One of the first things people do when faced with the loss of a job or the threat of being laid off is to stop spending on items that have a sales tax, such as cars and large appliances. On a large scale, this can cause an immediate drop in tax revenues that were projected to fund the present budget. Most of the time this fluctuation cannot be predicted, such as the pandemic that hit the world in 2020. Therefore, financial managers should always keep a personal list containing items in the budget that could be reduced in case of emergencies. This list should be the result of an internal analysis of each line item and its importance to the mission.

Some states greatly benefit from the sales tax, especially those that have a lot of tourists. Tourists can pay a substantial amount of tax to help finance state and local governments. For example, state and local governments may aggressively pursue tourism to counteract the cost of public schooling, which is a major cost for the state and local governments. There are many examples of state or local governments reducing or eliminating real estate taxes on new amusement parks, tourist attractions, or manufacturing developments in anticipation of sales tax receipts and new jobs.

There are several other types of taxes, such as **fire service taxes**, **user taxes**, and **government fees for services**. A fire service tax can be a set value for all residences or can vary depending on the square footage of the dwelling or the fire flow formula, which calculates the approximate amount of water needed to extinguish a fire based on the combustibility and area of the structure.

A user tax applies only to the use of specific goods or services, such as taxes on motel rooms, rental cars, and tourist attractions; these funds are applied to offset the cost of providing public services to tourists. One of the most common use taxes is the road tax, which is added to the price of fuels used by motor vehicles. Consumers see this as a fairer tax because it is directly based on the use of the roads. The more miles driven or the poorer the fuel economy of the vehicle, the more tax that is paid. However, this tax is not collected when vehicles use electricity to power their vehicles either partially or totally. This is a complex problem because the amount and cost of charging an electric vehicle is not kept. And there are some imbedded taxes in the electric utility bills but they are not a road tax. This is a new problem being studied by policy makers.

Tax Considerations

"Taxes are what we pay for civilized society," said Supreme Court Justice Oliver Wendell Holmes, Jr., in 1887 (https://www.irs.gov/newsroom/tax-quotes, 2020). But what is a tax? What types of taxes are there? When is a tax fair? What impact does the tax have on the local economy?

Because taxes are the major source of revenue for most FES departments, it is worthwhile for administrators to have a basic understanding of important tax considerations that can affect their agencies. The major concern is that the tax base or service fees on which FES agencies depend may not be adequate to fund increases in costs or other growing demands. Administrators may see increases in taxes as the best solution, but state and local governments are limited legally and practically in their ability to increase these taxes. Even when increases are possible, citizens and government officials are often not supportive of tax increases unless they are convinced about the possible consequences of not taking action, such as reductions in vital emergency service and response times. Administrators must recognize that businesses, special interest groups, and the taxpaying public are all affected by changes in taxes and also recognize that tax changes have a significant effect on the market economy.

It should also be noted that increases in taxes do not always lead to the forecasted increase in revenue. Every product and service has a relationship with supply and demand **FIGURE 5-2**, which is affected by its overall cost, including taxes or fees. When the price rises, the demand falls. Therefore, if the government increased taxes significantly, the demand for the taxed good or service would decrease and tax revenues might even fall. For example, when the government increases a tax on cigarettes, fewer people may purchase cigarettes, thus leading to decreased tax revenue and a decrease in smoking. This will reduce medical costs, because smoking is the cause of many serious medical diseases. However, nonsmokers on the average live longer and will require pension payments (including Social Security) for additional years. These are very complicated calculations that require a sophisticated cost-benefit analysis to determine the fiscal impact on government revenues.

Similarly, if the government significantly raises taxes on new housing by impact fees or new regulations such as earthquake and hurricane protection, rates of construction and home purchases may fall, leading to a loss in potential increases in tax revenue. Still, the overwhelming reasons for purchasing a particular home are based on the characteristics of the neighborhood and the schools; therefore, unless the new total cost is substantially increased, the increase in taxes or costs does not have much of an impact on home purchases.

It is important to consider who is being taxed (e.g., the consumer or business). In 1996, Congress reinstated the 10% federal excise tax on airline tickets. Even though the price of airline tickets should have gone up 10% to cover the new tax, the airline tickets initially rose only 4%. Businesses typically have to absorb some

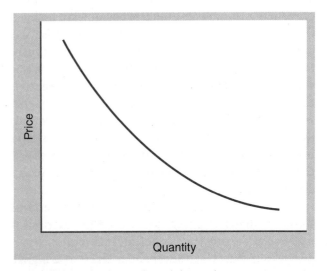

FIGURE 5-2 Classic supply and demand curve.

of the tax levied on their products and services. In this case, the airline companies lost 6% in revenue as a result of the airline excise tax. However, this result normally does not hold true over a longer period of time. Eventually consumers see prices increase.

In this case, and for all taxing issues, businesses do not actually pay the tax. The loss in revenue is passed on either to the investors in reduced dividends or to employees through reduced salaries. If you follow money to its source, you find out who actually pays the tax. For example, when buying a car from a large auto manufacturer, federal taxes, which are estimated to be about 23%, are actually included in the cost of the car. Therefore, in this case, the buyer pays the tax.

Similarly to taxes on goods or services, the government cannot simply increase property taxes significantly without backlash from businesses and home owners. Property tax rates affect commercial developers' decisions as to where to construct a new building or business. This can be of great concern to the appointed and elected officials who prefer to attract new development and additional tax revenue to the municipality. Some jurisdictions reduce, suspend, or eliminate property taxes to attract a major industry. Sometimes other economic factors, such as the availability of public water and sewer and costs of local labor, are of greater concern.

There are situations in which higher property taxes may seem worthwhile. For example, parents with school-age children may be willing to pay more taxes to be in a city with an outstanding public school system or lower crime rates. Similarly, retirees and senior citizens may be willing to pay more taxes to be in areas with convenient and quality medical facilities. The quality of FES services is normally not a consideration in selecting an area to buy a home.

In some areas, a dedicated property tax is used exclusively to fund FES, eliminating the cross-agency competition for available funds. In some emergency service districts, an elected board of commissioners is responsible for managing the tax revenues. Often, these commissioners may be firefighters or EMS providers, which can create a conflict of interest with regard to financial oversight and decision-making.

Administrators should be aware that property taxes are the major source of revenue for state and local governments, and for their FES departments. Therefore, if additional funds are needed to support existing programs or start a new program, the property tax and user fees probably are the primary sources of funding. Taxes that require minimum enforcement effort, such as the property tax, are those that work best at the local and state level. Increases in taxes or fees are more likely to be approved when the program they will fund is shown to increase the quality or quantity of fire or EMS service and the tax or fee is viewed as fair, equitable, and easy to enforce.

Tax Affects

There are flat (or regressive) taxes (e.g., sales tax, applied to all buyers in a certain area equally) and progressive taxes (e.g., income tax, applied at a varying scale with higher wage-earners paying a higher percentage). When recommending that a tax be adopted or increased, keep in mind that the people in the highest 10% of the income bracket may have significant influence in the local political process and pay a major portion of taxes already. However, flat taxes may face opposition from those with fixed incomes who may not be able to afford the additional expense. Some jurisdictions have revised taxing ordinances to reduce or limit the burden on the truly needy, such as exempting homes under a certain assessed value from property taxes or the application of a fixed fee tax per dwelling unit for FES. Remember that a tax proposal is still evaluated for its fairness. As part of this analysis, the resulting benefits or public good from the increase are compared with its additional cost. If the benefits derived create an increase in service or an improvement in the quality of the service, then there may be stronger support from taxpayers.

There are some taxpayers in this country who have a distrust of government and believe there is never a good justification for an increase in taxes. However, remember that the fire service is trusted more than any other public institution and may be able to overcome the public's apprehensions. If there is a chance of overcoming these negative government feelings, it is with a strong justification of improving service. There must be a clear cause-and-effect relationship between the new program spending and increases in service or the quality of the service. Arguments that are ambiguous or subjective often fail.

Interstate Considerations

States cannot legally tax anything involved in interstate commerce. They also cannot discriminate against non-residents or out-of-state businesses in their application of taxes. If taxes in one area become excessive, citizens or business owners may elect to move to another state. Therefore, there is some competition to keep taxes low to attract and keep businesses and residents. Areas with large tourist industries often have low taxes for their residents, because taxes on hotels, entertainment, and rental cars are paid by nonresidents.

Global and National Economy Effects on Local Taxes

Most FES organizations are funded at the city or county level. The tax revenues at the local level can be impacted not just by the local markets, but also by state, national, and global markets and economic trends. Anything that hurts the local economy impacts the collection of these taxes. The exact timing of any economic impact on the local budget can vary depending on the local tax type and collection process.

More and more of the U.S. economy is affected by global influences and financial trends. Problems with the stock markets in Indonesia, Japan, Russia, Brazil, parts of Europe, and other countries have had substantial impact on the U.S. economy. Recently, terrorist attacks by organizations and individuals from other countries have had a substantial negative worldwide economic impact. Trade agreements open up U.S. markets to competition with foreign products.

Most of the international trade agreements are being renegotiated to reflect the Trump administration's goal to make them fairer and level the trade playing field. The goal of these renegotiations is to help the U.S. economy, including increasing jobs and salaries. However, in the short term there may be some wide variances in the U.S. economy until all of these trade agreements are modified.

The United States is the world's largest debtor nation. Although this situation does not seem to have any real impact on the economy at the present time, keep this potential adverse economic situation in mind. A recession occurred between December 2007 and June 2009 mainly because of the practice of approving mortgages to buyers who were not financially qualified to make the payments. When the price of homes started falling, these buyers could not sell their homes, resulting in many foreclosures **FIGURE 5-3**. To compound the

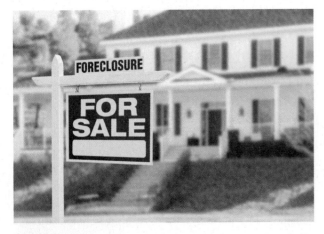

FIGURE 5-3 Foreclosure.
© Andy Dean Photography/ShutterStock, Inc.

Facts and Figures

The U.S. Financial Problem

Here are some U.S. financial statistics from fiscal year 2018:

- U.S. tax revenue: $3,330,000,000,000
- Federal budget: $4,411,000,000,000
- New debt: $779,000,000,000*
- 2018 national debt: $21,460,000,000,000

To give this a personal view, compare these statistics to a household budget, by removing eight zeroes from the above numbers:

- Annual family income: $33,300
- Money the family spent: $44,110
- New [annual] debt on the credit card: $7,790
- Outstanding balance on the credit card: $214,600

How would you feel about having a family income of $33,300 and having a personal debt of $214,600? Another way to look at this debt is to consider the national debt as the federal government using your personnel credit card. Each year you pay the interest on the debt when you pay your federal income tax, and it continues to grow. If all you do is pay the interest, nothing goes to pay down the debt.

*Not including stimulus federal funds to offset the economic impact of COVID-19.

Source: www.usgovernmentspending.com/federal_budget

© Smeby Jr, Charles L/Jones & Bartlett Learning

---

problem, these mortgages were bundled into investment packages that were sold to banks and other investment companies. The financial companies caught holding mortgage-based investments at the time of the subprime mortgage crisis lost so much value that some needed a bailout from the federal government; others went into bankruptcy. Numerous risky investment strategies and some outright dishonest practices added to the magnitude of the financial crisis.

If the local community relies on industry or manufacturing for a good portion of its tax base, there may be trouble for these companies in the future and trouble for taxes. The Second Edition of this textbook reported that "The decline in manufacturing jobs is a natural evolution of the labor market in this country. As the costs for labor have risen in the United States, the costs of labor for manufacturing in other countries have stayed at a lower level." However, more recently this paradigm is changing, and manufacturing jobs are now being created or returned to the United States.

This is the result of the newly negotiated trade treaties. In 2018, 264,000 new manufacturing jobs were added, representing the highest number of new manufacturing workers since 1988.

Other globalization factors to consider include the reality that many U.S. companies have been downsizing and re-engineering themselves. Many of the jobs lost are in the middle management ranks. Simultaneously, the American educational system is receiving low marks for its ability to produce graduates that excel in high technology and the skilled, educated workers needed to compete in the global market.

The U.S. economy is highly regionalized. An older city that contains a large number of heavy manufacturing firms may experience a greater reduction in local taxes than the suburban area that surrounds the city. Older cities in the United States have already seen reductions in city services as a result of declining tax revenues. Social Security and Medicare are projected to run out of money in the future. Some of the options to maintain them are to either raise taxes or reduce benefits. Because most federal elected representatives are not anxious to take either option, these problems may not be resolved until a crisis occurs, and then change will be unavoidable.

Be aware that major changes are always possible in the economy at the local, state, and national level and may happen without notice. For example, these changes are very evident as a result of the 2020 coronavirus lockdowns. This is a huge problem. As a fire or EMS administrator, you must be prepared to defend your agency's budget to the municipal administration and other department heads if a downturn in the economy occurs. Use of sound financial practices and strategies puts FES agencies in the best position to provide their valuable services, regardless of local, economic, or political factors.

Non-ad Valorem Assessments (NVA)

A few jurisdictions charge what looks like a fee per parcel in residential developments. NVAs are not calculated on a percentage of the property and building's value but equally for all parcels. One justification for these fees is that in these jurisdictions the real estate tax has been reduced or eliminated by homestead exemptions on many low value properties. For example, the value of a mobile home and property may be as low as $50,000. This same jurisdiction may allow the first $50,000 of value to be exempted when calculating the real estate tax. Therefore, they pay no property tax. NVAs can be used for many varied services such as drainage, landscape services, police and fire response

COVID-19 Financial Consequences

The 2020 COVID-19 virus caused a worldwide fatal disease that resulted in many revenue issues for local, state, and national governments. First and foremost was the lockdown of most of the economy of the country. Therefore, no economy, big decrease of taxes except real-estate. For your information, there is a growing number of the public policy professionals and scientists that are considering the premise that this was unnecessary, unconstitutional, and ineffective.

Much of this pessimistic notion of unsubstantiated justification for the lockdown comes from the realization that this virus was not as fatal as originally predicted. Computer models predicted up to 2.2 million deaths in the United States. The actual numbers are over 200,000 deaths, certainly not trivial and about twice the normal seasonal flu, but not as extremely devastating as first predicted. In addition, the majority of people who came down with the virus were indoors at the time and those cases that were fatal were older patients with morbidity issues caused primarily by obesity that resulted in diabetics, heart conditions, and many other diseases.

As far as tax revenue effects, with 13.5 million unemployed (August 2020), there will be a substantial reduction in federal and state income taxes along with the taxes supporting Social Security and Medicare funds. The one tax that will be felt at the local level is the real estate tax, which may not be as drastic in the short term. There will be some homeowners who will not be able to pay mortgage payments (including payment of real estate tax) and eventually be foreclosed on. If the economy does not rebound quickly, this will be a big problem. A few examples are listed below:

- By JUDITH PRIEVE | jprieve@bayareanewsgroup.com | Bay Area News Group

PUBLISHED: June 11, 2020 at 1:01 p.m.

"Faced with continued financial shortfalls, the East Contra Costa Fire Protection District has announced that, beginning on July 1, fire fighters won't enter burning structures unless lives are at stake."

"In a news release earlier in the day, the chief said having only three fire stations instead of the six needed to cover 250 square miles is 'pushing

FIGURE 5-4 Some fire prevention agencies charge fees for plan review and inspections.
© Glen E. Ellman/Jones & Bartlett Learning.

or garbage collection. This assures that all home owners will pay something toward the service. These non-ad valorem assessments add to the revenue for the FES but in many cases do not cover the complete costs.

Fees

Government fees may be applied to offset the cost of specific services when they benefit the user directly, such as road tolls, driver's license fees, and professional licenses. Specifically with regard to providing EMS services, an ambulance transport fee can be a substantial source of revenue. Some fire departments have agreed to take over EMS transport after calculating the amount of revenue that can be realized from transport fees. Insurance companies, including Medicaid and Medicare, usually pay most of the transport fees. However, there are many examples of these fees not covering the full cost of EMS transport. Fire organizations must factor in all of these elements and then decide whether ambulance transportation fees are a significant revenue generator or just another financial and liability burden. In busy departments, there simply may not be enough time to attend to all FES duties. Mandatory paramedic re-certification can be very time-consuming and should be a significant consideration before committing to providing either patient transport or paramedic services.

In fire departments, a fee for a fire prevention plan review and fire inspection may be an alternative revenue source **FIGURE 5-4**. When applied, the fire department may serve as the government's enforcement authority for required inspection of fire protection equipment. In these instances, inspection fees should not be used as a way to acquire additional funding for the agency's overall needs, but just to cover the actual costs of the inspections. It is a conflict of interest to use the government's authority to require inspections and mandate a fee at the same time.

Government Bonds and Short-Term Borrowing

Although most of a fire or EMS agency's operating costs are covered by taxes and fees, additional funds, such as those necessary for major capital improvements, are generally provided by loans through the sale of government bonds. A capital item is typically an item that has a long life expectancy, such as a fire station. Fire stations have a 30- to 50-year life expectancy and are very expensive; they are a perfect example of a capital item purchase that must be funded by loans from the sale of bonds. State and local governments sell bonds for a low interest rate at a set interval for up to 30 years with tax-free interest.

Several governments have suffered financial crises because of inappropriately using bonds to bridge the gap between necessary expenditures and incoming tax revenues. For example, if the city has only enough funding to cover fire services through June and tax revenues are distributed in July, the city could get a one-month government loan to cover its expenses for that month. This can end up being a troublesome practice, because the 1-month loan extends to 2 months, then 3, and so on. This was one of the questionable financial practices that brought New York City to the brink of bankruptcy in 1975. In cases such as this, the government is borrowing to cover operational costs such as salaries. This is a poor practice, and the options for the elected officials are to either cut costs or raise taxes. Neither is a pleasant option for a politician. To mitigate this and related issues, many local and state governments have either an official or a recommended debt ceiling. Often this is expressed as a percentage of the total revenues (e.g., bond debt payments should not be more than 10% of tax revenues).

In fiscal year 2019, the federal government had to borrow almost 24% of the total budget to cover excess expenditures, most of which included operating expenses and progressively greater interest payments on the debt, not capital improvements. Further, the interest payment was 8.3% of the budget and will continue to grow each year unless the federal budget is balanced, which would result in no new debt. This federal debt is funded by Treasury bills, Treasury notes, Treasury bonds, and Treasury Inflation Protected Securities. Theoretically, federal debt never has to be repaid; only the interest payments must be paid each year. Therefore, the federal government and its elected officials can vote to spend money, but do not need to vote for taxes to pay for the spending or cut spending to balance the yearly budget. However, this situation is not sustainable long term. State and local governments receive federal tax dollars for numerous programs and any interruption may be offset by cutting FES budgets.

At the state and local level there may be legal restrictions on borrowing, and 39 states require the budget to be balanced. However, this does not include actuarial fiscal future demands that are required for pensions and benefits, for example. Again, any interruption in either revenue or shortage of assets to pay pension benefits may cause elected or appointed officials to look at the FES budget to make up the difference.

A few larger departments have used a bond sale to replace all (or a large portion) of their units at one time. In these situations, the department has all brand new equipment in the first year, but as time goes on, all of the apparatus ages simultaneously, increasing the probability of repairs, breakdowns, and out-of-service time. Because most apparatus divisions are staffed and funded at a fixed rate, there is no repair work in the early years, because the apparatus is new and under warranty. However, with all the apparatus aging at the same time, the maintenance cost is substantial in later years, with many vehicles in the shop for repair.

A replacement plan that budgets for the replacement of apparatus on a scheduled basis is easier to justify. After approval for this type of funding is gained, it is easier to achieve approval for the next fiscal year. For example, if last year's budget contained $250,000 for a new engine, then next year's budget request would just include an incremental increase to cover the increased cost for 1 year. It is far easier to gain a 3% increase in the budget than to go back and justify each new apparatus purchase.

However, smaller FES agencies may also use money from the sale of bonds to purchase fire rescue apparatus. Fire and rescue apparatus can cost more than $1 million, so it is very difficult for small departments to be able to save the amount of money needed to pay for them. Because these departments buy only one

new fire rescue apparatus every 5–20 years, a bond may be a good method to raise the needed capital. This spreads the costs over many years. Larger departments generally have purchase plans to buy a certain number of apparatuses each year.

It is important for FES administrators to remember that bonds are a form of borrowing and that any time money is borrowed, there is an additional cost associated with the loan. It also is important for administrators to recognize that in many jurisdictions, voter approval by ballot is necessary to sell government bonds. This process of voter approval can take a long time and is not guaranteed, although fire bond referendums historically have high approval rates among voters.

Whenever revenue is generated by the sale of bonds for a specific capital improvement project, these funds cannot be used for any other purpose. In some cases, dependent on local or state laws, funds may be transferred from one project to another project as long as both are capital improvement projects. For example, if excess funds remain at the end of the construction of a new station, these monies may be used for a rehabilitation project at a different station. Funds for a turnkey project, which means the project is immediately ready for use, can be used toward station construction, furnishings, and, in some cases, new apparatus. However, funds from bond sales cannot be used to cover salaries because these costs are part of the operating budget.

Investments

Investing funds is another source of revenue for governments. Governments have the ability to earn interest on their bank accounts. Many sources of revenue for governments, such as taxes, are due to be paid by a certain date. For example, the federal income tax is due by April 15 each year. Most of this revenue is needed to pay salaries and other fixed costs throughout the year. Because these funds are not needed immediately, the government can invest in short-term investment instruments. Many larger governments hire financial managers to invest these funds.

Some investments can be very risky, especially those that report high returns. One notorious example occurred on December 6, 1994, when Orange County, California became the largest municipality in U.S. history to declare bankruptcy after its risky financial investments unexpectedly lost most of their value. Generally, however, governments tend to stay with low-risk investments. For example, the federal Social Security Trust Fund can invest only in federal government bonds (T bills). Even in times when some investment funds were reporting up to 30% returns, the Social Security Trust Fund was only making 2.9% on its surplus funds.

Caution is always necessary with government revenues. Although the economy has improved since the 2009 financial crisis, states have cut their budgets and dipped into emergency funds to make ends meet. In some situations, this is compounded by the reduction of investment income in retirement accounts. During the first few years of the Trump presidency, the increases in stock values were substantial, although as a result of the world pandemic most of this increased value disappeared in the short term. While stocks took a beating in the early days of the pandemic, the stock market has seen a good rebound, which is very impressive since their lows in March 2020 were around 20%. However, when states have to contribute more into retirement accounts to offset losses in stock investments, they usually have less tax revenues to support the yearly budgets of agencies, including FES. This is problematic for FES administrators who may have to cut their budgets.

Other Forms of Borrowing

If government bonds are not available because of the debt ceiling, an agency may lease apparatus, with the option to purchase it later. Lease-purchase contracts are an option when the need for new equipment or apparatus is critical. In addition, a lease-purchase contract might have a better chance of approval from the legislature because the costs are recorded as a line item in the operating budget instead of a capital purchase. This concept is similar to purchasing a new vehicle; an individual may purchase it by using a loan with a monthly payment, leasing it with a monthly payment, or paying cash for it up front.

Techniques for Increasing Funding

Once an administrator has a basic understanding of the agency's sources of revenues, it may be useful to consider various approaches to increase available funding. These techniques include providing supportive evidence from national standards, performing a thorough cost-benefit analysis, and defining potential sources of funding.

Provide Justification from National Standards

The FES administrator must recognize that the FES organization is part of the government bureaucracy and is in competition with other agencies for tax revenues. Most of these other agencies have very few standards

for the services they provide to the public, because output from these agencies is very difficult to measure. However, FES agencies have had standards to benchmark services provided and to justify their budgets since the 1970s and 1980s. Therefore, FES administrators are at an advantage when they can support their requests for additional funding with evidence about the actual and legitimate benefits it will bring.

Several relatively recent national standards such as NFPA 1500, *Standard on Fire Department Occupational Safety and Health Program*; NFPA 1710, *Standard for the Organization and Deployment of Fire Suppression Operations, Emergency Medical Operations, and Special Operations to the Public by Career Fire Departments*; NHTSA, *National EMS Education Standards*; and OSHA, *Fire Brigade, Hazardous Materials, and Respiratory Protection*, can be used to focus budget requests on resulting effective and safe service to the public. These standards address training, professional qualifications, personnel safety, and the deployment and staffing of FES units. When chief officers use these national consensus standards as part of a formal budget planning process, they increase government and public support for new and existing programs.

Performing a Cost-Benefit Analysis

If a survey was performed about the value of FES, there probably would be a great variance in the valuations. For example, a family enjoying a nice quiet evening in front of their television, with no fire or smoke in their home, would probably express a low value for having an FES organization ready to respond and rescue them in a fire emergency. Take this same family and place them into a 13th floor apartment with a fire blocking their exit to the stairway, and there would be a new higher value for FES.

In most cases, it is very difficult to measure benefits or direct outcomes from FES programs. What makes this cost-benefit analysis difficult is that it is hard to assign a value to an individual's life or quality of life. This type of financial analysis attempts to define benefits from the input of tax dollars. Moreover, the outcome (benefit)—saving lives and property—is often the result of a complex set of circumstances. The following is a hypothetical set of circumstances that led to a death in a house fire:

1. Fire prevention regulations did not require smoke alarms to be installed in existing dwellings.
2. The local FES organization did not offer any home fire safety inspection services or the free installation of smoke alarms.

CHIEF OFFICER TIP

FES as Free Public Services

In most communities, FES services are provided as a free public service. Unfortunately, because there is no direct charge to the consumer, these public services are often requested more times than they are needed. This results in an excessive demand, causing delayed response times when units are not available. The higher the call volume, the greater the chance of the closest unit being busy on another call. As a result, FES organizations struggle with false alarms caused by faulty equipment and excessive EMS calls caused by requests for nonemergency medical problems. To try to reduce excessive use of fire services for nonemergency situations, some jurisdictions have adopted an ordinance that fines businesses that have multiple false alarms to encourage them to fix malfunctioning alarm systems.

Many departments that provide EMS transport are struggling with heavy demands for service. This service is essentially free to the patient in most cases. Even where there is a fee, it is either paid by medical insurance or Medicaid/Medicare, or not paid at all. A free public service can create a market inefficiency when people request the public good (emergency ambulance service) when it is not really needed.

Most funding for FES services is provided by the property tax. However, many properties do not pay any property tax, because it is normal to exempt federal, state, and local government properties along with nonprofit organizations, such as churches. Because FES are a public service afforded to all property owners, visitors, and residents regardless of their tax contribution, property owners who do pay property taxes essentially subsidize free services for those properties that are tax exempt. In jurisdictions where governments or nonprofit organizations have large holdings of property and buildings, there is a concern that these organizations unfairly do not pay for the emergency services they receive.

There are some solutions to generate revenues while still providing a public service. In some cases, agreements have been achieved that require an annual fee in lieu of property taxes. This situation helps fund a substantial portion of the Washington, DC, government, including the FES department. In other cases, a donation of property for the construction of a station has occurred, for example, at some large universities. In addition to these efforts to generate revenue, government regulations that encourage owners to purchase and maintain smoke alarms and fire sprinkler systems may be necessary to keep demand for fire services down.

---

3. The occupants of the dwelling did not see any value in purchasing and installing smoke alarms.
4. The occupants of the dwelling did not recognize the hazard of careless smoking habits in the family room.
5. The first arriving fire unit was understaffed or personnel were not adequately trained.
6. When the victim was eventually located and brought to the front yard, the FES organization was not equipped to provide EMS.

It could be argued that correcting any one of these six items could have potentially saved this fire death victim. However, anything less than 100% effort in all these key areas can lead to loss of life and property.

Still, there are some benefit evaluation measurements that may be helpful. For example, in this specific case, the costs associated with correcting all six of the items can be listed. The benefit is described as saving victims and property from a structural fire. Since it is impossible to measure something that did not happen (i.e., fire deaths), an outcome cannot be measured directly. Thus, this should justify the measurement of correcting all items listed previously.

In the business world, the benefit is new or increased profit; however, the vital piece to this analysis is to have benefits that are measurable or can be described in common sense consequences. For example, an administrator is considering equipping the next fire engine purchased with a compressed air foam system (CAFS). The costs are easily obtained from the manufacturer. The benefits need to be described in subjective terms because sound numbers are impossible to calculate. Some of the advantages of CAFS are lighter weight hose and increased firefighting efficiency of the foam over plain water. It is not possible to quantify these advantages as a result of the many variables that are typical in any hostile fire situation. However, CAFS has many positive benefits to firefighting that can be described as potential benefits rather than hard numbers.

Define Sources of Funding

To increase the likelihood of a spending request being approved, the fire or EMS administrator can point to specific areas of available funding. Sources of funding may include increases in tax revenues or fees.

The United States Fire Administration guide, *Funding Alternatives for Emergency Medical and Fire Services*, provides information on locating and implementing traditional and nontraditional methods of funding.

In addition, several federal grant programs offer financial assistance to fire departments in the United States. All of these programs are under close scrutiny each fiscal year, especially now that the federal debt is a major issue with voters and elected representatives. There are four separate grant programs. The Assistance to Fire Fighters Grant Program is a 1-year grant provided directly to fire departments and non-affiliated EMS organizations to enhance their firefighting and emergency response needs. Fire Prevention and Safety Grants provide assistance for local, regional, state, or national organizations for fire prevention and safety, primarily directed at high-risk target groups, including children, seniors, and firefighters. Staffing for Adequate Fire and Emergency Response Grants are awarded directly to volunteer, combination, and career fire departments to help the departments attain 24-hour staffing by hiring career firefighters or to help recruit and retain volunteer firefighters.

Finally, the Department of Homeland Security offers fire departments Assistance to Fire Fighters Fire Station Construction Grants to build new or modify existing fire stations. Initially designed as a stand-alone program, they are now folded into the U.S. Department of Homeland Security's consolidated grant program. The U.S. Fire Administration has a webpage (https://www.fema.gov/welcome-assistance-firefighters-grant-program) that helps departments with their grant requests. Evaluation of grant applications is done by fire service peers.

EMS grants may be obtained for equipment, apparatus, and education through the U.S. Health Resources and Services Administration and other state or local organizations. Insight into how to be a successful grant applicant can be obtained by contacting other departments that have been successful and asking them for advice.

Another possible source of funding discussed previously in this chapter is cost recovery programs that bill insurance companies for services provided. Ambulance companies and fire departments may bill for transport fees, or hazardous materials teams can file for reimbursement through the transport vehicle's insurance coverage or OSHA's *Emergency Response Operations for Releases of, or Substantial Threats of Releases of, Hazardous Substances Without Regard to the Location of the Hazard*. Some departments charge a fee for FES services at motor vehicle crashes to the at-fault driver's insurance company.

Administrators should be prepared to recommend a funding source from a new tax or an increase in an existing tax or fee if asking for substantial additional funds. Proposing a tax increase should be saved for those occasions when attempting to make substantial improvements in the organization and its services or when implementing a master plan. For example, in a department that needs three new stations for adequate coverage, an implementation plan may detail the opening of one new fire/rescue station a year for the next 3 years. This also requires the hiring of new personnel each year. The proposal for funding may include step increases in a tax to generate an additional number of dollars each year.

Along with a proposed funding source that is a tax or fee, be prepared to explain all the ramifications and who is affected. In general, if the taxpayers are convinced that there is added value or service from the increase in the tax or fee, they will probably support the increase.

Expenditures
Salaries and Benefits

In challenging economic times, a steady stream of stories in the media point out some of the most outrageous examples of salaries and benefits of public sector employees. This news can be very troubling for the public, especially those without jobs or generous benefits. However, this was not always the case.

Many years ago it was common for most government workers to make very poor salaries and, in the case of firefighters, work very long hours. As a reaction to pressure from employees and the ever-increasing use of public sector unions, it was popular for elected officials to grant generous increases in benefits, especially those that did not have to be paid in the near future such as pensions, medical insurance, and cash-outs for vacation and sick leave. At the same time, public sector employees were gaining increases in wages and, over many years, their salaries ended up equal to or above those in private industry.

Elected officials and the public are now examining some private sector practices to reduce the cost of employees' salaries and benefits. An example of reducing the costs of a benefit, while actually providing a healthier employee, is a policy that Safeway Inc., one of North America's largest supermarket chains, instituted in 2005. This plan rewards employees for meeting certain health wellness criteria. These criteria are based on the notion that 74% of all healthcare costs are confined to four chronic, mostly preventable, conditions: (1) cardiovascular disease, (2) cancer, (3) diabetes, and (4) obesity (Burd, 2009). To determine whether an employee receives the incentive, Safeway tests employees' tobacco usage, healthy weight, blood pressure, and cholesterol levels. A plan like Safeway's could "ultimately reduce our nation's healthcare bill by 40%" (Burd, 2009). In fact, during the first 4 years after the plan was implemented Safeway was able to keep its per capita healthcare costs flat, whereas other American companies' costs increased 38% (Burd, 2009). For example, wellness programs have been included in the Affordable Care Act (ACA) in 2013 and contain provisions for rewards. Similar programs could be very valuable in providing increased health and safety for firefighters and EMS personnel. This also has the advantage for FES organizations that require strength and cardiovascular fitness to provide enhanced physical performance by members.

Tracking Expenditures

FES administrators should be able to spot mistakes or unplanned expenses as soon as possible so that spending can be adjusted. A system should be set up that keeps track of expenditures on a monthly basis. At the beginning of the fiscal year, the administrator should calculate monthly anticipated expenditures. Some expenses, such as salaries, are predictable, whereas others may be one-time or variable expenditures, such as overtime.

TABLE 5-2 shows a sample line-item budget for 4 months of a fiscal year. Because this is only one-third of the annual budget, the benchmark for its expenditures is one-third of the total (or 33%). On the bottom line, the report points out that 33.01% of the department's budget has been spent or encumbered, which indicates a healthy budget position. However, several line items are over the 33% mark (e.g., printing, electricity, and travel). If the bulk of these areas of expenditure had been anticipated in the first 4 months, this would not be a problem. However, because a substantial amount of travel is projected for the upcoming months, the administrator should closely monitor the situation and may need to cancel travel plans or transfer funds from other line items. Other expenditures

TABLE 5-2 FES Department Budget April Expenditure Report							
Object	Description	Budget Plan	Current Month	Year-to-Date	Encumbrances	Balance Available	%
102	Salaries, permanent	$695,271.00	$57,939.25	$230,975.90	$0.00	$464,295.10	33.22%
103	Salaries, overtime	$48,668.97	$4,055.75	$13,294.27	$0.00	$35,374.70	27.32%
104	Health/welfare insurance	$75,228.32	$6,269.03	$24,999.00	$0.00	$50,229.32	33.23%
105	Dental insurance	$11,402.44	$950.20	$3,907.65	$0.00	$7,494.79	34.27%
106	Retirement	$108,462.28	$9,038.52	$36,795.00	$0.00	$71,667.28	33.92%
107	Vision care	$2,781.08	$231.76	$939.27	$0.00	$1,841.81	33.77%
120	Medicare	$2,363.92	$196.99	$786.46	$0.00	$1,577.46	33.27%
Total Category 1		$944,178.01	$78,681.50	$311,697.55	$0.00	$632,480.46	33.01%
201	General expense	$3,500.00	$605.00	$421.00	$0.00	$3,079.00	12.03%
241	Printing	$1,300.00	$137.50	$550.00	$0.00	$750.00	42.31%
251	Communication	$8,000.00	$608.00	$2,505.00	$0.00	$5,495.00	31.31%

(continues)

TABLE 5-2 FES Department Budget April Expenditure Report (*Continued*)

Object	Description	Budget Plan	Current Month	Year-to-Date	Encumbrances	Balance Available	%
261	Postage	$2,400.00	$177.00	$792.00	$0.00	$1,608.00	33.00%
291	Travel	$500.00	$0.00	$0.00	$225.00	$275.00	45.00%
331	Training	$800.00	$0.00	$0.00	$275.00	$525.00	34.38%
341	Facilities operations	$4,500.00	$0.00	$996.00	$0.00	$3,504.00	22.13%
361	Electricity	$16,000.00	$2,310.00	$9,240.00	$0.00	$6,760.00	57.75%
371	Heating oil	$18,000.00	$591.00	$2,110.00	$0.00	$15,890.00	11.72%
410	Professional and special services	$45,000.00	$3,712.53	$4,418.27	$12,000.00	$28,681.73	36.49%
431	Data processing	$1,800.00	$148.50	$594.00	$0.00	$1,206.00	33.00%
503	Uniforms	$500.00	$0.00	$0.00	$500.00	$0.00	100.00%
524	Vehicle operations	$49,000.00	$3,791.16	$15,348.66	$0.00	$33,651.34	31.32%
569	Equipment rental	$12,000.00	$0.00	$3,960.00	$0.00	$8,040.00	33.00%
571	Unallocated	$0.00	$0.00	$0.00	$0.00	$0.00	0.00%
Total Category 2		$163,300.00	$12,080.69	$40,934.93	$13,000.00	$109,465.07	33.03%
Total Org/Prog		$1,107,478.01	$90,762.19	$352,632.48	$13,000.00	$741,945.53	33.01%

that are below 33% (e.g., heating, which is not yet in season) may take place later in the year, so this surplus is only temporary.

At the end of each month, the administrator should review the actual expenditures for the month and compare them to the expected amounts. A running total should be calculated along with any variance to be sure the agency is not spending more than budget allocations allow. To ensure that adequate funds are available for the purchase of goods or services, fire administrator can use an **encumbrance**, which sets aside designated funds until the invoice is received and payment is issued. In this way, funds are earmarked for a particular expenditure and are not confused with other available funds. An encumbrance is an accounting procedure that indicates that the money has already been committed to be spent even though the actual payment may be made sometime later in the fiscal year. In some cases, the encumbrance can be paid in the following fiscal year, allowing the use of those funds from the current year. For example, if an administrator encumbered funds for CAFS but there was a delay in the manufacture and delivery, the contract may be extended at the same price, allowing delivery and invoicing to occur in the next fiscal year.

Keeping real-time spending records for each month facilitates the planning of any future and actual seasonal peaks (e.g., winter heating oil) or one-time costs occurring in a later month. If, for example, the purchase of protective clothing for wildland firefighting was planned for the eighth month of the fiscal year (2 months before the next wildfire season), these funds could not be encumbered until the order is actually placed. The administrator would note this future purchase in a personal spreadsheet.

It is also a good idea to keep personal records of expenditures to double check computer reports from the budget department. Data entry mistakes are always a possibility. When dealing with number codes to designate the department and line-item category, hitting the wrong number key can put expenditures into the wrong agency's budget **FIGURE 5-5**. A procedure of double keying, where two different operators enter the same data, helps to eliminate most of these errors. The administrator should also verify computer reports for timeliness and accuracy to ensure that the budget department is not late in documenting expenditures or revenues so there is no confusion about the actual amount of available funds. For example, if the department has just received an order of fire hose costing several thousand dollars, but it has not shown up on the latest budget summary report, the administrator needs to ensure records are accurate.

Finally, it is a good idea to keep a list of the previous year's total expenditures for each month to identify any significant changes that may have caused variances in the present budget cycle. Any unexpected costs that were the result of new mandates, labor contract changes, severe weather, or large-scale emergencies may require the administrator to recalculate the budget or request additional funds.

At the end of the fiscal year, many government agencies attempt to spend any unspent funds. In many cases, these purchases are not needed but are used to assure that there is not a surplus to be given back to the general fund. To avoid this practice, budget department personnel at the direction of the city or county administrator may temporarily suspend capital purchases around this time of the year.

Purchasing Policies

Because purchasing policies vary widely among different governments and their agencies, the administrator must become familiar with local rules and procedures; failure to follow these rules can lead to disciplinary procedures. Policies may determine the following:

- Limits on spending for various services and commodities
- Limits on purchases made without a formal bidding process
- Standard procurement of expendable items
- Contract items on a master purchase order list
- Policies for emergency procurement during a disaster incident
- Policies regarding petty cash
- Procedures for inventory of property or equipment
- Required approval for capital items or travel
- Standard service contracts
- Lists of approved vendors

If there is no formal policy for a particular purchase, it is prudent to find out what the past practice has been and to document the details of the purchase with the budget analyst assigned to the department. It is also wise to consider how the transaction would look to outside observers (e.g., local media) before making a purchase.

Purchase policies may be based on the useful life span prediction for an item. For example, if the department has four fire engines, and the chief can convince the governing body that the useful life of a fire engine is 12 years, then a good policy dictates purchasing a new fire engine every 3 years to ensure that no engine is over its useful life.

The quality and completeness of the "Request for Proposals (RFPs)" for major capital items is extremely important. RFPs from other similar sized departments should be sought and requested for use. Statewide cooperative purchases can be a good place to find a quality and very competitive price for these costly items. This is especially valid for small to midsize departments. This is not a good place to reinvent the wheel.

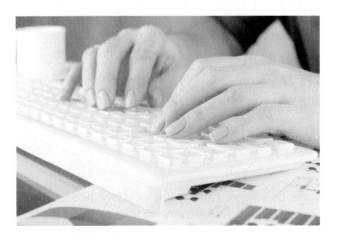

FIGURE 5-5 Double key entry is important to catch mistakes in keyboard entry.
© Yeamake/Shutterstock.

Total Cost Purchasing

Another calculation that the administrator should consider is total cost purchasing. The total cost of an item considers both the cost of purchase and any other associated costs, such as maintenance and downtime (including the cost to rent or reserve associated equipment), and any surplus value the item may have at the end of its useful life. The total cost of an item is especially useful to compare bids for major purchases.

When comparing bids from different vendors, it is not always best to go with the lowest bidder. If the low bid vendor's equipment has high maintenance costs or worse than average down time, the actual total cost of the purchased item may be more than that of another bidder with a slightly higher purchase price. Remember, the business of emergency services is about always being ready to respond. An unreliable piece of equipment should not be acceptable. It may be useful to contact peers to obtain a list of other agencies that have purchased the same equipment and ask about their experiences with a particular manufacturer.

Cooperative Purchasing

One way to reduce the total cost of an item is to find opportunities for cooperative (or joint) purchasing of equipment or supplies. In some areas, items such as fire hoses, personal protective equipment (PPE) and gear, and even apparatus may be on a regional or state purchasing bid list. In some cases, purchases may also be made from the U.S. Government Services Administration's Cooperative Purchasing Program. These joint arrangements consolidate purchasing power across several departments, resulting in lower total costs. Cooperative purchasing can save money because companies typically give a discount on bulk purchases. However, cooperative purchasing may also result in a few disadvantages, such as the need for compromise regarding local preferences of design or manufacture, a potentially longer delivery time, and a lack of control over the contract process. Most of these disadvantages can be overcome by detailed review of the specifications, noting where they differ from the department's previous purchases. If the variance is truly significant, a request for a modification to the supplier or manufacturer may be obtainable. The administrator should be involved in the process for the development of the consolidated bid specifications to get better results for their agency.

Legal Considerations

There are many legal considerations to the purchasing process, and the administrator needs to understand the

CHIEF OFFICER TIP

Disaster Purchasing Plans

Agency plans and budgets are designed to cover routine call loads, and may include extra funds for unanticipated larger incidents; but in general, FES budgets cannot afford to set aside funds for major disasters. When Oklahoma was hit with a catastrophic tornado in 1999, EMS agencies formed ad hoc emergency rooms to handle the increased demand **FIGURE 5-6A**. When massive wildfires hit San Diego in 2007, responders tried to battle the flames and care for 500,000 evacuees in an already overloaded EMS system **FIGURE 5-6B**. In these types of situations, agencies do whatever they can to acquire supplies, including going to noncontract vendors or pulling from other non-affected areas. At some point, a state or federal disaster declaration can be made to use government supply caches. Disaster purchasing plans are essential to provide guidelines in these emergency situations. These plans should be done prior to any possible disaster and involve other municipal agencies.

For example, the federal government had stockpiled what was believed to be an adequate number of ventilators and personal protective equipment (PPE) to help hospitals respond to an epidemic of seriously ill patients. The Federal government's role is designed to back up state governments. This is by design because the United States has a federalist form of government that limits the central government power. But as a result of state governments, including hospital and medical facilities, having little or no stockpile of critical ventilators and PPE in the 2020 coronavirus national emergency, the federal government has taken a leading role. In the past, emergency planners learned this lesson in our response to devastating major hurricanes the hard way, by failures. This will play a major consideration in future planning for major disasters and should be a coordinated effort of federal, state, and medical facilities.

rules, policies, and laws for the local jurisdiction. For example, in some jurisdictions, an affirmative action plan requires a certain percentage of the purchasing dollars be awarded to minority- and female-owned businesses. In other instances, relationships with previous contractors,

A

B

FIGURE 5-6 Purchases for major disasters may need to occur during the disaster, but preferably before.
A. Oklahoma tornado, 1999. **B.** San Diego wildfire, 2007.

suppliers, or manufacturers may affect the bidding and awarding of contracts. Unfortunately, some of these practices may cause additional dollars to be spent from the budget and inferior equipment to be purchased.

When discussing these issues with government officials, argue from the perspective that the department needs the best equipment or supplies that can be purchased within reason. Stress that emergency services to the public require the most reliable equipment and supplies possible. In some cases, the U.S. Supreme Court has found these set-asides unconstitutional; however, some administration and elected officials strongly support these practices. Therefore, it might be wise to avoid discussing the legal aspects or ethical considerations and instead review these practices on a case-by-case basis with an emphasis on quality and reliability. Whenever there is a question, the administrator should consult with municipal attorneys for legal counsel before awarding a contract.

Making Budget Cuts

If forced to cut the department's budget, it is a good idea for the chief to have a plan that protects the highest priority items. It may be very tempting to choose items that are the least controversial to cut, but in the long run these types of cuts may have a distinctly adverse effect on quality of service. For example, when faced with mandatory cuts in their budget, many chiefs of FES organizations cut training or prevention items to keep all existing fire stations open. This can be short sighted; the long-term effect of cutting in these areas can be substantial. It may take a year or more, but cuts in training impact the quality of services to the public, as firefighters lose their level of firefighting skill proficiency and injury rates increase.

An article in *Firehouse* magazine explains the typical budget cut situation (https://www .firehouse.com/safety-health/article/10468568/ telling-it-like-it-is-february-2010):

> We have never encountered, and we don't know anyone who ever has encountered, a budget meeting where funds were reduced while raising performance expectations. We typically impose that upon ourselves. We want to do better every year…In the face of budget cuts, most fire departments do everything possible to maintain the same or similar levels of service. What they ought to be doing is reducing services every time the budget is reduced.

In the FES, "Boy Scout syndrome" is sometimes used to refer to personnel who try to help others under any circumstance, even when the situations are dangerous. Others call this the "can do" attitude. It is the optimistic view that goals can be accomplished even when resources and members are not adequate or safe for the job. However, if cutting services is unavoidable, how are budget-cutting decisions ultimately made and consequences evaluated?

Adaptive Policies

Budget reductions may be disguised by dangerous and ill-advised adaptive policies, such as brownouts or reduced staffing. Brownouts are a strategy in which fire companies go unstaffed on a rotating basis. In most cases the object is to meet a budget reduction without closing any fire stations. This type of cut is almost invisible to the public unless the closest company to a tragic fire happens to be the one out of service. This practice is dangerous because it is impossible to anticipate if a fire will occur in the first due area of a temporarily closed station. An alternative measure may be to reduce staffing. Fewer firefighters per company may be able to keep costs down. However, mini-pumpers or fast attack units can never adequately replace fully staffed fire companies. For safety and efficiency, national standards and rigorous scientific research have substantiated that a minimum of one officer and three firefighters are needed on duty at all times; this minimum is also in compliance with NFPA standards and OSHA regulations. Therefore, neither brownouts nor reduced staffing are recommended practices.

One safe alternative for nonmetropolitan departments is shared specialized services with other surrounding departments. For example, if three departments are bordering each other, one could staff a ladder truck, one a heavy rescue unit, and the third a hazardous materials company. Some states have organized and funded hazardous material companies to cover large areas protected by smaller departments. These services must be covered with governmental automatic aid agreements.

Another approach, zero-based budgeting (ZBB), is a financial decision-making technique that reverses traditional budgeting. In traditional incremental budgeting, only variances from past years are justified, based on the assumption that the "baseline" is automatically validated. In ZBB, every line item must be approved as if it was the first time it was requested.

ZBB is based on the assumption that the present level of funding may not be justified. Funding levels evolve historically by each agency fighting for its fair share and may not match present-day needs. In addition, existing funding may not reflect the relative desires of the public and its elected officials. It may be argued that if last year's budget was a fair distribution of tax revenues, the same percentage split should also be fair for the following year. ZBB challenges this assumption and requires agencies to justify budgets from the ground up. This can be a cumbersome and time-consuming process.

Many times a modified ZBB is a more reasonable goal by identifying a smaller percentage of the total budget to provide new updated justifications. This technique could be used inside the agency as an "across the board" percentage cut that is used in a simulated evaluation. Depending on the size of the department, distributions vary. The following is an example of a department that has to make a 10% budget cut:

- Fire prevention division has 10 inspectors: eliminate one inspector.
- Training and safety division has 20 members: eliminate two members.
- Operations division has 82 members: eliminate 8 members. This elimination would probably require the closing of one company along with its station to keep staffing at safe levels for the other companies.

The screams can be heard throughout the fire department when a fire station is proposed to be closed.

For the safety of the public, a sign should be placed on the station that says CLOSED. That would keep any citizen from stopping by with a heart attack or other emergency, expecting to get help. It would also be a daily reminder to anyone passing the station that it was closed for lack of funding. If the station was truly justified by minimum response time criteria, funding for staffing would most likely be found in the near future. However, if the department deeply cuts non-emergency services or staffing per company, it may never get these resources back.

Common Areas for Budget Cuts

The most common areas cut in FES budgets are staffing, stations, salaries, and retirement benefits. For the FES administrator, it may seem unfathomable to make cuts; the thought of reducing services may seem totally unacceptable. Yet, at some point, all administrators may be faced with the difficult task of finding ways to reduce the agency's budget. Although closing stations or reducing services may seem like the last option from a personnel or community perspective, these options may be considered along with other cuts. The chapter, *Ethics,* discusses these dilemmas further.

Cutting retirement benefits is a very controversial endeavor to reduce the budget. Often these retirement plans are already underfunded because of unrealistic union contract agreements or pensions that were negatively affected by stock market downturns. Even though experts project that individuals should have 80% of their working income to maintain a comparable lifestyle after retirement, this seems like an impossible dream to many firefighters and EMS responders. Lower retirement ages and longer life spans post-retirement add to the current financial pressures.

A 2016 Forbes' article warned, "…U.S. state and local government pensions have 'unfunded liabilities' estimated at more than $5 trillion and funding ratios of just 39%. There are two glaring reasons for our pension crisis: poor investment decisions and greedy [and maybe dishonest] assumptions" (Maurer, 2016). The market crash and the economic fallout from the coronavirus have been reported to result in nearly $1 trillion in investment losses for U.S. public pension funds. This was reported by Moody's in a March 2020 analysis. It is not clear at this time if the market will recover these losses over time.

Indeed, some governments have now resorted to cutting retirement benefits, increasing the retirement age, making employees pay more into the retirement system, and providing more robust oversight and investment rules. Some have forced employees to switch to 401K and 457B retirement plans, relieving the cities, counties, and states of any future defined pension obligation.

As part of the 2013 bankruptcy settlement, the City of Detroit retirees accepted a 4.5% cut in their pensions. They lost cost-of-living adjustments and saw healthcare benefits cut. A contributing factor was not the pension plan itself, but rather the deterioration in the city's finances. In response to decreases in tax revenue to finance the budget, instead of cutting services or raising taxes, municipalities reduced their contributions to pension funds. They say they will catch up later when times are better. Many never do.

Wrap-Up

CHAPTER SUMMARY

- FES financial managers are responsible for the creation, monitoring, and real time adjusting of the use of public funds to achieve the department's goals.
- A comprehensive understanding of the budget process for the municipality is essential to achieving a fair share of public funds.
- A thorough knowledge of the government's revenue sources is imperative, especially when making a new proposal or proposing substantial increases in the FES budget.
- Monitoring the budget by tracking expenditures and personnel costs must be ongoing and accurate.
- For new programs and substantial increase in the FES budget, techniques to justify the increase in funding must be fully comprehended and applied to influence the elected officials.

KEY TERMS

Brownouts A cost-cutting strategy in which fire companies go unstaffed on a rotating basis.

Budget reports Periodic documents created by the administrator to monitor budget expenditures and detect any deviation.

Capital item An operating cost that has a long life expectancy, such as a fire station.

Cooperative purchasing Joint buying arrangement that consolidates purchasing power across several departments, resulting in lower total costs.

Compliance audit Documentation review that checks for adherence to a legal or policy mandate, which can include elements of fiscal management.

Cost-benefit analysis A cost-benefit analysis is used to measure the benefits of a decision or taking action versus the costs. It is commonly used in the business world because the benefits and profits can be more easily identified. A simple FES example would be the justification of a new fire station. The benefit would be reduced response time to a known number of citizens. This benefit in and of itself is not easily defined. Studies that have tried to correlate the relationship of reduced response time to either life or property loss have not been successful. Although, on a common sense level, reduced response time is always beneficial.

Debt ceiling In the federal government, this is the maximum national debt that cannot be exceeded by law. It is typically increased by Congress as needed with some posturing and even some compromising because the two political parties have opposing views on public policies.

Encumbrance Accounting procedure that sets aside designated funds for a particular expenditure until an invoice is received and payment is issued.

Expenditures This is the cost in dollars used to acquire some merchandise or service.

Fire service tax A government tax that can be a set value for all residences or can vary depending on the square footage of the dwelling or the fire flow formula.

Financial audits Accounting procedure usually performed on an annual basis to ensure that financial records are accurate and complete.

Financial management The process of translating financial resources into services while providing fiscal accountability.

Financial manager A person (s) responsible for creating, justifying, and monitoring the department's budget.

Fiscal year A full year budgetary period starting on July 1 for local and state governments and October 1 for the federal government.

Government bonds A form of money lending that has a low interest rate at a set interval for up to 30 years with tax-free interest.

Government fee for service These are fees charged by local, state, and federal governments for various services such as admission to national or state parks, drivers licenses, and toll roads.

Income tax This is a tax based on a person's income and can be either progressive or flat. Most are based of the federal income tax that is progressive, requiring a larger percentage the more income is earned. Many earners at the low end of the pay scale are not taxed.

Lease-purchase contracts An agreement to lease a capital item with the option to purchase it later.

Line-item budgets A statement of used and available funds broken down into categories.

Procedural audit A review designed to evaluate management policies and procedures, which also can include financial policies and procedures.

Property tax Government tax assessed as a percentage of the total value of property and improvements, or in some cases based on income for a commercial business.

Revenue This is the money received by the political jurisdiction from taxes, fees, or borrowing.

Sales tax Government tax applied to consumer purchases, such as electronics and vehicles.

Total cost purchasing A purchasing strategy that considers both the cost of purchase and any other associated costs, such as maintenance.

Unfunded mandates Regulations requiring expenditures that are not combined with any government funding.

User tax A government tax that applies only to the use of specific goods or services, such as a road tax.

Zero-based budgeting (ZBB) Financial decision-making technique in which every line item must be approved as if it was the first time it was requested.

REVIEW QUESTIONS

1. Identify an existing need in an FES department.

2. Quantify the resources, both staff and equipment, needed to implement a program to address this need.

3. Provide a list of all costs.

4. Provide an overview of potential support or opposition to the program, including key players.

5. Describe a hypothetical or real situation that could support an emergency request to fund the program.

CHAPTER ACTIVITY #1: BUDGET INCREASES

In the article "A Master at His Art," a unique technique for gaining budget increases is exemplified in a story involving former Los Angeles County fire chief, Keith E. Klinger.

According to the article, a rapid wildfire in Bel Air, California in 1961 destroyed nearly 600 homes within 72 hours. Throughout that night, Chief Klinger actively led the response efforts, never taking a break or slowing down. The very next morning, he drove straight over to the Los Angeles County Board of Supervisors' weekly meeting. Although he wasn't on the meeting's agenda, Chief Klinger walked right up to the podium and, still covered with soot from the night before, gave the board a descriptive report on the catastrophic wildfire that was taking over the mountains.

Without hesitation, Chief Klinger then proceeded to describe a 10-year plan for fire protection improvement. Although the cost of the plan was expensive, the board could not refuse it due to the current predicament. The chief had developed the plan prior to the Bel Air fire, but his effectiveness came from waiting for the best time to present it to the powerful elected officials (Page, 1998).

Discussion Questions

1. Can you describe a large emergency in your jurisdiction that would rise to this level of importance?

2. What is the schedule for meetings of the elected officials in your jurisdiction? Would there be an opportunity to make this type of presentation?

3. Discuss one item that you think is needed to improve the fire protection or emergency medical service along with a short justification.

CHAPTER ACTIVITY #2: BUDGET ADJUSTMENTS

Central City is an older city that has grown slowly over the years by annexing several pieces of adjoining land **FIGURE 5-7**. The city covers approximately 60 square miles and is a typical medium-sized city with diversified light industry, a busy commercial area outside the central core, a community college, one hospital, and several medium-sized assembly buildings in the downtown area. More recently, the city annexed an area primarily consisting of expensive homes in the northwest side in order to bring in more tax revenues to cover existing shortfalls.

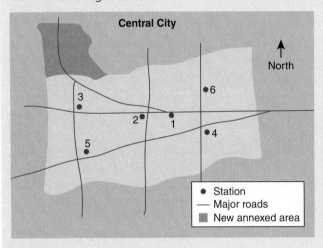

FIGURE 5-7 Central City.

The Central City Fire Department operates six engines (one from each station) and two ladders (Stations 1 and 5) under the supervision of two district chiefs. The department responds to life-threatening medical calls, but it does not transport patients. Firefighters work a three-platoon system. Review the Central City Fire Rescue Department organizational chart **FIGURE 5-8** and table containing the characteristics of the fire stations **TABLE 5-3**.

The firefighter's union contract calls for a 53-hour work week with an additional 3 hours of overtime pay. Therefore, each firefighter on shift works 56 hours per week. These costs are already built into the present budget. In addition, the budget contains enough funds to pay 5% of vacant positions due to sick and annual leave.

A master plan was completed recently and the following major recommendations were made to the department: (1) increase the staffing of all companies to four on-duty persons to meet national safety standards; (2) construct a new fire rescue station in the northwest section of the city to reduce response times to the city average of about 4 minutes; and (3) start a new fire safety education program by increasing the staffing of the Fire Prevention Division by two personnel.

FIGURE 5-8 Central City Fire Rescue Department.

TABLE 5-3 Central City Fire Stations

Station	Date Constructed	Average Response Time	Population Served	Square Miles in First Due
1	1937	2:45	13,000	7
2	1943	2:33	9,000	4
3	1957	5:50	36,000	16
4	1963	4:10	22,000	11
5	1971	4:35	22,000	13
6	1982	4:05	20,000	9
Total		3:54	122,000	60

Discussion Questions

Consider the following two scenarios and answer the discussion questions that follow each. It is helpful if you can approach this activity without any biases. Many FES members prefer not to cut areas where they are presently working. Where appropriate, use reference material to justify any cuts or additions.

Scenario #1: You have been instructed to cut your budget by 10% and specifically advised that you must cut 13 positions from the 121 total positions.

1. Where would you cut positions and how would you justify your proposal?

2. How would you select and measure "outcomes" or consequences of your plan?

3. What type of reaction would you expect from: (1) the union president, (2) the public, (3) the press, (4) the city manager, and (5) the city council?

Scenario #2: Starting with the beginning level of staffing, you have been instructed that your budget will be increased by 10% and specifically advised that you must hire 13 new employees, binging the total personnel to 134.

1. Where would you add positions and how would you justify your proposal?

2. How would you select and measure "outcomes" or consequences of your plan?

3. What type of reaction would you expect from: (1) the union president, (2) the public, (3) the press, (4) the city manager, and (5) the city council?

CHAPTER ACTIVITY #3: DISASTER PLANNING

Some of the most expensive types of fire incidents are those that require both fire suppression and patient care. In the San Diego wildland fires of 2007, 500,000 people were required to evacuate, many of whom needed acute medical and trauma care. Even though local departments often know what types of natural disasters they should expect, they can't plan for any specific event. In the case of San Diego, the EMS system was quickly overloaded. Convalescent home evacuations brought patients with special needs into ad hoc evacuation sites like schools, churches, and civic centers.

In a more recent disaster, review the issues that surfaced during the 2020 pandemic in the United States. These lessons are many times all the same in that there is a critical need for either equipment or personnel.

Discussion Questions

1. What are some of the things that make disaster planning difficult?

2. Does your FES district have a cache of disaster supplies or a disaster purchasing plan?

3. What other resources are available in a larger incident?

4. Describe any existing mutual aid agreements your department has with adjoining departments. Would this be sufficient for a major disaster? If there are none, describe a proposal for a regional agreement.

REFERENCES

Burd, S. A. 2009. How Safeway is cutting healthcare costs. *Wall Street Journal.* Retrieved from http://online.wsj.com/article/SB124476804026308603.html

Cebula, R., and E. Feige. 2008. *America's underground economy: measuring the size, growth and determinants of income tax evasion in the U.S.* Retrieved from http://ideas .repec.org/p/pra/mprapa/29672.html

KTVU.com. 2001. *Drowning death pushes Alameda officials to address water rescue training.* Retrieved from http://www .ktvu.com/news/news /drowning-death-pushes-alameda -officials-to-address /nD63s/

Internal Revenue Service. 2015. *Individual Income Tax Returns.* Washington, DC: Statistics of Income Division, Unpublished Statistics.

Maurer, Tim. 2016. *American Pensions Crises: How We Got There.* Forbes Media LLC. Retrieved from https://www .forbes.com/sites/forbes-personal-shopper/2020/08/19 /nordstrom-anniversary-sale-sneakers/#3d7843f96e69

Page, J. O. 1998. A Master at His Art. *FireRescue.*

Tharpe, Wesley. 2019. *Raising state income tax rates at the top a sensible way to fund key investments. Center for Budget and Policy Priorities.* Retrieved from https://www.cbpp.org /research/state-budget-and-tax/raising-state-income-tax -rates-at-the-top-a-sensible-way-to-fund-key

CHAPTER 6

Human Resources Management

LEARNING OBJECTIVES

After studying this chapter, the student will be able to:

- Understand the importance of well-trained and well-equipped professional fire and emergency services (FES) members (p 104).
- Understand the function and operation of human resources personnel (pp 104–105).
- Recognize ways of providing diversity in the department (pp 109–113).
- Examine the legal issues relevant in hiring FES members and recruiting volunteers (pp 105–107).
- Understand and review the disciplinary process, along with legal issues that may arise (pp 121–124).
- Examine the process of job analysis and validation (pp 106–108).
- Recognize the influence and understand the operation of public sector unions (p 124).
- Understand the concept of motivation and review ways to retain members (pp 126–128).

Case Study

The Fair Labor Standards Act and Overtime

In an appeals court decision, *Benshoff v. City of Virginia Beach* (1999), the court dealt with the issue of Virginia Beach career firefighters volunteering at independent volunteer rescue squads (ambulances) in their jurisdiction. These stations are completely staffed by volunteers and govern themselves. They also do not perform firefighting duties. The court decided that the Fair Labor Standards Act (FLSA) did not grant overtime to career firefighters volunteering at these independent volunteer rescue squads. This decision holds only in the U.S. Fourth Circuit Court of Appeals: Maryland, Virginia, West Virginia, North Carolina, and South Carolina.

This points out the frequent involvement of paid firefighters living in the suburbs of a large city volunteering their free time as volunteer firefighters. Even though this may not violate the FLSA, if the firefighter is injured volunteering, who pays the costs?

Discussion Questions

1. Does your local paid fire department have a policy to handle this issue?

2. Does the local volunteer fire department have an insurance plan that covers medical and the loss of compensation that is of equal worth as their benefits from their paid firefighter's position?

3. Some paid fire departments have a policy against volunteering off duty. Is this a beneficial policy or should it be changed?

The Most Valuable Resource

The staff of a fire and emergency services (FES) organization are its most valuable resource in terms of costs and results **FIGURE 6-1**. Without well-trained, well-equipped, and well-prepared members, the essential tasks at an emergency scene cannot be accomplished effectively, safely, or competently. Fortunately, most FES members are highly motivated and dedicated to the profession. This makes the FES administrator's job relatively trouble-free most of the time. However, as discussed in this chapter, some goals may be difficult to implement and/or monitor. In addition to a network of supporting personnel, resources—including stations, emergency apparatus, safety equipment, and specialized tools—are another component critical to success on the emergency scene.

The following depicts the cost of one engine company amortized for 1 year:

- Staff: Five firefighters (minimum crew of four plus one to cover leave impact) at $42,000 salary each plus fringe benefits (~40%) for three shifts (53 hours/week) = $882,000

- Facilities: One fire station ($3 million using a 30-year bond [6% interest] = $217,947 annually) and one engine ($300,000 using a 20-year bond [6% interest] = $26,115 annually) = $244,062 per year

The personnel costs for this engine company are more than three times the costs of the capital items needed to respond to emergency incidents. These costs do not include the overhead expenses of staff and administrative personnel. Behind each piece of emergency equipment are people who need to be hired or recruited, trained, evaluated, and motivated to do the job effectively and professionally. FES personnel management prepares individual members to do their jobs well.

The Function and Operation of Human Resources

The human resources (HR) department is the agency within the local government responsible for recruiting, hiring, training, evaluating, and paying employees. The employees that make up the HR department have

FIGURE 6-1 Firefighting is a demanding physical job that includes team work developed from considerable training and ongoing practice.

© Glen E. Ellman/Jones & Bartlett Learning.

specialized training and education, but generally have limited knowledge regarding FES personnel issues, including the skills and aptitudes required for emergency services operation. Instead, these employees are support staff for line agencies, such as the police, fire other government agencies, and schools. HR uses merit rules and regulations along with union contracts to provide direction for individuals carrying out the job duties. Merit rules are legally adopted by the jurisdiction and contain provisions for employee pay, benefits, discipline, and job descriptions. HR can also play a major role in acquiring additional funding for personnel and selecting the best people for various jobs.

It is very important to have a close relationship with the director of the HR office of the municipality. Make an appointment to meet the director and get to know him or her personally. The first step in forming this relationship is to familiarize HR personnel with the unique nature of FES operations. The FES organization needs new members who:

- Can learn complex skills
- Work independently to acquire knowledge
- Have courage
- Are physically fit
- Are mentally stable
- Do not mind getting extremely soiled and sweaty
- Can endure temperature extremes (hot and cold)
- Tolerate working with or around critically injured or deceased victims
- Enjoy operating in a family-like situation for up to 24 hours
- Are tolerant of diverse cultural backgrounds of other members
- Are team players

If HR employees are not fully aware of the need for these unique abilities, they may unintentionally hire individuals who cannot perform the essential job functions when faced with other policy goals dictated by the municipal administration.

Properly educated HR employees will have a better understanding of FES personnel issues. One way of accomplishing this education is to take them to an FES station or the training academy so they can experience the working environment (minus live fire or medical emergencies) and job tasks. HR employees should don the protective clothing and self-contained breathing apparatus, carry the trauma bag, and perform various job functions. Some of the same multimedia materials, lectures, and demonstrations that are used in public relations or risk reduction education programs can be very helpful in educating these employees.

The chief officer should have as much influence over personnel management functions as possible. For smaller organizations, an FES officer should be assigned part time to coordinate with the HR office; a larger department should have permanent staff assigned full time. These officers should have some formal education in personnel management. A civilian employee with the appropriate guidance and knowledge from the chief can carry out this function in some departments.

State and Local Hiring Laws

Many states have statutes referencing the hiring and, in some cases, operations of FES. For example, in the state of Florida, an applicant must be a state-certified firefighter to become an employee of a municipality. The state specifies the curriculum using National Fire Protection Association (NFPA) Firefighter I and II and a minimum of 360 hours of training. In actual practice, because the State Board of Education must approve the curriculum, these training programs are a minimum of 480 hours. These programs are delivered throughout the state by community colleges, vocational technology centers, and a few larger FES organizations. The state is legally responsible for approving these training programs and their facilities. In addition, each graduate of these firefighter training programs must take a state-administered skills test and a written examination for final certification. This process provides a pool of certified firefighters who become applicants for firefighter positions.

A minimum hiring requirement for emergency medical technicians (EMTs) and paramedics is possession of a basic license or certification and a valid cardiopulmonary resuscitation card. Some states require a current National Registry of EMT certification, but other states use the National Registry of EMTs only as a first-time certification assessment. Hiring requirements for paramedics may also include locally required certifications, such as advanced cardiac life support, pediatrics, and trauma training certificates. Candidates must also be tested for competency on optional scope knowledge and skills practices required in their local area. In some counties or regions, the local emergency medical services (EMS) agency requires an orientation on county or jurisdictional medical treatment protocols.

Many merit or personnel laws regulate employment at the local level. In addition, union contracts in most cases have stipulations affecting salary and benefits. These are too numerous to discuss here, but are all legally enforceable.

Job Analysis

A job analysis is a common method of defining job requirements. A good example of the outcome of a job analysis can be found in **FIGURE 6-2.** In fire and EMS, very detailed training and testing criteria are readily available. There are three distinct parts to a job

---

FIRE DEPARTMENT

I. Position Title: Volunteer or Career Firefighter

II. Summary Statement of Overall Purpose/Goal of Position:

Under the direction of the Fire Chief and general supervision of the station officer; respond to emergencies involving fire, medical, and environmental concerns. Responsible for the care, operation, and condition of fire and rescue apparatus and the station.

III. Duties:

- Respond promptly and efficiently to fire, rescue, hazardous materials, and medical alarms.

- Operate at emergency fire and rescue incidents efficiently and safely.

- Maintain proficiency in fire and emergency medical knowledge and skills.

- Conduct tours, lectures, and video presentations. Display fire apparatus. Participate in public demonstrations at local school programs and various civic and city functions.

- Inspect business, public, and private properties for hazards and code violations.

- Drive and operate department apparatus in emergency and nonemergency situations.

- Maintain equipment on apparatus to include daily, monthly, and annual testing. Test and rotate hose.

- Complete daily, monthly, and annual reports on the testing of fire and medical equipment.

- Complete 8 hours of department training each month.

IV. Qualifications:

At time of application screening:

- Must be at least 18 years old

- Must live within 20 miles of the municipality

- Must not have been convicted of a crime for which the applicant could have been punished by imprisonment

- Must not have been convicted of an offense involving dishonesty; unlawful sexual conduct; physical violence; or the unlawful use, sale, or possession of a controlled substance

Before being accepted:

- Complete entry-level written test

- Pass physical agility test

- Complete and pass drug test

- Successful driver's license check

- Check for convictions and outstanding warrants

- Enter initial training program

Education: High school diploma or GED.

Medical and Physical Fitness: Must meet department's medical and physical agility requirements.

V. Working Conditions

This job entails regular exposure to dangerous situations under disagreeable conditions, including smoke, heights, fire, fumes, heat, cold, emergency driving; frequent exposure to dangerous situations with medical emergencies; must meet the Fire Department's physical agility standards; must be able to wear and work in fire department breathing apparatus.

FIGURE 6-2 Example of firefighter job requirements.

analysis process: (1) hiring criteria, (2) recruit training and testing, and (3) incumbent evaluation. All three should be consistent with the job analysis.

There have been very few legal challenges to new employee requirements put in place by most FES organizations. For example, a very detailed and comprehensive basic skills test is generally required of a new recruit. However, incumbents are not tested periodically to ensure that they can still perform the basic skills that a recruit must complete. Because a new recruit is on probation, he or she does not have the same legal rights that an incumbent possesses. There may come a time when a future court decision requires equal treatment. Although it seems like a fair and ethical policy to treat all firefighters and EMS personnel equally, it is common to have different requirements for incumbents. Practically speaking, it is a good idea to upgrade new employee and recruit requirements before upgrading incumbents, and then phase in requirements for incumbents. They may need time to relearn some skills.

Many HR professionals recommend using incumbents, supervisors, and personnel analysts as sources to develop a job analysis. A good strategy includes using each of these sources in the process; however, one thing to keep in mind is that incumbents may not be able to perform all job functions or not in a timely performance level, thereby skewing the results.

The effort to create a job analysis should start with review of a standard, such as NFPA 1001, *Standard for Firefighter Professional Qualifications*. The administration should also do some research and contact several FES agencies to get copies of what peers are using, asking if they have had any problems with their job descriptions and whether they are happy with the abilities of their new recruits.

After this basic research is complete, the information should be consolidated into one comprehensive job description. This draft should be circulated to the incumbents and supervisors in the department. In addition, input should be actively solicited from labor organizations and the personnel office.

Because of the possibility that incumbents and their supervisors may not encounter critical job functions on a regular basis, they may not identify some of the job functions as being essential. Advise those questioned that a critical job function may have to be performed only once in a career. It becomes essential not because of daily use or need, but because it is how the public and the national standards expect a firefighter or EMS provider to perform.

To use an example from the Americans with Disabilities ACT (ADA) regulation, the expectation that a lifeguard who works at a pool must be able to bring a heavy person back to safety from deep-water depths is described in detail in the job analysis. This job function is one of the most essential functions because of the life-saving potential, which is the ultimate purpose of a lifeguard. Lifeguards may never face this situation, but they must still meet this essential job requirement.

Incumbents can also be valuable when planning an implementation strategy. Testing incumbents provides a realization of where the incumbent's skill levels are to successfully complete the tests. An implementation plan can be proposed that should allow a phased-in compliance for existing members.

Job Classification

Most members in any FES organization are at the firefighter or EMT level. The firefighters and EMTs are the people who get the job done at the emergency scene, and the organization should show its respect and acknowledgment of these special qualities, especially as each member gains additional training and experience. The department should continually emphasize that the FES profession is a noble pursuit, requiring high levels of skill, courage, and physical ability.

Some FES organizations have an annual awards ceremony that singles out individuals for heroic achievements. Most members are capable of heroic acts, but only a few firefighters or EMTs end up at the right place at the right time. Therefore, rewarding all members for their potential to perform at these heroic levels of service may be warranted and should be acknowledged at these annual awards ceremonies.

To reward competent members in a job that is very skill-, knowledge-, and physically-oriented, including the requirement to act instantaneously, the FES organization could use a three-tier promotion system for firefighters: (1) recruit firefighter, (2) firefighter, and (3) senior firefighter. Each new level would have its own set of training requirements, testing, and experience on the job. For example, to gain the firefighter level, the member would have to successfully complete recruit training, a 1-year probation program, and 3–4 years' experience. The senior firefighter would require an additional 3–5 years of experience, demonstrating competency of all of the basic skills and knowledge tests, and completion of an apprentice program. Each new level would be a promotion with an appropriate increase in salary.

Similarly, the National EMS Education and Practice Blueprint has established four levels of EMS providers: (1) emergency responder, (2) EMT-basic, (3) Advanced EMT, and (4) EMT-paramedic.

Emergency Medical Responder

Emergency medical responders usually have the minimum training and certification to be firefighters and many are often police officers who are trained to carry and use automated external defibrillators at the scene of an emergency medical incident. In some communities, especially senior neighborhoods, groups of qualified residents are organized into response units and alerted when a 911 call is received for their community. They perform lifesaving interventions such as CPR and AED application while awaiting additional EMS response, then transfer responsibility, ensuring a progressive increase in the level of assessment and care. They can assist EMTs and paramedics in the field.

Emergency Medical Technician

EMTs provide basic emergency medical care interventions for patients who access the emergency medical system. In rural jurisdictions, EMTs may provide the highest level of prehospital care. The EMT is the minimum licensure level for personnel transporting patients in ambulances. This job requires 150–190 hours of training, culminating in certification after rigorous testing of skills and knowledge.

Advanced EMT

Advanced EMTs provide limited advanced emergency medical care and transportation for patients who access the emergency medical system. The advanced EMT scope of practice includes emergency medical responder and EMT skills and competencies and limited advanced skills, including some pharmacologic interventions. Advanced EMTs may function as part of a tiered response system but are not generally used in systems that have paramedic-level care. This job requires 150–190 hours of training beyond EMT.

Paramedic

The paramedic's primary focus is to provide advanced emergency medical care for patients who access the EMS. The paramedic's scope of practice includes basic and advanced skills with equipment typically found on an ambulance as well as invasive and pharmacologic interventions. In many communities, paramedics provide a large portion, and often the highest level, of prehospital care. This job requires 1,000–1,200 hours of training, leading to a rigorous certification procedure.

Community Paramedic

A higher proficient EMS professional that has completed an appropriate educational program and has demonstrated competence in the provision of health education, monitoring, and services beyond the roles of traditional EMS. The specific roles and services are determined by community health needs and in collaboration with public health and medical direction.

Recruitment

Selection of new members is one of the most important decisions for the quality and effectiveness of the agency. Hiring decisions can and often do have long-range impacts regarding the competency and effectiveness of the department. People who do not meet the minimum job requirements should not be hired or accepted as members.

The Selection Process

Some of the legal ramifications of selecting new members will be discussed later in this chapter. Consideration should be given to cutoff scores and pass-fail requirements for cognitive abilities (speaking, reading, math, listening, and writing skills); medical tests; drug and alcohol tests; criminal background checks; physical fitness; and cigarette smoking. For example, in some cases, firefighter job applicants must sign an affidavit that they have not used tobacco products for 1 year before being hired. Many departments further require firefighters to sign a preemployment agreement to not use tobacco at any time during their employment. For EMS, this is usually a recommendation rather than a requirement, although company policies may vary on this issue.

Setting acceptable minimum scores high enough so that the department selects those who will be competent members in their ability to learn and perform the job is a good idea. Selecting marginally functioning members can cause unacceptable performance in the future. Remember, these are selections for new FES members for a minimum 20-year career.

Another issue is the option of hiring either trained or untrained certified firefighters. Each option has its merits. Selecting trained firefighters reduces training costs to a brief orientation. This is a substantial cost savings because a new firefighter recruit might be in training for 12–16 weeks. Also, many smaller departments cannot hire enough new firefighters at one time to justify a rookie school. In addition, it is common for larger departments to find that their applicants are trained firefighters from smaller departments looking for better salaries and benefits, or certified volunteers trying to obtain paid positions.

Some prospective firefighters may apply to a certified firefighter training center to become a student. The student pays his or her own tuition and other costs for the typically 12- to 16-week training programs. With successful completion of the Firefighter I and II course, applicants are eligible to take the state firefighter certification examination. After they successfully pass the practical and written examinations, they are state certified and can apply for a paid position in the state.

Some larger jurisdictions hire personnel and pay them to attend training programs. These departments prefer to train their own recruits so they can impart their organizational culture and the specifics of their department to the recruit firefighters.

In addition, some departments also prefer paramedic certification. Many potential career firefighters/medics must have both certifications to be selected for an opening. This trend is becoming more common throughout the nation.

Private EMS is simpler to get into. Although private schools may offer shorter EMT classes, many personnel attend semester-long courses offered at community colleges. With the naturally high rate of attrition experienced by private ambulance companies, they will hire right-out-of-class EMTs. Many companies structure their training program with the understanding that this may be the employee's first job.

Additionally, some ambulance providers have their own paramedic schools. It is an advantage to an EMT to seek out these companies, because they may provide employee discounts, or in some cases a scholarship to the company's program. There is also a likelihood that he or she will be placed in the company as a paramedic after successfully passing the course. This arrangement is mutually beneficial. The EMT gets a discounted rate on training and the hard-to-find internship, and then a job. The company retains a well-trained (and grateful) employee who already knows the policies, procedures, employees, and the response area. In some cases, as part of the paramedic school agreement, the new paramedic may be bound to the company for a period of time.

The quality of training programs varies from one company to another, but they generally include an orientation, classroom and/or self-study of EMS knowledge, field training time, emergency vehicle operator training, and a probationary period. After the employee passes initial training, the company may offer further training opportunities or continuing education classes.

Diversity in the Department

Complete diversity reflects a blend of gender, race, age, education, country of origin, and other characteristics, such as personal values. In the past, the typical FES organization was composed of people with similar characteristics—those who joined were expected to conform. In today's world, each member wants and deserves to be valued as a distinct individual. People from diverse backgrounds bring diverse values and experiences to their roles.

Starting in the 1970s, coalition groups began to support the rights of particular categories of people. These groups are very vocal in their identification of perceived or real discrimination and demands for corrections. Today, most FES organizations have actively recruited minorities and women to compensate for memberships that had been predominately white and male. In a few cases, these organizations were under court orders or consent decrees to provide affirmative action programs to increase the numbers of minorities and women.

There is disagreement about the use of affirmative action, preferences, set-asides, and quotas—strategies used to hire minorities and women in many FES organizations. New legislation and contemporary court rulings have changed the laws of the past. For example, judges used to rule discrimination cases based on a lack of a specific percentage of the underrepresented group compared with the general population served. Now, discrimination in the FES organization cannot be proved solely based on these percentages.

Many FES departments use the diversity of the municipality as their benchmark for diversity goals. This method may skew the diversity statistic because the local city may have large numbers of one group. It may be better to use a larger area to represent the region where potential new members are recruited. The minimum area probably should be a county; larger departments might use several counties.

Affirmative Action Cases
The Boston Fire Department Affirmative Action Case

Some argue that an affirmative action program may lead to widespread lying on applications to schools and jobs. This was the case in 1990 for the Boston Fire Department. Six firefighters from the department were found to have lied on their applications, claiming false minority statuses.

Two of these firefighters, white twin brothers Phillip and Paul Malone, had initially failed their qualifying examinations for the department. However, on subsequent applications, the brothers claimed that their great-grandmother was black, thereby qualifying for minority outreach standards. It was not until 10 years later that their lie was discovered (Vlahos, 2003).

The University of Michigan's Affirmative Action Case

This was a 2003 U.S. Supreme Court case that upheld an affirmative action policy at the University of Michigan law school. However, the Court also ruled that the undergraduate policy at the University of Michigan was not legally correct and must be changed. "In that admissions process, which the university now has to revamp, the school literally gave extra points for applicants who were black, Hispanic or Native American" (data from Vlahos, 2003). In fact, the diversity points had the potential to outweigh points given for academic achievement. However, a more recent affirmative action case may affect the decision made in 2016. In *Fisher v. University of Texas*, Abigail Fisher claims she was denied admission to the school because she was white. The U.S. Supreme Court found that the University of Texas affirmative action program was constitutional on a 5–4 vote. The swing vote was Justice Kennedy and since then he has retired and been replaced by a conservative Justice. This same case would probably have a different outcome today. Although, the court did warn other universities that not all affirmative action programs will pass constitutional muster. For an update, "The Justice Department sued Yale University on Thursday, weeks after prosecutors found the university was illegally discriminating against Asian American and white applicants, in violation of federal civil rights law". (Ref: https://www.sfgate.com/news/article/Feds-sue-Yale-allege-discrimination-against-15632358.php, October 2020).

The New Haven, Connecticut Affirmative Action Case

On April 22, 2009, the U.S. Supreme Court sided with firefighters claiming they should have been promoted based on examination results for lieutenant and captain positions. The results of the promotional examination were discarded when the city noticed that no African-Americans scored high enough to be promoted.

The firefighters that should have been promoted filed suit, alleging that the city violated Title VII of the Civil Rights Act of 1964, the Equal Protection Clause of the Fourteenth Amendment, and the First Amendment of the U.S. Constitution. The firefighters argued that the examinations should have been race neutral and solely based on job performance and duties. They claimed to have been punished for their race. In June 2009, the U.S. Supreme Court sided with the firefighters (Ref: https://www.law.cornell.edu/supct/html/07-1428.ZO.html).

This specific legal issue, affirmative action, is undergoing many new challenges throughout the country. It is extremely complicated. When faced with political pressure to hire or promote members based on something other than merit, attempt to get the municipal attorney(s) to research the issue and provide guidance.

Diversity Selection in Practice

What does all this mean for the typical department? The following hiring scenarios comply with the latest U.S. Supreme Court decisions and other legal precedents:

- A policy can be used to rate each applicant based on scores or ratings from written and other cognitive examinations. The person with the highest score is chosen first, followed by others, and strictly based on their ranking. The U.S. Supreme Court did not say that this traditional way of selecting new members could not be used legally, but it must use job-related criteria.

- Court orders must be followed if it has been determined the institution has discriminated. Affirmative action programs must be provided when order by a court, resulting in a mandatory plan and quota. Most have a time table for compliance and termination of the program.

Banding

A compromise policy called banding takes into consideration both points of view. In a banding program, each applicant is put into a band and each band of applicants must be exhausted before going to the next one. After the applicants take the examinations, they are put on a traditional list and ranked from highest to lowest. One common cutoff for the top band is 85%. Then new members are selected from the top band first.

In actual practice, the department may offer jobs to women or minorities first and then offer jobs to white males in a ratio to achieve diversity of the total new hires. All members of the band must be offered jobs (promotions) before going to the next lower band, even if this results in a greater percentage of white males. This helps prevent what some would characterize as reverse discrimination. Legally, there is no reverse discrimination, only discrimination. For example, if there are 100 people in the band, selecting the top 20–30 people by numerical scores often may not provide a diverse group.

Banding with a cutoff of 85% to 100% uses a rationale identical to the typical grading schemes in school. For example, an "A" grade is actually a numerical grade of 90% to 100%. This approach takes into account the fact that human beings are consistently inconsistent to some degree in test taking within a small band variance. On any particular day, the numerical grade received and the position on a list can change for each individual within a predicted range. For example, one day's performance may see a person at the top of the list, but another day that same person may have a lower score, although both scores may be in the top band.

To ensure a diverse outcome, the department first has to recruit and encourage minorities and women to apply for the positions on a continuing basis. An affirmative action officer normally handles this job.

If the department has vacant positions and the ideal diverse mix cannot be achieved, the positions still have to be filled. There are numerous and varied situations in which this may occur, such as with smaller departments that typically have fewer applicants. The administration must try its best to meet the goal, and if it cannot, the positions must be filled with qualified applicants from the band even if this does not comply with the diversity goals.

It is typical in many departments for the chief to have the option of choosing from the top three on a list for a promotion. This option allows the chief to completely skip over someone who did well on the promotion process but who is not acceptable to the chief. It is a compromise with unions who often want the first on the list to be the first promoted. This chief's option is generally not used to support diversity but to acknowledge the chief's professional judgment on the qualifications for promotion. Regardless, all individuals in the band must be promoted before promoting candidates on the next lower band.

It is never a good idea to lower the job requirements to select applicants outside the band, or to select applicants who cannot meet minimum requirements, such as physical fitness standards.

Minimum Job Performance Requirements

Diversity should be accepted and encouraged. Society shows that different cultures can peacefully coexist, and the workplace should attempt to reflect that as much as possible. The public can more easily relate to a workforce that is representative of their society. Furthermore, diversity provides a broader perspective in organizational problem solving. However, while attempting to achieve this diversity, the department should not lower standards of performance. Acceptance of the new diversity of employees must be supported by minimum standards for job competencies. For new employees to become part of the family, they must be able to perform the minimum job requirements. In the emergency services profession,

members must rely on each other to work in teams and, more importantly, if one member needs help, other members must be able to come to their aid.

It is unethical to hire or accept members into FES organizations who are not capable of performing critical job competencies. The public trusts that FES personnel can perform emergency functions safely and efficiently. If the personnel cannot perform to professional standards, they will fail the public and fellow members. There is strong support for this viewpoint as expressed in the International Association of Firefighters/International Association of Fire Chiefs Candidate Physical Ability Test: "Diversity should never come by lowering validated entry standards. Rather, it should come from actively recruiting qualified men and women candidates from all racial and ethnic backgrounds for careers in the fire service" (data from International Association of Firefighters, 2011). Although the private ambulance industry does not have a standard physical agility test like the Candidate (firefighter) Physical Ability Test, physical agility tests may be required by an employer.

Some FES organizations, in their efforts to do the right thing regarding diversity, have hired or accepted members who cannot meet minimum job performance criteria. These situations are very challenging to resolve, requiring an administrator with a good understanding of leading change (see the chapter *Leading Change*).

There are three reasons to turn down applicants who do not meet minimum firefighting job requirements:

1. Firefighting is a very hazardous job; skills and physical strength can be a matter of life or death. It is also a team sport and any one member may prevent the team from being successful and/or safe in emergency operations. This is critical to serve the public with competent professional service.
2. If firefighters are not able to adequately perform essential job functions, the public receives inferior service.
3. It is frustrating and demoralizing to the firefighters who do meet the requirements. They may need to make up for the inability of others to perform safely and effectively.

EMS jobs may have similar, but slightly different, minimum standards. The following are three reasons to turn down applicants who do not meet minimum EMS job requirements:

1. Candidates need to be in good physical condition and able to lift at least 125 pounds, or a weight required by the company. Even with advanced

equipment such as hydraulic gurneys, the EMT or paramedic may have to move a patient out of a cramped bedroom or down a flight of stairs without injuring the patient or themselves.

2. Candidates must show competence with a basic written didactic knowledge test and a skills return practical scenario. The company must confirm that candidates have an adequate job-specific knowledge base and can adequately perform patient assessment skills. This is generally achieved by completing a specific certification process.

3. Candidates must display an attitude indicating they would be well accepted as a partner and patient care technician. They must respond to questions with honesty and without ego, suggestion of bias, or display of negative attitude. Compassionate and respectful interaction with partners, public safety personnel, emergency room employees, and the patient and family are critical.

Diversity Sensitivity Training

Many agencies have started offering **sensitivity training** to educate employees about personal and cultural differences. This training works to achieve social harmony in a group that may have little in common. For example, one FES department started an affirmative action program to recruit minorities for firefighter positions. The affirmative action officer decided to go to some local churches where the membership was mostly minorities. The officer found many applicants who successfully passed the written and physical fitness examinations. The new firefighters were assigned to permanent fire stations after completing recruit training. Because the new firefighters were recruited from churches, they had strong religious values and personal beliefs that some existing members did not share, in many cases. The result was a distinct lack of socializing for most of the shift, including when off duty. These situations can be discovered and corrected with professional sensitivity training.

Sensitivity training can go a long way in melding diverse personalities to work together in harmony. For some minorities and women, this training can make the difference between feeling welcome and feeling alienated. Individuals who feel alienated could become less motivated on the job, and may show evidence of carelessness, resulting in accidents. In addition, they may resign because they feel out of place, which becomes a big loss to the organization that spent many hours training the employee.

"The underlying idea behind sensitivity training is developing respect for each person as an equal and as a human being. Just as we have wants and needs, so do the others around us. And just as we want people to show respect to our feelings and ways of living, so too must we show respect to others. Sensitivity training teaches people to engage in perspective-taking so that it is easier for them to appreciate another point of view and come to a mutual understanding [without being judgmental]. Sensitivity training also involves developing an eye towards what is more and less likely to be offensive, and likewise developing the ability to backpedal and apologize sincerely if feelings are hurt" (Reproduced from Chaubey, Dhani & Subramanian, Kalpathy. (2014). Role of Sensitivity Training and its impact on Organization Behavior in a world of cascading Demographic divide. Management. Volume2. Retrieved from https://study.com/academy/lesson/what-is-workplace-sensitivity-training-for-employees-definition-lesson-quiz.html).

Recruitment and Selection of Firefighters to Achieve Diversity

It is a good idea to ensure that the company is committed to providing diversity in the organization as new employees are hired or new volunteer members are accepted. It is rare that a formal quota system can be legally used. Quotas are allowed only when a court has ordered a remedy to a proved past discrimination practice. Therefore, the department has to set up a legal recruitment and selection process to attract a diverse mix of applicants. For example, because minorities and women tend to be underrepresented in most FES organizations, the chief officer should assign a member the responsibility of recruiting minorities and women to apply for jobs or membership. There are many techniques for accomplishing this goal, but the department should consider only ethical, legal, and easily implemented techniques. The chief officer should consult with the HR office staff for more options.

Additionally, a professional presentation and web site should be created or acquired to help educate and attract applicants. Locations or specific communication channels, including social media, should be identified to reach the selected audience. For example, because firefighting is physically demanding work and generally women are not as strong as men, visits to women's athletic programs at local colleges and universities can locate good prospects who are physically qualified to be members.

Fire and Emergency Services Workforce Issues

A common workforce issue concerns the understanding that the typical FES organization is different

Hiring and Recruiting

Hiring and recruiting are important issues to the human resources of both career and volunteer departments. Not only do the policies and procedures regarding hiring and recruiting need to comply with all applicable legal and ethical issues, they also must ensure that potential members of the department are well qualified and fit the job description of the position to be filled. NFPA standards such as NFPA 1001, *Standard for FireFighter Professional Qualifications*, and NFPA 1582, *Standard on Comprehensive Occupational Medical Programs for Fire Departments*, can be helpful references in the hiring and recruiting using minimum job performance selection criteria.

FIGURE 6-3 The U.S Supreme Court building.
© Gary Blakeley/ShutterStock, Inc.

from other occupations. Most of those differences are because FES work is extremely time-sensitive, risk-oriented, and team-based, with members in close contact with one another for long periods of time. The duration of the typical FES shift and the closeness of the club-like surroundings of the career and volunteer FES organization make it imperative to manage diversity effectively.

Legal Issues

This section contains up-to-date information from recent Supreme Court and legislative actions **FIGURE 6-3**. The law is constantly changing. Many of the recent Supreme Court cases have been decided by a slim majority (5–4), so replacements for retiring justices may swing the Supreme Court in a different direction. Recent Supreme Court justice appointments are inclined toward "constitutional" or "conservative" viewpoints during President Trump's term, making it more likely that the Court will follow the original interpretation of the U.S. Constitution in future decisions.

Two aspects of legal issues affect FES: the recruitment and retention of members, and emergency operations. This chapter discusses only personnel management issues. Legal representation should be consulted for any changes in local, state, or national laws or interpretations that may make the legal opinions in this section inaccurate.

The Screening Process

One of the most important labor functions is the hiring of new personnel or the acceptance of a new

volunteer member. Any mistake in this process can leave the department with an undesirable member who must either be tolerated or separated. Separation is never an easy or pleasant process.

The screening process checks learning potential, knowledge, and physical abilities that are needed to master essential job performance functions. For firefighters, the necessary skills for most job functions are not mastered until recruit training is completed. This basic training may occur before or after formal acceptance. It is essential to test for the basic skills that are needed to master the training program. Pass/fail screening tests that indicate a minimum ability in reading, writing, math, and cognitive abilities along with physical fitness should be administered before the start of basic training.

This issue is illustrated by the experience of St. Louis in 2007, where "more than 70 percent of about 1,350 applicants failed the test [reading and math]" (data from Wagman, 2007). Three years earlier, in 2004, the city was sued by a group of black firefighters who challenged a promotional test for captain and battalion chief, claiming discrimination against them. A judge ruled that the city did not discriminate; however, the experience left the city wary of future legal battles. Therefore, after receiving the examination results in 2007, the city decided to scrap the test to avoid more controversy. However, Chris Molitor, president of the International Association of Firefighters Local 73, argued, "I don't think it's asking too much for someone to read, write, and do basic math. This test was designed to eliminate those people who could not perform those basic functions. It sounds to me as if the test was doing its job" (data from Wagman, 2007).

Psychological testing is another area that deserves consideration in light of incidents of destructive behavior and arson by FES members. To keep the public respect and trust, firefighters and EMS personnel need to be upstanding role models. Some basic testing categories to use are:

- Psychological
- Aptitude and achievement
- Minimum cognitive abilities
- Personality
- Integrity and honesty
- Physical fitness

To reduce the likelihood of liability, FES organizations should use a standard cutoff score on screening examinations that indicates the minimum qualifications necessary for successful job performance. To compensate for normal variances in all testing results and to guarantee that the applicants selected can perform at minimum levels of competency physically and mentally over a 20- to 30-year career, the minimum cutoff should be above the theoretical lowest scores. It is not a good idea to hire or accept marginally qualified applicants; it is not fair to the applicants, other members, or the public they serve.

Reference Checks

Many companies have adopted a strict "name, rank, and serial number" approach to requests for information about past and present employees. The reluctance to give negative information about employees' performance and disciplinary employment records has backfired in some cases, causing liability and negative exposure for these organizations. Furthermore, this practice can lead to increased risk of negligent hiring and retention of employees, which is especially concerning in organizations with public trust. For example, a new employee was hired as a paramedic after a complete background check. A past employer failed to provide information that the individual had stolen jewelry at the scene of a medical emergency incident. The previous employer was held liable for withholding information that was relevant to the employee's behavior on the job.

In response to this type of scenario, the state of Florida recently enacted a statute providing legal protection to employers who furnish information about present or former employees. This protection provides immunity from civil liability when sharing information about employees. The state attorney general can provide guidance about similar statutes in other states. The employer is covered unless it can be shown that the information was knowingly false or deliberately misleading, which would be a violation of the employee's civil rights, and the employee could seek appropriate legal action. The following are guidelines for providing reference information:

- Designate one individual in the personnel department as the only contact for requests.
- Require that a written request be made on company letterhead. Telephone the requesting agency to confirm all requests.
- Disclose only documented job performance information, not subjective evaluations.
- Do not disclose any information regarding discrimination complaints or medical/disability information that may reflect on equal employment opportunity status or a protected category.

First Amendment: Freedom of Speech

In 2006, in the case *Garcetti v. Ceballos*, the Supreme Court placed an entire category of speech outside the protection of the First Amendment: statements made by government employees in the course of their official duties. Justice Kennedy, when writing for the majority, said, "[W]hen public employees make statements pursuant to their official duties, the employees are not speaking as citizens for First Amendment purposes, and the Constitution does not insulate their communications from employer discipline" (data from Garcetti v. Ceballos, 2006). Fire officials need to be especially careful of this restriction during poor economic times when they may be tempted to blame a fire death on inadequate funding. It should be noted that this restriction does not cover whistle-blowers or union representatives acting for their organization.

Civil Rights

Several acts of Congress have been based on the 13th and 14th Amendments to the U.S. Constitution guaranteeing equal treatment or protection against discrimination based on race, religion, gender, or national origin. These acts are generally referred to as civil rights acts. In the FES profession, violations of these federal laws have resulted in monetary awards and court-ordered remedies such as affirmative action plans. This area of law has been slowly evolving and is different today than it was 10 or 20 years ago. This subject is also complex, and legal advice should be obtained when considering a change in policy. A brief summary of the present situation follows.

A statistical imbalance between the percentage of a protected class in the workplace and the percentage in the general population is an indicator of a possible problem. The organization must study this imbalance carefully to see if any discrimination has taken place.

In the past, this type of imbalance was prima facie evidence that discrimination had occurred. Now it is only a possible symptom, not conclusive evidence.

If a department enters into an affirmative action plan voluntarily (in writing or not) without a court order, and that plan contains specific percentage goals to hire minorities and women, the department may be open to civil court action based on reverse discrimination (which is actually just discrimination). Affirmative action plans can be used if they are designed to encourage and attract underrepresented groups to apply for employment or membership, including pursuing promotional opportunities.

The Americans with Disabilities Act

The ADA provides civil rights protection to people with disabilities and guarantees equal opportunity in employment, public accommodations, transportation, government services, and telecommunications. The act, which is enforced by the **Equal Employment Opportunity Commission** (EEOC), is designed to protect individuals from discrimination based on a disability that substantially limits a major life activity. Employers must make a reasonable accommodation for a mental or physical disability of an individual, unless it would pose undue hardship on the employer's operation. In most cases to date, reasonable accommodation for disabilities has been shown to create undue hardship if applied to public safety jobs. Therefore, it is very important to document job performance requirements along with the requisite medical, physical, and mental abilities.

Because of the intimidating nature of a federal regulation and the different analyses by many lawyers, some FES organizations have hired or accepted new members who physically or medically were not capable of performing the job of firefighter or EMT. Organizations should be very careful when getting advice about the ADA and research the many articles that have been written specifically about FES workers. Very simply, if standards for job performance are based on verifiable job requirements, they are acceptable and legally defensible. Even though job standards may eliminate some people with disabilities, the standards are valid and meet the exceptions stated in the ADA.

Furthermore, organizations should be careful not to fall into the trap of lowering the minimum requirements to accommodate a perceived exposure to legal action based on the ADA, civil rights, or another federal law designed to protect against discrimination. The goal of all FES organizations is to give the best service possible within financial and practical limits of human performance, both mental and physical.

The ADA and Hiring Practices

During the hiring process, applicants cannot be given a medical examination or asked any specific questions concerning any medical or physical disabilities. However, applicants can be screened using mental and physical ability tests as long as they are validated as needed for essential job functions. For example, for EMS jobs requiring an ambulance driver's certificate, a physical examination may be required to determine adequate hearing, visual acuity, and basic mobility for driving purposes. In addition, drug and alcohol testing is allowed before a job offer.

After an employee has accepted a job offer, he or she may be required to take a medical examination that could be used to rescind the job offer. In addition, it is very important, especially in FES organizations, to have an extended **probationary period** during which the employee can be separated without cause. It may take many months to evaluate a firefighter, EMT, or paramedic in actual emergency situations or those situations that occur infrequently, such as a large structural fire. The probationary period should run at least 1 year after the successful graduation from a recruit firefighter school, completion of an EMS certification program, or hiring of an employee already trained. The department then has a whole year to evaluate the new employees and be able to separate any who do not meet minimum expectations. The private sector is less standardized; successful orientation and field evaluation generally follows a probationary period lasting about 6–12 months.

Any standards for separation should be validated for essential job functions and each employee should be made very familiar with the rules. Legally, probationary employees have very few rights. It is better to make sure everyone, especially the probationary members, knows what the rules are, the justification for them, and how the members are evaluated.

It is also typical and preferable to have a 1-year **apprentice phase** in which the new member learns additional knowledge and skills. For example, in the private ambulance arena, the candidate, after successful completion of pre-employment tests and orientation, may be assigned to a Field Training Officer. Successful completion of this phase is based on competency rather than time, and it is generally shorter than in the fire service. The candidate must successfully meet required training goals, such as competent patient care and equipment handling, understanding of local policies and protocols, and driving skills.

Injuries on the Job

Policies should be created to ensure documentation and verification of injuries that are reported on the job. Standardized forms should be submitted to the appropriate municipal official for workers' compensation.

High-ranking FES officials and the employee's immediate supervisors should visit any employee in the hospital or make personal visits to the home, especially in cases of serious injuries on the job (IOJ). Employees heal faster if they believe they are valued and missed. The following was reported in a paper for the *Harvard Business Review:* "a large number of empirical studies confirm that positive social connections at work produce highly desirable results. For example, people get sick less often, recover twice as fast from surgery, experience less depression, learn faster and remember longer, tolerate pain and discomfort better, display more mental acuity, and perform better on the job" (Cameron and Seppälä, 2015). Supervisors should send a get-well card, make appropriate visits during recuperation, and provide the employee with the paperwork necessary for the workers' compensation claims and the payment of medical bills **FIGURE 6-4**.

Supervisors should stay in close contact with the IOJ members until they are released to either light duty or full duty, keeping a constant dialogue to inform the employee of what is going on at work. If the department's official cannot visit in person, notes should be sent often. Finally, a small welcome-back gathering should be planned with the employee's close peers.

FIGURE 6-4 Injuries are a hazard of the job and can be expected. They should of course be prevented when safety solutions are known.
© Glen E. Ellman/Jones & Bartlett Learning.

Light Duty

To lower workers' compensation costs, FES departments often bring individuals back to a light-duty position if they are still recovering and are not ready for full duty. The temporary duty assignment should not consist of clerical work, such as answering telephones or filing. Instead, for those firefighters and EMS members who are exemplary, temporary assignments may be provided in such areas as fire inspection, fire safety education, or 911 dispatching. Use caution because these positions are highly technical and require a great deal of knowledge and in some cases certifications, such as fire inspections. A mentoring position with an experienced member is a good way to initiate a light-duty member to their new duties. Basic knowledge, skills, and abilities training can also be provided.

This technique seems especially effective for employees who are on shift work and are assigned to light-duty day work. If the administrator brings the injured worker back on light-duty day work, 9 to 5, there seems to be a strong correlation to the injured employee wanting to return to full duty faster than expected. In fact, this policy is effective even when the department can bring the employee back to duty only for short periods at the beginning of recuperation. The important point is to get the employee back in uniform and back to work as soon as possible. Otherwise, employees may get used to a sedentary lifestyle and lose the motivation to get back to work.

While the member is on light duty, provide him or her with written guidelines covering work hours, tardiness and absenteeism, rest periods, and personal appearance. Specify the procedure for requesting leave, including IOJ leave for medical appointments and physical therapy.

Some employees in the FES organization have attempted to use ADA and its reasonable accommodation feature to claim that an employer has the responsibility to create a light-duty job. However, this is not a requirement of ADA; light-duty positions do not have to be offered to employees who cannot perform the essential functions of their job.

Light Duty Requests

For example, consider an employee IOJ who is covered by workers' compensation. After a recovery period off work, medical evaluation states that this person will never be able to fully recover to perform the essential functions of a firefighter. The employee may want to work in another position in the FES department that does not require strenuous physical work, such as in the dispatch office. However, this would require the department to remove one full-time firefighter position from the staffing of fire and EMS units and transfer the position to the dispatch office. Instead, the employer (government agency) may offer a job in another agency that the employee could manage with his or her permanent physical limitations. In this case, the same rules as for a new hire apply to the new position.

Discussion Questions

1. What is the policy of your agency on light duty?
2. What is the policy of your agency on disability benefits?
3. Does the agency owe the employee a duty to provide a job when they are injured and cannot perform the duties of a firefighter? Or should the employee be discharged with a disability pension and allowed to find employment on their own?

Pregnancy Issues

If a pregnant employee cannot perform the essential functions of her job, should the department provide an alternative light-duty job? In most cases the answer is yes, this is a good policy. Although there is no absolute requirement to provide a light-duty assignment, if the general practice is to provide other employees (especially males) who have temporary disabilities with light-duty positions, then it is advisable to do the same for a pregnant female employee.

A pregnancy creates extra weight, changes in equilibrium, and loss of agility. Any of these situations can cause injuries and inability to complete essential job functions. Generally, the department can order a pregnant employee to take a job performance test or be evaluated by the department's physician at the beginning of the third trimester and earlier if there are any medical complications or observations of substandard job performance. The department should adopt a written policy with input from the physician, the union, and the department's legal advisor.

Previously, employers have prevented pregnant workers from doing their jobs simply because of a concern for the well-being of the fetus. There is some evidence that elevated temperatures and the high concentration of carbon monoxide typically found in firefighting can cause medical harm to the fetus. In EMS, the biggest risk to a pregnant woman comes from lifting patients (especially obese and overweight patients), dealing with combative or assaultive patients, or working at potentially unsafe scenes, such as motor vehicle accidents. There is also concern for communicable disease exposure, especially with diseases such as rubella and of course the unknown COVID-19 consequences, which can harm a developing fetus. However, the Supreme Court has held that the well-being of the fetus is the concern only of the woman. At this time, several states are adopting legislation that restricts abortion and this may also affect these concerns for pregnant workers. In addition, many of these laws are being challenged and new laws are adopted in each state, some pro-choice, some pro-life.

Interestingly, in March 2015, the Supreme Court sided with a former United Postal Service (UPS) worker, Ms. Young, who claimed pregnancy bias after she requested light duty and was denied. The request was documented by her doctor who did not want her to lift over 20 pounds. UPS requires their employees to lift up to 70 pounds, thus the conflict. UPS has since changed its policy, and now says it will try to accommodate pregnant workers for light-duty positions. Subsequently, the EEOC has updated guidance to employers to make clear that they should accommodate people in Young's situation. However, the U.S. Postal Service said it will make no changes to its policy. Again, it is probably helpful to request the advice of the city/county attorney if you encounter this issue as an administrator.

Family and Medical Leave Act

The Family and Medical Leave Act (FMLA) of 1993 has affected many FES agencies. Those with minimum staffing levels commonly must hire additional personnel or provide overtime for leave impact. Studies have shown that up to 20% of additional personnel must be hired to compensate for leave in traditional sick, annual, and IOJ leave provisions; that is, to maintain four people per company on duty without overtime, the department needs to hire five firefighters.

The FMLA mandates that employers with more than 50 employees must offer employees up to 12 weeks of unpaid leave per year for family responsibilities, such as the birth of a child; care of a seriously ill child, spouse, or parent; or a serious illness of the employee. The employer must give workers their previous position back when they return to work. Therefore, during the absence, the FES agency may have to cover the vacant position with overtime. Larger organizations may hire additional personnel to cover these vacancies.

The additional cost to cover the shift with overtime may actually cost about the same as hiring additional personnel. This is because a firefighter's normal pay includes a base salary and benefits, and benefits may run up to 40% or more of the salary. Therefore, it costs the difference between regular pay with benefits and overtime, which is calculated at time and one half or 50% more. Clearly, this can be a substantial unexpected cost. However, the one big advantage to overtime is filling the position with an experienced FES employee on an as-needed basis. This can be a big drawback if the employee becomes fatigued by too many hours or long shifts of overtime.

If an employee needs IOJ leave to recover from a worker's compensation injury or is granted sick or annual leave for family obligations, the employer should notify the employee that the leave of absence is counting against FMLA leave. If this notice is not given, the employee may be able to apply for an additional 12 unpaid weeks off after he or she is found fit for duty from their worker's compensation injury or when sick or annual leave runs out.

In addition, the employer should have a written policy that combines FMLA leave with other paid leave when appropriate. For example, if an employee is off on sick leave for a serious medical condition, the FMLA leave can run concurrently. It is important to have this type of policy to reduce the chance of employees abusing time-off potential.

Drug and Alcohol Testing

The FES organization should have a written policy prohibiting members from using, consuming, or being impaired by (legal and illegal) drugs or alcohol. In some organizations, this is a policy that needs to be part of a union contract, although the requirement that a member be not under the influence of a chemical substance is normally deemed an employer right because it is necessary for the critical professional service to the public and the safety of the individual and peers. When the policy is first implemented, all existing members and any new applicants should sign a consent form for testing that may be implemented in the following circumstances: preemployment or premembership; at random; after an accident **FIGURE 6-5** or injury (as soon as possible); and suspicious activity or behavior.

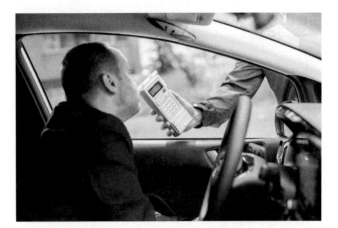

FIGURE 6-5 Governments must be prepared to test immediately for drugs or alcohol since they have a limited time in the body.
© Jones & Bartlett Learning/Photodisc.

If a member refuses to take a drug or alcohol test, a copy of the signed consent form should be produced and the member reminded that rejecting the test will result in disciplinary action. The alcohol test could use the same pass-fail criteria used for determining an impaired driver under the state's motor vehicle statutes. Refusal to submit to a drug or alcohol test or a positive result should be considered legal grounds for immediate dismissal.

Sexual Harassment

The administrator should be aware of two types of sexual harassment: quid pro quo and hostile work environment. Both of these expose an employer to substantial liability that can result in large monetary awards. Therefore, it is important for an employer to set up policies, training, and procedures to prevent and eliminate this conduct.

One key to whether sexual harassment has occurred revolves around the issue of whether or not the sexual advances were welcomed. For example, at work, a member becomes friendly with another employee, and eventually asks that person out on a date. At that time, there is no knowledge of whether the advances are welcome. If turned down for a date and the other person makes it clear that there is no romantic interest, any future request for a date may constitute sexual harassment. This is especially true if the person proposing the date is the supervisor of the other employee.

Quid pro quo sexual harassment occurs when a person in authority controls an employee's future in the organization. This type of sexual harassment may occur when a supervisor offers the employee a promotion, raise, or other benefit in exchange for a sexual favor; it may also occur when a supervisor threatens demotion, transfer, or termination. Whether the consequences for the employee being harassed are positive or negative is irrelevant.

Hostile environment refers to employees' conduct of a sexual nature that may be offensive or intimidating. The simplest example of a hostile environment is allowing male employees to post pictures of women either naked or in sexually suggestive poses. The telling of sexually explicit jokes is another good example of a hostile environment. The EEOC has outlined the following factors for determining if a hostile environment is present:

- Conduct was verbal, physical, or both
- Conduct was frequently repeated
- Conduct was hostile or patently offensive
- Alleged harasser was a coworker or supervisor
- Others joined in perpetrating the harassment
- Harassment was directed at more than one individual

The following guidelines are helpful in preventing and defending the organization in sexual harassment situations:

- Implement a written policy of antiharassment with examples of prohibited actions.
- Provide training for all members, including a description of conduct that is illegal or unacceptable.
- Have all members sign a receipt that they have taken the training and understand the policy.
- Clearly post the policy in work areas.
- Clearly post and issue standard sleeping attire for the dorm room.
- Identify the persons or office to be contacted if harassment occurs.
- Encourage members to report any potential situations that may lead to a violation.
- Vigorously investigate and discipline any violations.

Dating Policies

Some organizations have created formal policies covering potential dating relationships at work. An outright ban is usually neither possible nor preferable. In an FES organization where employees are assigned to a shift crew and sleep together in a dorm, it may be preferable to have a policy that requires the transfer of one of the partners in a consensual relationship. This also solves the problem of a possible quid pro quo situation where a supervisor is one of the partners. In the typical FES station, employees spend long shifts together and sleep at the employer's facility, requiring specific considerations such as sleep clothing. A different policy may need to be used for staff employees working in an office.

Even in a volunteer department, dating can be disruptive to the relationships of the other members if there is any hint of favoritism. A strong chief officer and an oversight board can be very effective at preventing any adverse consequences.

Fair Labor Standards Act

The core requirements of the Fair Labor Standards Act (FLSA) are fairly simple. Employees must receive overtime pay for hours worked over 40 per workweek at a rate not less than one and one-half times the regular rate of pay. Once covered, employees working more than a designated number of hours in a pay period must be paid overtime. However, public employees have some special rules. First, the employee can accept compensatory time at a rate of 1.5 times instead of overtime. Also, firefighters' work period is defined as a 28-day cycle, and the average hours per week before overtime are a maximum of 53 (212 in 28 days). These regulations are meant to accommodate the common 24-hour on and 48-hour off shift that many firefighters work.

This act has caused controversy in two areas. First, the definition of firefighter contains the provision that a majority of on-duty time be spent in fire protection duties. Some fire departments providing EMS assign firefighters to staff these units, causing them to work more than 40 hours per week. This qualifies as overtime for all other FLSA-covered workers (other than police, who can work up to 171 hours in a 28-day cycle) (U.S. Department of Labor, 2012).

In the past, there have been a number of firefighters assigned to full-time EMS duties who have filed lawsuits and won their cases. These were very costly to the jurisdictions, which had to fund large back-pay awards. With more and more fire departments offering EMS, this issue became a major problem. In response to this conflict, Congress passed a new law defining a firefighter's job requirements as including EMS, fire prevention, and other duties where life, property, or the environment is at risk. This definition now allows FES departments to use the 53-hour workweek as long as these workers are dual-trained in fire and EMS.

The second issue surfaced when it was noted that overtime could be awarded if employees volunteer in the municipality where they work. For example, in a large county fire department, an employee may work at Station 1 and be a volunteer member of Station 15. Because of the overtime interpretation, paid employees have been instructed not to volunteer at any FES station in their jurisdiction. This restriction has caused some hardships in combination FES departments.

The Financial Impact of Lawsuits

Administrators should be keenly aware of the financial consequences of legal improprieties. Although very difficult to quantify in dollars, many of these complaints end up in your local and sometimes national media. In addition, these incidents do not endear you to the city/county mayor or legislators. At best, it may make your job to get your budget approved more difficult or worst case, you may be asked to resign or be fired.

Relevant Supreme Court Cases

Sutton et al. v. United Air Lines, Inc. (1999)

This case concerned two pilots who were refused employment by United Airlines. Even though both pilots met the Federal Aviation Administration vision requirements for the position, they failed to meet United's more stringent requirements. The two pilots had uncorrected vision of 20/200 but had corrected vision identical to United's more stringent requirements. They sued United for discrimination under the ADA. Not only did the Court support United's stringent requirements, but the Court also did not find the pilots substantially limited in any major life activity. Therefore, they were not regarded as disabled under the ADA.

This ruling has application for those departments holding medical examinations to specific standards, such as the NFPA Standard 1582, *Standard on Comprehensive Occupational Medical Program for Fire Departments*. This standard contains several items that a potential member may argue permits an accommodation under ADA. Provided the department's medical and physical requirements are based on job performance standards and are essential to the job, they can be used even if they discriminate.

Also, this Supreme Court decision supports higher (but not unrealistic) benchmarks, especially for physical pass-fail levels. For example, a department may have a requirement for hiring using the pass-fail mark of running 1.5 miles in 10 minutes. This exceeds the minimum requirement supported by the International Association of Firefighters (IAFF) and the International Association of Fire Chiefs (IAFC), but has been recommended by other experts historically. Therefore, it is legal.

Kimel et al. v. Florida Board of Regents (2000)

In this case the Supreme Court found the Age Discrimination in Employment Act not enforceable in state and local governments. Therefore, unless a state law prohibits age discrimination (and some have an exemption for police and fire), an FES organization is allowed to have a mandatory minimum or maximum hiring and retirement age. For example, the federal government requires its firefighters to retire at 57 years old. Because of this requirement, and because their retirement plan requires 20 years of service, the maximum hiring age is 37. In another example, a metropolitan department had a maximum age of 29 for hiring and a mandatory retirement age of 55 years old. These types of age requirements were viewed as discriminatory in the past, but are now allowed in state and local governments.

In explaining its decision to repeal the Age Discrimination in Employment Act, the Court stated, "Old age does not define a discrete and insular minority because all persons, if they live out their normal life spans, will experience it" (data from Kimel et al. v. Florida Board of Regents, 2000, IV C).

If the department is using a mandatory retirement age, the cutoff should be justified using hard data and studies. For example, the 2017 Firefighter Fatalities report from the NFPA states, "The [mortality] rate for firefighters aged 60 and over was two-and-a-half times the average. Firefighters aged 50 and over accounted for half of all firefighter deaths over the five-year period, although they represent only one-quarter of all career and volunteer firefighters in the U.S" (NFPA, 2020). This evidence seems to support a mandatory retirement in the 55–60 year age range. However, these statistics have not been normalized for criteria such as medical and physical fitness levels. It is likely that a firefighter who is medically and physically fit would have a different death risk than the general firefighter population. Therefore, a yearly extension based on a comprehensive medical examination and physical fitness test could be allowed.

A comprehensive research paper authored by the Ontario Association of Fire Chiefs in 2009 looked at the ramifications of ending the mandatory retirement age of 65 in Canada. The Canadian study concluded that physical fitness declines with age, but that exercising regularly can slow the process. Therefore, the Ontario Association of Fire Chiefs suggests that "prior to the age of 40 firefighters take a physical fitness test once every five years, between the ages of 40 and 50 once every two years and over 50 once a year" (data from Ontario Association of Fire Chiefs, 2009, p. 44). This issue has been controversial but acceptable because many retirement systems have generous pensions for firefighters at relatively young ages. The argument for physical fitness tests can be furthered by an NFPA report that states that in 2020

"Heart disease and other cardiac issues have long been recognized as significant health risks in the fire service. Sudden cardiac death has consistently accounted for the largest share of on duty firefighter deaths since NFPA began this study in 1977" (NFPA, 2020). These medical problems should have resulted in treatment or forced retirement before the death occurred.

Janus v. American Federation of State, County, and Municipal Employees, Council 31 (2018)

This was a landmark U.S. labor law case concerning the power of labor unions to collect fees from non-union members. This is common policy in many states for FES employees. The Supreme Court (5–4 vote) ruled that such union fees in the public sector violate the First Amendment right to free speech. Succinctly, government non-union members won the right to cease automatic union fee payments.

Discipline

Generally, some studies indicate disciplining bad behavior is more effective than rewarding good behavior both at work and home. Although there are recommendations from many behavioral experts that say that rewarding good behavior is a good technique, especially for pre-adolescent children, most people correct bad behavior successfully after experiencing adverse consequences or discipline they receive. These negative observations of inappropriate behavior are best delivered privately. These are remembered vividly by the individual and are labeled "Significant Emotional Events" (NeuroAlchemist, 2016).

It is also very difficult for most supervisors (parents) to notice good behavior, so it goes unnoticed and not rewarded in most cases. For example, a station officer does not verbally congratulate each firefighter arriving on time to work. The best technique for encouraging good behavior is to lead/manage by setting a good example. Your example is also noticed when another member of the group commits a violation of a rule or regulation and the supervisor does not discipline the individual.

Insubordination

To improve morale and workplace harmony, employers may attempt to silence employees who regularly spread unsubstantiated rumors, vociferously challenge management, or publicly attack an organization's practices and policies. In this endeavor, the administration should be careful not to violate a person's First Amendment rights (freedom of speech).

The government has a legitimate interest in promoting the trust of the public in the services it provides. Therefore, unsubstantiated public statements can be handled as **insubordination**. Also, any high-ranking officer can be ordered not to communicate with the public.

The preferable way to handle this situation is to assign a public information officer who normally gives all official statements to the media. Then, by written order, all members should be directed to refer any inquiries to the public information officer. This prevents confusion by the public when an official statement is released.

Silencing Complaints without Violating the Law

First and foremost, psychologists recommend that the best method of reducing worker complaints is to create a safe, respectful, and productive workplace. The administration should listen to and take action, when justified, regarding the concerns and views of members. A formal process is for employees to submit, anonymously if preferred, comments, and constructive criticism should be provided. A post office box and an e-mail account should be used to receive this feedback. In addition, some departments use a telephone number connected to a voice message recorder.

It is also important that these concerns be acknowledged and attempts be made to correct unsafe situations. Some type of feedback should be given to employees, such as through an employee newsletter indicating the department's position and any attempts to solve the situation. If the members see attempts to solve the problem, they are more than likely to support the department's efforts.

For those complaints that are completely unfounded, the chief officer may want to follow this five-step procedure:

1. Remain objective, even in the realization that some members are chronic complainers and there is always a group that is opposed to any changes.
2. Focus on conduct when the complaints are disruptive in nature.
3. Determine if the member needs or could use some professional mental health services for truly disruptive behavior. Privately suggest the member contact the employee assistance program, because it can be helpful for managing his or her dissatisfaction.

4. Evaluate observations and supporting evidence to see if the behavior calls for disciplinary action.
5. Select an appropriate level of discipline.

Public Sector Discipline

There is a general perception that after a public sector employee is hired and finishes probation, it is impossible to terminate the person. This is not true. Public sector employees have **property interests** (Cleveland Board of Education v. Loudermill, 470 U.S. 532 [1985] (https://supreme.justia.com/cases/federal/us/470/532/)) in their jobs, requiring the employer to use due process in disciplinary procedures to separate an employee. Property interest is a legal term indicating that rights to a job are similar to and as strong as the rights the owner of a piece of real estate has. Before an adverse action can be applied, due process should be followed. The administration should follow these general guidelines:

1. Provide a written notice of charges. This preferably involves a three-step process starting with a verbal reprimand unless the charge is of great consequence to the organization.
2. Include a complete explanation of the evidence and reasoning on which the charge is based.
3. Provide a meaningful opportunity for the employee to be heard by an impartial decision-maker, preferably with a representative of the employee present, such as a union shop steward.

Additionally, consider using the following checklist to help guide the administration through the separation or serious disciplinary action:

- Is there sufficient basis for discipline?
- Is there appropriate documentation?
- Is the inappropriate behavior related to conduct, such as insubordination or tardiness, or performance, such as failure to perform a job requirement?
- Were all of the employee's rights respected during the investigation?
- What is the employee's previous disciplinary record?
- Was the violation serious, and were there any mitigating circumstances?
- Have other employees been disciplined for the same actions; if so, what were their consequences?
- Were all the appropriate steps in the disciplinary process followed?

- Did all of the notices clearly detail the observed actions and the violation of the law or policy?
- Did the employee have adequate time, representation, and opportunity to respond to the charges?
- Was the agency's attorney consulted before proceeding with the notice of disciplinary action?

Fair, Reasonable, and Evenly Enforced Discipline

The first step to ensure that discipline is fair is to make sure written policies and standard operating procedures (SOPs) include a listing of potential inappropriate behaviors and performance standards. In some cases, a general statement and explanation may be sufficient, such as in the case of insubordination. In others, as in illegal drug and alcohol use while on duty, the standards should be specific. Generally, it is a good idea for the administration to limit these policies and SOPs to as few as possible. In the emergency services arena, it is always better to keep rules and regulations simple because many of the critical decisions have to be made in the first few minutes.

An ongoing training program to inform all members of inappropriate behaviors and minimum performance standards is mandatory. Members should be given adequate notice and time for compliance, or members may win appeals of the adverse action in many cases, allowing another chance and additional time to comply.

The department's disciplinary actions must also show a consistent and nondiscriminatory record of enforcement; it is very important to have an active enforcement policy. For example, a department has an SOP requiring drivers responding to emergency incidents to come to a complete stop at all stop signs and red traffic lights. During the 3 years after the SOP was issued, the department made no attempt at enforcement. Then, a driver was charged with failure to follow the department's SOP (this is the first written enforcement record), and a penalty of 1 month's pay was proposed. Between the severity of the adverse action and the lack of any prior history of enforcement, the department will have difficulty sustaining this action if the employee appeals to an arbitrator or the courts.

Also, a department should have a policy of due process that outlines a disciplinary procedure starting with a verbal warning and concluding with written warnings and, with continued violations, substantial consequences. Appropriate notices of hearings and the ability to acquire legal or union representation need to be fair and realistic.

Probationary Period

As stated earlier, a newly hired FES employee should be put through a vigorous probationary program. Many potential employees present extensive qualifications, education, and experience to gain a job offer. This probationary period is a chance to gauge the effectiveness of previous education and experience. The probationary period and its requirements must be documented and presented to the employee. This will assure the employee cannot argue ignorance of the policy or Knowledge, Skills, and Abilities expectations.

An organization should have a structured evaluation in place reviewing all of the essential job performance criteria and standards. The assessments should be standardized at prescribed periods, including a formal written evaluation completed by the supervisor. All supervisors should be trained in the use of the evaluation systems and encouraged to make specific comments and observations in the report.

The probationary period should emphasize feedback to the employee to aid the individual in becoming a better employee, but should also be designed to weed out those who are not qualified. A formal testing procedure for the basic essential skills and knowledge that the employee should have mastered in recruit training should be used.

Supervisors should be reminded of the grave consequences of providing charitable probation evaluations. Retaining an FES member who is not qualified has double-edged consequences. The public does not get the best service and the member cannot operate safely, creating a hazard to himself or herself and the team.

A probationary member need not be told the reason for termination; the administration can simply state that the new member has not satisfactorily completed the probationary period. In some cases, it is better not to tell the probationary member the reason for termination, especially in writing. Although the employee has no winnable legal appeal, a written reason may unintentionally encourage a lawsuit. Anyone can instigate a lawsuit, even if there is no chance of winning. Probationary firefighters or EMS personnel should be aware that they have not satisfactorily completed the basic fundamental skills or behavior expectations.

Terminations

This section discusses appropriate methods for conducting terminations in the hopes of lowering exposure to legal challenges. Even with adequate evidence and support for a performance or disciplinary termination, if handled inappropriately, it can lead to legal disputes that can be costly and disruptive to the FES agency.

Members may be motivated to sue simply because they feel they were not treated fairly or with respect. The administration should approach the separation as a situation where there was not a good fit between the member and the job, rather than labeling the member as incompetent, lazy, or another derogatory characterization. Not everyone has the aptitude or mental and physical fitness to perform the demanding FES duties. They are not bad people; they are just in the wrong occupation.

Everyone (except the probationary member) has a right to respond to their accusers and have competent representation in the administrative process leading up to termination. Even when it may seem like a clear case, the agency should follow all of the policies that provide due process for the member. Specifics on termination policies vary, but all state and local employees have substantial job protections after their probation period. This protects employees against arbitrary dismissals. In all municipalities, public employees have legal rights to receive notice and the opportunity to discuss the inadequate job performance before they can be fired. Things are more complicated in civil service systems, which provide specific protections such as the right to appeals. In union environments, the process is complicated further. Nobody wants to be fired, and it's easier to find a new job when still employed, so the counseling approach to a dismissal situation can often be effective, resulting in the employee leaving voluntarily for another job. Sometimes there could be an opportunity to find another position within the municipal government where there is a good fit.

Reductions in force (RIF) is another situation that calls for employee termination. It is absolutely imperative to have an established RIF policy and to make all members aware of the selections process. RIF decisions that may be discriminatory, such as selecting all members older than 50 years old, should be avoided. From an accountant's perception, terminating one chief officer may create a cost savings equivalent in salary and benefits to terminating two or three firefighters or EMTs. A task force of members should provide recommendations and adopt a written policy.

In the situation of a voluntary termination (the employee quits), the supervisor should gather as much information about the member's motivation as possible through an exit interview. The objective is to uncover any issues that could later support a legal claim of discrimination by the member. The supervisor should attempt to find out if any underlying reasons for the resignation were stated previously by the member, and whether these complaints should have been addressed by the agency, such as would be appropriate for discrimination. Ask the following questions:

- Why is the member leaving?
- Would he or she consider returning to the organization?
- What plans does the member have in the near future?
- Does he or she believe that the FES should consider changes? If so, what changes and why?

Constructive Discharge

Constructive discharge is a legal term indicating that the member was somehow convinced to resign because of intolerable working conditions. Some common example scenarios are:

- Hostile work environment (e.g., sexual harassment)
- Unsafe working conditions (check Occupational Safety and Health Administration and NFPA safety standards for compliance)
- Insufficient information about alternatives in disciplinary or performance actions
- Not given the option of continuing employment in lieu of an early retirement (sometimes used with RIF to reduce the payroll). This is sometimes a preferred technique during reduction of funding in FES. The employees who have reached their minimum retirement benefit are good candidates because they have higher salaries in merit systems. Even though they may benefit in several more years of service to increase pension pay, they may be encouraged to retire early, possibly with incentives.

Unions

Some states are known as "right-to-work states" because they do not provide for the organization of public unions. In these states, it is common to not have a union contract or formal discussion with unions or other employee organizations. Even if not required, it is a good idea to meet with employee representatives on a regular basis and discuss topics such as salary, benefits, working conditions, and safety issues. This practice generally raises the morale of members.

Public Sector Unions

Unions are groups of workers who have formed an association to discuss issues with management. Unions protect jobs and negotiate to improve salary and benefits, including safety issues from local governments. They also promote companionship and respect for fellow members. This creates a strong loyalty to the union that sometimes can conflict with the main mission or goal of the FES organization—namely, quality service to the public.

The relationship between unions (and the representatives of the public, e.g., chief officers, mayors, county executives) is a system of checks and balances—the administration is watching out for the best interest of the public at a reasonable cost, and the unions are looking out for the rights of their members. This relationship can be especially challenging for a chief hired in from outside. This can be a difficult situation for the new chief because they do not know about any politically influential individuals or groups in the municipality (see Chapter 2, *Power Elite*). However, this is an oversimplification; there are many situations in which unions have supported changes that were primarily for the benefit of the public.

Unions come in all sizes and types. Some are extremely strong and influential; others are mere social clubs. In right-to-work states, unions tend to be weak and more advisory in their actions. As an administrator, never underestimate the power of the local union. You do not have to give in to all of their demands simply because they have influence; however, the chief officer must respect their opinions and give them a chance to voice their concerns and beliefs.

Union Dues for Non-Members

In a February 2018 decision, the Supreme Court decided that forcing union non-member public employees to pay union fees violates free speech. This affects every employee who decides not to join the union. In many political jurisdictions, the non-union employee is required to pay the union dues, even though they are not a member. The union in most cases was obligated to provide protection for the non-members , rights, including bargaining for salary and benefits. The rationale for this decision is that non-union members support the union's political endeavors even if they disagree with the politician, policy, or party being financially supported.

Private Sector Unions

The National Labor Relations Act was enacted by Congress in 1935 to protect the rights of employees and employers, to encourage collective bargaining, and to curtail private sector labor and management practices that could harm the welfare of workers. There are many varied opinions concerning the pros and cons of union affiliation. Whether a union is effective depends on the union itself and the agreements made with the company. When employers deny the rights of employees to organize and refuse to accept the procedure of collective bargaining, this can lead to strikes. Generally, strikes are not an option for most fire service organizations and are in most cases not allowed by state or local legislation.

Strikes and Job Actions

In most labor unions, the strike is the ultimate tool to gain agreement for an improved salary or benefit. Without this negotiating tool, public safety unions are at a great disadvantage in discussions with the dominant governments.

Some states allow binding arbitration for public safety employees. This permits a fair and equal status for labor and management. However, in right-to-work states, binding arbitration is rare. In these situations, it is common to lack a written contract or for negotiations to continue endlessly. This can be very bad for the morale of members, resulting in safety concerns, which may have a negative impact on service to the public.

Bargaining Units

It is very common for there to be two bargaining units—one for workers and one for officers. In many cases, both units may be represented by the same union president or are in some way combined into the same organization. Representing two classes that may be in opposition can cause a conflict of interest for the union. For example, in some departments, only the fire chief is not a member of the union. If both the worker and supervisor are members of the same labor organization, they may become friends in the social atmosphere that surrounds these organizations. When differences surface, it can become a one-person administration versus everyone else. This can isolate administration and make it very difficult to negotiate solutions to problems.

It can be very difficult for supervisory or management personnel to separate their loyalty to the firefighters from the duties they must perform as a supervisor. In the day-to-day running of the department, and in the negotiation process, supervisory personnel may not know where their loyalty should stand. This situation is exacerbated by the nature of FES work. Personnel perform their duties in small teams that become very close professionally and personally. Firefighters often become very close friends. Often, one of these friends may be promoted and become the

others' supervisor. The problem with this may not revolve around being a part of the same bargaining unit but rather from being a part of the same social organization. Workers and supervisors have several needs that are the same, including pay and working conditions. It could be argued by a union that this is clearly a case of equal goals and, therefore, supervisors should be part of the union.

One way to address this predicament is to set up a policy that awards the same working conditions to both groups but insists on separate bargaining units. For example, when bargaining for pay raises, work hours, and other monetary benefits, the city could bargain with both groups with the understanding that the final outcome would be equal for both groups. For nonmonetary or management prerogative items, separate discussions with each union president would suffice.

Many times, high-ranking officials from the FES agency do not sit in negotiations because they are seen as the connection between the administration and the union. Their attendance and views during negotiations can be detrimental to the relationships with the union and its members. Many administrations have officials from FES agencies meet with them in private to learn what they want to have changed in the contract during negotiations. Finally, personnel who have middle management positions (above the company officer) should not be in any bargaining unit. These individuals should be a part of administration and treated accordingly.

Local Government Representatives

When formal negotiations start, the management is represented by several people from the government along with high-ranking officials from the FES agency. These government officials may have a narrow point of view or goal, such as a target budget figure. They do not have the knowledge or experience to foresee any possible adverse effects on FES that could result from a new union contract that saves the city money. This same lack of knowledge is evident if using arbitration or fact-finding services.

An education program should be provided to inform these officials about the actual functioning aspects of FES. For example, they may not understand why there is a need for four firefighters per company when 61% of the department's calls are EMS calls that require only two people (a cardiac incident may require five or six). Emergency services are very technical and require a high level of skills and knowledge. Only when members have been providing emergency services for many years do they get an intuitive feel for what constitutes quality service, including the safety aspects of the profession.

Grievances

This is an important aspect of any labor agreement. This process allows for the fair resolution of misunderstandings of the contract or an appeal process for disciplinary actions. The union is obligated to represent the member to guarantee that the determination of any adverse action proposed or taken was fair. This process also keeps many issues from being litigated in the courts. Most courts will not overturn a grievance procedure, especially if the process ends at binding arbitration. Therefore, when complying with the provisions in the labor contract for settling grievances, the decision by an arbitrator is most often final.

When processing a grievance, make sure the administration has all the facts and is well prepared for the hearing. If these facts indicate that the administration made an oversight or error, this should be corrected immediately. If not, include a full review and representation by the city or county attorney. The results of an arbitrated grievance can set a precedent that is very difficult to change in the future. Do not take these cases lightly. They can have long-term effects on the department's operations and administration.

Progressive Labor Relations

It is now more common to see cordial relationships between labor and management. In the past, each viewed the other as the enemy and would do anything to win in negotiations and grievance proceedings. Today, although each has to represent its own constituents, where there is a mutual interest, each helps the other.

A lot of the previous mistrust was the result of confrontational behavior that was common in the private sector during the Industrial Revolution. Even violence by management and labor was not uncommon. As a result of federal laws and regulations, and the revelation that competing in the new global economy requires well-motivated workers, labor and management are finding ways to reasonably settle their differences without hurting the competitiveness of the industry or business.

This type of new cooperation has also resulted in improved communications. The chief officer should schedule meetings with union officials on a regular basis. In addition, visits to work locations to talk and listen to all members are very important.

There have been cases where the union or volunteer leaders have lost touch with their own people. In these cases, internal communications can break down. Labor and volunteer leaders may be using old paradigms or their own agendas as a basis for representing the membership. Although this is not a good situation, it is not uncommon. The technique of Managing by Walking Around would assure the administrator has unfiltered communications with all the members because it provides a link directly to the members. The larger the organization, the greater the likelihood of this type of impediment to communication.

Getting to Yes

The book *Getting to Yes: Negotiating Agreement Without Giving In* by Roger Fisher and William Ury outlines a successful type of negotiation process that attempts to separate facts and issues from people and their emotions. "Principled negotiation methods focus on basic interests, mutually satisfying options, and fair standards that typically result in a wise and acceptable agreement" (data from Fisher, Patton, and Ury, 2011). The techniques described in the book can also be used to problem solve differences between spouses, parents and their children, and many other interpersonal relationships.

One great method to use in negotiations is to brainstorm for possible solutions. The ground rules should allow any suggestion to be recorded without judgment. The group then needs to go back to each idea and analyze its cost, complexity, and reasonableness. After a set of realistic possible solutions has been agreed on, the separate parties can then concentrate on them in their sometimes contentious discussion over a union contract or other personnel disputes.

Motivation

In every organization there is the potential for members who can pull the entire organization toward mediocrity if the administration fails to deal with the situation. Motivation is an important concept that can be used to create a satisfied member who will be a top performer for the organization. The following two areas, when combined, seem to produce a motivated member: benefits (hygiene factors) and pride (motivators); this is also known as Herzberg's motivation-hygiene theory. Examples of benefits include fair salary and work hours, health insurance, workers' compensation, and length-of-service awards (volunteers) or pensions. Examples of pride factors include modern and well-maintained fire and rescue

FIGURE 6-6 Recognizing an individual for heroic or exemplary service is a great way to improve motivation.
© Pazargic Liviu/Shutterstock.

FIGURE 6-7 Volunteer firefighters.
© Ariel Skelley/DigitalVision/Getty Images.

equipment; professional qualifications for membership; a rigorous training program, including physical fitness; and, in general, anything that leads a member to feel good about his or her contribution to the FES organization. These factors can be reinforced by recognition from the organization **FIGURE 6-6**.

Although relative, pride factors tend to be more important to the individual than benefits. For example, volunteers will serve without any pay if properly motivated. Some would argue that the additional time required for a volunteer to complete a firefighter comprehensive training program would be a disincentive to retaining members. However, in an article titled, "Breaking Bad Habits: Recruitment and Retention of Volunteer Firefighters" (IAFC, 2020), "One of the main reasons why members quit is because they do not feel appreciated."

In addition, a rigorous training program will create loyal and motivated members leading to fewer disciplinary problems and retention of volunteers. Members who feel good about their ability to provide professional emergency service are those that have been trained in the knowledge and skills necessary to perform proficient professional services. Never underestimate a person's pride—it is a very strong human emotion and is a major contributor to high motivation.

Recruit and Retention within Volunteer Departments

In past years, members were free to commit many hours of their time to the local volunteer fire company **FIGURE 6-7**. However, more recently, economic

pressures have reduced the availability of members. This is especially true for older members who find themselves with new demands on their time as they start families. Furthermore, changing social values have influenced many men to take a more active part in the raising of children. Between the increases in demand and newer values, men are no longer staying as active in FES organizations after taking on the responsibility of a family. Some of this downward trend has been offset, however, by an increasing number of women volunteering.

For additional insight, the U.S. Fire Administration report "Recruitment and Retention of Volunteer Fire Fighters" discusses the core problems of recruitment and retention in the volunteer fire service, and provides suggestions for solutions to these problems from volunteer fire chiefs and firefighters from across the United States. This free publication can be ordered from the U.S. Fire Administration Web site (https://www.usfa.fema.gov/). In addition, more recent information can be found on the Internet using https://duckduckgo.com/ and searching for "recruit and retention of volunteer fire fighters."

People who volunteer have a basic need for recognition, belonging, and the satisfaction that comes from helping people and serving the community. Retaining volunteers is dependent on acknowledging these needs. Some helpful suggestions follow:

- Cut unnecessary bureaucracy in terms of application and retention.
- Give volunteers clear expectations and guidelines.
- Provide flexible schedules to accommodate complex work and family schedules.
- Use non-FES people for administrative tasks.
- Create training programs to defray the expenses of certification and re-licensure.

- Consider grant funding for obtaining and replacing equipment.
- Investigate community groups that may donate or fund equipment.
- Encourage all volunteers to be recruiters and take part in training teammates.
- Continually show all members they are valued.

Retention within Combination Departments

According to the NFPA, combination departments make up 24% of all departments and protect 33% of the U.S. population. These departments come in all sizes and, therefore, generalizations fail to describe many of them. The NFPA attempts to separate combination departments into categories of "Mostly Volunteer" or "Mostly Career."

Many of these departments report conflicts between volunteer and career firefighters, the origin of which often stems from issues of group loyalty and job security. For example, volunteer members may be concerned that if more career members are hired, fewer volunteers will be needed. It is because of this reasoning that many volunteer departments wait long periods of time without adequate volunteer staffing before venturing into the combination solution.

The first career member of a combination department is generally the driver, who works weekdays and is expected to maintain the station and apparatus. This career member often cannot advance in his or her profession to an officer rank (reserved for volunteers only) but is relegated to remain a firefighter. There are, however, many firefighters who are satisfied with these duties, and the volunteer department should be very careful to hire such an individual.

Still, there may be times when the combination department needs a full career crew, such as for a weekday shift. For example, in the past, the New York City Fire Department has staffed some ladder companies for the high incident hours of 3:00 PM to 1:00 AM. This arrangement is also commonly seen for EMS units. At this point, all members should be required to meet the same training, medical, and physical fitness standards. If the career officer must be certified as an NFPA Fire Officer I, then this same requirement should pertain to the volunteer officer.

One organizational technique is to keep the two factions separated physically by staffing separate stations. An example can be found in Baltimore County's Fire Department. This department is made up of 25 career stations and 33 volunteer stations. At fires and other emergency scenes, volunteers work side by side with career personnel under a single chain of command and set of operating procedures. This department protects over 800,000 citizens in 610 square miles. Interviewed fire officials, both career and volunteer, report smooth operations without major conflicts.

Another important issue related to combination departments revolves around pride. Companies can be represented through the uniforms they wear and the markings seen on their fire apparatus. Even when away from the station and off duty, members often wear T-shirts proclaiming their pride in their company. This holds true for the career firefighter and is even commonly seen at the station level. Just like volunteer companies, many career stations are allowed to create a logo for their company, which increases the pride and morale of the members.

Bearing in mind these issues of pride and loyalty, the following guidelines can be very helpful in reducing personnel conflicts in combination departments:

1. Maintain equal professional standards of training, education, and physical fitness among all members.
2. Maintain stations and apparatus for each group. If that is not realistic, organize members in like groups with their own identification. When either group is staffing the apparatus, it should be clearly marked as such. This can be done by permanently marking assigned apparatus or using temporary removable signs.
3. Remember, Pride = Loyalty = Good Morale = Retention.

Wrap-Up

CHAPTER SUMMARY

- Understanding that members are most critical to being able to provide quality service and the most expensive share of the budget is important.
- Understanding legal mandates with hiring and member continued participation is critical to prevent legal complications and public perception problems.

- Knowing how to motivate and, when there is noncompliance, discipline members is essential to an effective and efficient organization.
- Understanding how to negotiate with public sector unions is critical to come to mutual agreements on pay, benefits, and safety disputes.

KEY TERMS

Affirmative action A hiring program used to recruit minorities and women in many FES organizations.

Americans with Disabilities Act Federal act that provides civil rights protection to people with disabilities and guarantees equal opportunity in employment, public accommodations, transportation, government services, and telecommunications.

Apprentice phase 1-year period in which a new FES member learns additional knowledge and skills.

Banding This is a technique that replaces the traditional hiring or promotional lists used by many FES organizations. For example, for promotion a written exam is given and a list is created using the highest score first. This person is promoted first. However, in "banding" the list is separated and the top 15% are placed in the top band. Then either random choice or a diversity goal is used to choose the person for promotion.

Civil rights acts The 13th and 14th Amendments to the U.S. Constitution guarantee equal treatment or protection against discrimination based on race, religion, gender, or national origin.

Constructive discharge A legal term indicating that a member was somehow convinced to resign because of intolerable working conditions.

Diversity A mix of employees that reflects a blend of sex, race, age, education, country of origin, and other characteristics, such as personal values.

Equal Employment Opportunity Commission A federal agency that administers and enforces civil rights laws against workplace discrimination.

Fair Labor Standards Act The Fair Labor Standards Act of 1938 is a federal statute of the United States. The FLSA introduced the 40-hour work week, established a national minimum wage, guaranteed "time-and-a-half" for overtime in certain jobs, and prohibited most employment of minors in "oppressive child labor."

Family and Medical Leave Act Federal act that mandates employers with more than 50 employees must offer employees up to 12 weeks of unpaid leave per year for family responsibilities.

Hostile environment Repeated workplace conduct of an inappropriate and/or sexual nature that may be offensive or intimidating.

Human resources The division responsible for recruiting, hiring, training, evaluating, and paying employees.

Insubordination The act of willfully disobeying an order of one's superior.

Job analysis A common method of defining job requirements, including hiring criteria, recruit training and testing, and incumbent evaluation.

Light duty A temporary assignment for workers who have been injured on the job in such areas as fire inspection, fire safety education, or 911 dispatching.

National Labor Relations Act Enacted by Congress in 1935 to protect the rights of employees and employers, encourage collective bargaining, and curtail private sector labor and management practices that could harm the welfare of workers.

Probationary period A specified length of time after hiring (say, 1 year) used to evaluate a new hire, during which the employee can be separated without cause.

Property interests A legal term indicating that the employer is required to use due process in disciplinary procedures to separate an employee.

Quid pro quo Harassment that occurs when a person in authority takes advantage of their control over an employee's future in the organization.

Reductions in force A method to identify employees to be laid off when budget reductions dictate reducing the total budget.

Sensitivity training Education offered to employees about personal differences to achieve social harmony.

Unions Groups of workers who have formed an association to discuss issues with management.

Workers' compensation A form of insurance providing wage replacement and medical benefits to employees injured in the course of employment in exchange for mandatory relinquishment of the employee's right to sue his or her employer for the tort of negligence.

REVIEW QUESTIONS

1. What minimum job requirements/competencies were you expected to meet when you started out? How long was your probationary period (if applicable)? How have these standards changed since you were first hired, and do you think those changes were for the better?

2. Identify and read a recent news story about paid family leave; if possible, select one that is relevant to your state. How do you think the policy described in the article would or will affect your department?

3. Suppose one of your employees is caught drinking on the job. Describe the disciplinary process the employee would undergo from start to finish, being sure to highlight the role of the administrator and likely/possible outcomes for that employee.

4. Come up with one scenario in which an administration would need to work alongside the union to accomplish a shared goal, and another in which the two entities would be at odds. List at least two actions the administration could take in each instance to achieve a favorable outcome.

5. Do you think benefits or pride are more important when it comes to motivating and retaining employees? Defend your choice in a short essay of about 500 words.

CHAPTER ACTIVITY #1: NEW OFFICER

A newly promoted officer is assigned to a very busy engine company and on the first day is greeted by an energetic crew of three firefighters who seem pleased with their new officer. In reviewing personnel files and contacting the previous officer, the new officer has acquired some background information on the crew and is assured that they are aggressive firefighters who know how to put a fire out.

Before the morning coffee and apparatus check is finished, the company is dispatched to a reported fire in a single-family dwelling. The officer did not get a chance to explain any expectations or preferred procedures for emergency incidents and assumed that the crew would follow departmental SOPs. When the company arrives, it is clear that this is a working fire. Because it is a daytime incident, the officer is relatively confident that the dwelling is unoccupied.

Before the officer can make a brief size-up, the crew goes into action. Two firefighters pull the 150-ft (45.72-m), 1.75-in (4.375-cm) hose line and extend it to the side door (the fire can be seen coming out the windows near this door). Before the officer can join the two firefighters extending the hose line, they have forced the door and entered the house.

When the officer catches up, one of the firefighters has gone off on his own to search for victims. The fire was confined to a room and its contents and is extinguished using a few gallons of water. After the fire, the two firefighters discuss with pride their ability to get in fast and put the fire out before other companies arrived on the scene.

Discussion Questions

1. Describe any potential disciplinary problems in reference to safety or SOPs that would be apparent in this company.

2. Discuss your own personal thoughts about the situation.

3. In the next similar situation, how would you instruct your crew in their actions during emergency operations? Be specific. What are your expectations?

CHAPTER ACTIVITY #2: NONCOMPLIANCE WITH RULES AND REGULATIONS

Few things in EMS are as important as paperwork. The patient care report (PCR) is not only a part of the patient's medical record, but provides information required to bill the patient or his or her insurance. A record of every patient encounter is required by law, and certain data items are mandatory whether the PCR is paper-based or electronic.

In the past 6 months, an EMS agency that was using a paper-based PCR made the switch to an electronic format. The supervisor of the agency has faced a number of challenges ensuring that the SOP for completing PCRs is met. The SOP requires that PCRs be created on every call, that they are accurate and complete, and that they are submitted within 24 hours of the call.

Completion of PCRs is never an EMT or paramedic's favorite part of the job, and compliance can be a challenge. Specifically, the supervisor has one paramedic who is a good clinician, but hates doing PCRs. He is noncompliant with policy in terms of submitting PCRs on time, and sometimes does not do them at all. This issue has become worse since the agency switched to the electronic format.

Discussion Questions

1. How big of an issue is this?

2. What methods would you use to help this employee become compliant?

3. Is this a clinical or an operational issue?

REFERENCES

Benshoff v. City of Virginia Beach, 180 F.3d 752. 1999.

Cameron, Kim, and Emma Seppälä. 2015. *Managing People: Proof That Positive Work Cultures Are More Productive.* Harvard Business Review.

Fisher, R., Bruce Patton, and W. Ury. 2011. *Getting to Yes: Negotiating Agreement Without Giving In.* (3rd edition). New York: Penguin Books.

Garcetti v. Ceballos. 2006. U.S. Supreme Court. Retrieved from https://www.law.cornell.edu/supct/html/04-473.ZS.html

International Association of Fire Chiefs. 2020. *Breaking bad habits: recruitment and retention of volunteer firefighters.* Retrieved from https://www.iafc.org/topics-and-tools /resources/resource/breaking-bad-habits-recruitment -and-retention-of-volunteer-firefighters.

International Association of Firefighters. 2011. Fire Service Joint Labor Management. Wellness-Fitness Task Force. Candidate Physical Ability Test Program Summary. Retrieved from https://www.iafc.org/topics-and-tools/resources/resource /breaking-bad-habits-recruitment-and-retention-of -volunteer-firefighters

Kimel et al. v. Florida Board of Regents. 2000. 528 U.S. 62, 120 S. Ct. 631, 145 L. Ed. 2d 522, U.S.

National Fire Protection Association. 2020. *Firefighter Fatalities in the U.S.—2019.* Quincy, MA: National Fire Protection Association, Fire Analysis and Research.

NeuroAlchemist. 2016. *How significant emotional events can empower you; therapy & behavioral change.* Retrieved from http://neuroalchemist.com/category/therapy/

Ontario Association of Fire Chiefs. 2009. Discussion Paper: Managing the end of mandatory retirement [age] in the fire service. Retrieved from http://www.oafc .on.ca/lib/db2file .asp?fileid=1389

Sutton et al. v. United Air Lines, Inc. 1999. 527 U.S. 471. 1999. 130 F.3d 893, affirmed.

U.S. Department of Labor. 2012. *Police and firefighters under the Fair Labor Standards Act (FLSA).* Retrieved from http:// www.dol.gov/esa/regs/compliance/whd /whdfs8.htm

Vlahos, K. B. 2003. Race-based policies raise fraud concerns. Retrieved from https://www.foxnews.com/story/race -based-policies-raise-fraud-concerns

Wagman, J. 2007. City drops firefighters' entry test, *St. Louis-Post Dispatch.* Retrieved from http://mirroronamerica.blogspot .com/2007/09/st-louis-fire-department-entrance-exam.html

CHAPTER 7

Customer Service

LEARNING OBJECTIVES

After studying this chapter, the student will be able to:

- Understand the connection between market failures and public safety monopolies (p 135).
- Examine customer surveys and demographic changes (pp 135–136).
- List and describe the primary services that a fire and emergency services (FES) organization should focus on when planning customer service efforts (pp 136–141, 142–143).
- Identify issues relating to organizations that provide emergency medical services (EMS) (pp 143–145).
- Understand the importance of the customer's distress and grief after a fire or other emergency that has resulted in life or property loss (pp 145–146).

Case Study

Customer Service Considerations

A fire department responds to a house fire and very quickly extinguishes the fire. The officer at the site notices that the grass around the house is extremely high. After learning that the owner of the house, an elderly woman, has limited financial and physical abilities, the officer offers to cut the grass.

There have been influential fire department speakers that focus on identifying situations where FES members can provide a customer benefit that is not traditionally offered. The theory is that this is a good public relations policy to help the FES organization with citizens and elected officials to ultimately support and approve tax revenues for the department. This idea is borrowed from commercial marketing using personal services above and beyond what is customarily expected. These are positive experiences that the customer remembers and can last a lifetime.

The firefighters, after driving back to their station, load the station lawnmower onto the engine. They then return to the woman's house and cut the grass.

At all times they are available by radio for any emergency calls. The woman is extremely pleased with the firefighters, and the firefighters feel they have done a good deed in the spirit of customer service, which has been the subject of several recent training sessions. However, in the following days, headquarters receives several calls from other elderly and handicapped owners in the same neighborhood asking to have their grass cut by the fire department.

Discussion Questions

1. This story points out one pitfall of customer service that must be considered. Before volunteering a service, determine whether it is something that can be provided to anyone who requests or needs the same assistance.

2. Would you consider volunteering your own off-duty time to help this woman?

3. Are there other organizations or neighbors that could be contacted to help this woman?

Overview

In the business community, customer service equates directly to survival and increased profits. Those companies can identify what the customer wants, and then focus on these desires to increase sales and profit. Companies have gone out of business by failing to keep an eye on the customers and their needs.

Going out of business is rare for the typical FES organization. However, there are some exceptions, especially in the EMS business where there are a significant number of private ambulance companies. Generally speaking, organizations that practice quality service are those that provide the best service to the public and can secure adequate funding from elected officials. FES organizations need to provide the utmost in customer service and should be cognizant of this fact.

What Is Private? What Is Public?

Emergency services come in different forms. They may be provided by private or public organizations. The differences between these two raise a series of questions. Who should provide customers (the public) with emergency services? Should FES be privatized, or is this the rightful responsibility of public government agencies? What is the role of government regulation of FES providers? Should there be national standards for emergency services? Or is each local jurisdiction so unique that it should have its own tailor-made standards and customer service expectations?

Questions about the services governments can successfully and effectively provide for citizens need to be answered. Some Americans assume that the private capitalistic economy is the best place to find products and services. After all, this is what has made the United States great. Others believe government should provide for many of the needs of its citizens, especially the less fortunate. In general, most Americans prefer to rely on free markets, families, places of worship, and other private-sector associations whenever possible. Governments are called on mainly when private-sector institutions prove to be severely inefficient and when they fail to satisfy certain minimum standards of service, especially safety issues. Americans have built a strong economy based on competition and the free enterprise system that usually takes care of most basic needs.

Justification for Government Intervention

When can government legitimately intercede to provide services, goods, or safety to promote its citizens' (i.e., the customers') well-being? One type of justification occurs when there is a market failure, which validates regulations or services by government. For example, if education for young children was provided exclusively by private, for-profit institutions, the children whose parents could not afford the tuition would not get an education. In the United States, a high school education is a public benefit that is provided free to the student and supported by taxes. This does often bring up the question of quality because there is no direct competition. More recently, competition is being provided in many areas by private and/or charter schools.

Fire departments have a monopoly for the services they provide in the community. Improvements, service levels, and fiscal efficiencies are under the control of the department or the government officials that oversee the department. If service is below nationally recognized professional standards, there is no competing department for the customer to compare against or switch to. Unless the public demands better service or there is a regulatory body to oversee the operation, the FES organization sets its own level of service.

Fortunately, most FES organizations are dedicated to professional competent emergency service. However, the administration of these organizations is often at a disadvantage when it would like to make a change that will bring improved customer service. In the business world, if the manufacturer does not make the best product at the best price for the customer, it loses market share and profit, or even goes out of business. However, in the FES organization there is typically no measurable indication that a change is needed. Therefore, in many cases, it is difficult to gain the additional funding or cooperation of members that is needed to make the change.

Government regulation and elected officials, if properly educated, help to provide the balance needed for this monopolistic public service. They provide the FES administration with the independent oversight necessary to ensure the public that they are getting professional quality service.

Public Preference Surveys

For overall guidance, a survey of customers may provide some insight into their expectations **FIGURE 7-1**. If their expectations are supported by professional standards or expert advice, the administration can look for ways to implement the suggestions. If the public's understanding is uninformed, the agency can fashion an education program to correct any erroneous understanding of the department and its service to the community.

One useful technique to assess the needs of the customer and support budget requests is the public survey. These types of surveys, which require prior approval by the chief administrative officer and elected officials, can be used to recognize how the public perceives the department's present service, identify areas in which they would like improvement or expansion of service, and determine how much the public is willing to pay for improvements in service. A good time to do this is a month before the budget is announced publicly so the justification is fresh in everyone's mind.

For example, budget requests for additional personnel are always hard to gain approval for because personnel costs are always the most expensive items in the budget. After estimating the total costs for the proposal, including all personnel salary and benefit costs, the administrator would express the unit cost per person, per taxpayer, or per household for the community. Then, the survey question could be asked: "Would you be willing to pay X dollars more per year to bring all engine companies up to the staffing levels recommended by national safety and proficient training standards?" An effective survey would also provide detailed justification as part of the question or as an addendum.

Sharing the results of a public survey is one method of convincing the appointed and elected officials to buy into the proposal. This survey should be the ultimate test of whether the public is willing to spend tax dollars on the FES budget request. Although the results of this survey do not guarantee approval, the chief officer can argue that this is a fair indication of the public's (customer's) wishes.

FIGURE 7-1 Survey
© Ilker Celik/Alamy Stock Photo.

Fire and EMS organizations should look for opportunities throughout the year to communicate with citizen groups, the press, and elected officials to gain support for existing programs or enhancement programs that the administration would like to propose in the future. When these opportunities occur, they are often very short-lived. Organizations need to be prepared with all the information for justification and an accurate estimate of all costs.

However, for surveys to be meaningful, the persons completing them must have some idea of what constitutes quality service. In most cases, the public, press, and officials have little knowledge or background when it comes to fire, EMS, and emergency preparedness services. In a survey commissioned by the National Fire Protection Association (NFPA), it was noted that most people believe they are more at risk of dying in a fire in a hotel room than in their own homes. However, in 2017 NFPA records show that 97% of residential fire deaths occurred in homes and apartments. Over half of the respondents wrongly assumed they would have plenty of time to escape a typical living room fire. Most had no idea how fast smoke, toxic gases, and heat can fill a room, blocking escape routes and reducing visibility to zero (NFPA, 1996).

The first step to resolve this issue may be to educate the public. This should include fire safety education and education on how to evaluate the public FES. Although the latter tends to be complex, giving the public the knowledge of how to assess the public fire services and EMS is necessary if they are to have an informed opinion about the department and its services. This is especially true for the elderly and young.

Keeping Pace with Community Demographic Change

It is necessary that fire and emergency organizations monitor the community for demographic changes that may affect the services requested and tax revenues available. For example, older communities may see a decrease in population and an increase in the number of senior citizens, although this can bring developments of senior citizen communities that bring taxpayers without children—a good trend for the jurisdiction because typically over 50% of the local budget goes to K–12 education. Other suburban communities are experiencing growth, with strong demand for improved services in areas such as K–12 education. Typically, growth areas see an appreciable number of new families with school-aged children

whereas retirement areas see many senior citizens with no young children.

Demographic statistics are published yearly as estimates and can be found at state or local planning agencies. They are not 100% accurate on a yearly basis but are very close in most cases. The main source of this data is the U.S. Census Bureau and can be located at the web site: https://www.census.gov/data.html. The 10-year complete census is being collected in 2020 and should be available.

Expanded Services

The types of services a department can offer will affect its relationship with its customers. Many departments look at the addition of services outside of firefighting as a way to market the department to its customers. These organizations are quick to point out to citizens and elected officials that they provide extra value for the tax dollars spent.

The following is a list of services that an FES organization can provide to its customers. These are the primary areas on which to focus when planning customer service efforts. Several of these may be the primary responsibility of other agencies as dictated by local ordinances or historical evolution, but all could be provided in one fashion or another.

- Fire suppression
- EMS, including transport of patients
- Technical rescue service
- Major disaster mitigation and recovery
- Prevention and education (fire, medical, and major disasters)

Fire Suppression

As demographics and communities grow and change, fire service and EMS organizations need to revisit the issue of response time periodically. The capability of the fire department to respond with sufficient competent personnel and equipment in time to rescue any trapped occupants and confine the fire to the room of origin is a primary customer service goal. Until 2001 there was no consensus on measuring response times in the American fire service; however, it is now defined in NFPA 1710, *Standard for the Organization and Deployment of Fire Suppression Operations, Emergency Medical Operations, and Special Operations to the Public by Career Fire Departments*.

Another measure that can be used by fire service and EMS organizations is to look at existing average response times for all areas of the jurisdiction to see if their organization's times fall within this average for all areas. A fire or EMS department may want to benchmark its response times against a national survey that reports response times. Although not very common, there have been some national studies in this area, such as the 1999 Phoenix Fire Department National Survey on Fire Department Operations. The rationale behind this type of evaluation is that all citizens (the customers) deserve equal fire protection as measured by response time. This type of analysis can point out areas that need new stations and areas that could be served by fewer stations and still meet the jurisdiction's average response time.

However, response times are typically longer in rural areas that are sparsely populated. Response time becomes a cost-benefit calculation. Areas with few residents cannot collect the taxes needed to support shorter response times. For example, in built-up areas, the Insurance Service Office (ISO) Grading Schedule generally equates a 3-minute response time to a response distance of 1.5 road miles. In areas with predominantly single-family dwelling developments, ISO uses a 5-mile distance as a maximum for adequate protection (ISO, 2012).

According to the NFPA *Fire Protection Handbook* (NFPA, 2008), evaluation parameters for response times should be based on flashover **FIGURE 7-2**. Although these times vary based on a number of variables, NFPA notes, "[F]ire departments should operate on the assumption that prevention of flashover requires a response in less than 10 minutes." It has also been observed that flashover is correlated with fire deaths, including firefighters. This is a critical benchmark of a hostile building fire and should be one of the main goals of fire departments to attack the fire before this occurs.

The NFPA 1710 standard, the ISO Grading Schedule, and the *Fire Protection Handbook* give some guidance to the planner or administrator trying to determine an appropriate response time for a jurisdiction. The number and location of fire and rescue stations in a given area are controlled solely by the response time chosen as a service indicator for the local jurisdiction. In general, longer response times equate to fewer fire and rescue stations, and the opposite occurs when response times are shorter. For example, assume that a jurisdiction plans for a maximum response time of 4 minutes (NFPA 1710). If the response time is increased by 1 minute to 5 minutes, 36% fewer stations will be needed to cover the same area. If the response time is lowered to 3 minutes, then 66% more stations are needed. These are large variances.

When reviewing a department's equipment for compliance, the NFPA provides comprehensive standards for design and operation **FIGURE 7-3**. Most departments meet these standards and, unless limited by age, it is rare for this item to affect fire suppression.

The other key item that can critically affect fire suppression is competent personnel. In fire suppression and rescue service, competence has been equated to personnel who meet nationally recognized professional standards, such as NFPA 1001, *Standard for Firefighter Professional Qualifications*, through a training system that certifies competencies. In addition, larger metropolitan fire departments have developed "recruit" training programs that generally exceed the national standards.

However, in the United States, there is no consensus on the minimum time needed for a student to become proficient in the job performance items listed in the NFPA standards. There are some statewide certification systems, but no national programs. Generally, the length of the training program is directly proportional to the skills of the students. EMS, however, has a national curriculum and a set number of hours needed for completion of its several levels of competency.

FIGURE 7-2 Flashover is the point of an uncontrolled fire that causes major property damage and fire deaths, including firefighters.
© Christian.dk/Shutterstock.

FIGURE 7-3 Fire apparatus must be kept up to date because new innovations are continually discovered.
© Robert Wilder Jr/Shutterstock.

For customers to be able to make a rational judgment about the quality of their emergency services, they need to understand the previous discussion on response times, competent personnel (including minimum staffing), and equipment. It is very important that citizens understand that uncontrolled fires grow exponentially, meaning that every minute a fire grows faster than the previous minute. The NFPA has several excellent videos that visually illustrate this rapid growth. It is a good idea to purchase these videos and use them in public fire safety education programs.

What Is FES Customer Service Duty?

One of the customer services that typically is provided by FES agencies is enforcement of fire prevention codes and standards. An example of failing to provide this life-saving service can be seen from a February 20, 2003 tragic fire at The Station nightclub in West Warwick, Rhode Island that killed 100 persons **FIGURE 7-4**. The fire was started when pyrotechnics used during a Great White concert ignited flammable foam placed on the walls for soundproofing. There were many lessons to be learned from this fire, primarily regarding the effects from the absence of fire prevention. Patrick C. Lynch, the Rhode Island Attorney General at the time, was "criticized for not bringing charges against fire marshals [fire inspectors] or building inspectors for failing to cite the nightclub for code violations, or against members of Great White" (data from Laviue, 2006). In 2008, the State of Rhode Island and the town of West Warwick agreed to pay $10 million in damages to the victims as a settlement. Besides failing to prevent deaths, an FES agency and its officials could be personally liable or face criminal charges for similar incidents in the future.

FIGURE 7-4 Property where the Station nightclub tragic fire occurred. Original building has been demolished.
© Nicolaus Czarnecki/ZUMA Press, Inc./Alamy Stock Photo.

Fire prevention by FES agencies can be restricted by state and local regulations; however, some states allow the fire department to enforce the fire safety aspects of building and fire codes in new and existing construction. At a minimum, chief officers should determine who has the authority to enforce fire prevention codes in existing buildings. Through state fire organizations, officers should push for changes in the state law that allow this life-saving customer service to be provided by the fire department.

Fire Prevention

Fire prevention is a major customer service area for the public, specifically for building owners and occupants. The *NFPA Fire Protection Handbook* states, "One basic aspect of a comprehensive public fire protection plan is the concept that it is infinitely better for a community to prevent fires altogether…than to depend solely on the fire suppression capabilities of the community's fire department" (Courtesy of National Fire Protection Association; NFPA, 1996, pp. 7–31).

The customer may not have a good understanding of fire prevention and may not be as inclined to support funding for prevention programs. There is often a conflict between funding emergency operations and funding prevention programs, with advocates supporting emergency operations generally outweighing those who support prevention programs. The lack of support may be caused by the fact that in many cases the beneficiaries of a comprehensive prevention program do not even know that their life or property has been spared. Although this is customer service at its most efficient level, in most cases the outcome is invisible and therefore difficult to document life and property loss prevention.

In the International Society of Fire Service Instructors' publication *The Voice*, Vina Drennan further explains the difficulty of documenting the results of prevention (1998, p. 2):

> We do not know the name of a child that is spared the scars of fire because one of you [firefighters(s)] went into a school to teach the importance of developing an exit plan for their family. We do not know which elderly person you saved when you visited the senior citizen center in your neighborhood and gave a cooking safety demonstration. We do not know the names, but we do know the fatalities are decreasing.

There are a lot of customers who are alive today thanks to the efforts of modern fire prevention. Some of the major changes driving this trend are smoke detectors and the lower incidence of cigarette smoking. This is the ultimate goal in customer service.

Home Fire Sprinkler Systems

Home fire sprinkler systems may be the ultimate customer service for fire safety in the future. They are so effective that they are more highly recommended for life and property protection than smoke alarms. Still, the code changes needed to mandate fire sprinklers in homes have faced sustained and organized opposition on a national level, specifically from the National Association of Home Builders. When these safety devices were first proposed, the cost of installation per square foot was expensive; but now, partly because the devices are gradually being required by building and fire codes, installation in new homes can cost as low as $1.35 per square foot. Additionally, some of the installation costs can be recovered by reductions in fire insurance.

It is also common for the housing construction industry to argue that adding the cost of sprinklers to a home will put them at a price disadvantage compared with adjoining jurisdictions. This supposed disadvantage would hurt home sales in the jurisdiction adopting the fire safety requirement, resulting in loss of real estate taxes. However, a 2009 NFPA-sponsored study concluded that a sprinkler ordinance has no impact on the number of homes built. This study was conducted using the jurisdictions of Montgomery County and Prince George's County, Maryland, which have sprinkler requirements, and Fairfax County, Virginia, and Anne Arundel County, Maryland, which do not have these requirements. "The selected areas, all developmentally mature, cover a wide geographic area and contain a variety of housing stock and income levels, making them prime for comparing municipalities" (data from NFPA, 2009, p. 2). The results showed that such issues as schools and proximity to work have more influence on the decision to buy a home than the 1% to 2% cost differential represented by fire sprinklers.

The results of a 2011 study of Scottsdale, Arizona's 15-year experience with residential home sprinklers furthers the justification for installation:

- The average fire damage to homes with sprinklers was $2,166; the average for homes without sprinklers was $45,019.

CASE STUDY

Fire Sprinkler Requirements

In 1992, Prince George's County, Maryland, became the first county in the nation to require fire sprinklers in all newly constructed residential dwellings. During the previous years, Prince George's County studied the development of sprinkler technology, appointing a task force to review fire injury and death data during a 10-year period. Participants included representatives from the home building industry, multifamily developers, members from the Board of Trade, zoning attorneys, representatives from water and sewer authorities, local planning and zoning and subdivision commissioners, insurance industry, representatives from the International Association of Fire Fighters, P.G. County Volunteer Fire and Rescue Association, county executive's office, and city governments.

During the course of the task force's investigation, it reviewed fire statistics, scenarios, and visual evidence of actual tests. The task force agreed that residential sprinklers have the potential to save lives and property. Discussion then centered on resolving to what degree they should be mandated and what construction alternatives would be sufficient to offset costs while maintaining adequate fire protection.

During this time, the building industry voiced concerns regarding mandatory sprinkler installation in single-family dwellings, citing high costs and limited trade-offs. To counter these arguments, the task force noted that developers could apply for "density bonuses," which would allow them to create more building lots (up to 10%), thereby increasing property value by 10%.

With some potential cost savings identified and a mutual agreement that the cost of residential sprinkler systems would fall as demand increased, a compromise was worked out to implement the requirements.

Fifteen years after the adoption, a research paper from the Home Fire Sprinkler Coalition reported that because the ordinance was enacted, a reported 101 fire deaths occurred in single-family or townhouse structures in Prince Georges County that were not protected with fire sprinkler systems, whereas no fire deaths occurred in structures that had sprinkler systems (Weatherby, 2009).

For structures that did not have a sprinkler system installed, fire damages averaged $9,983 per incident, with that average going up to $49,503 when the fire resulted in a fatality. However, damages to structures with sprinklers averaged only $4,883. With 13,494 single-family or townhouse fires occurring during this time period, the County's fire damage totaled about $134 million. In contrast, installation of a sprinkler system costs under $2 per square foot (Weatherby, 2009).

- 90% of the fires were contained with one sprinkler.
- The cost of sprinkler installation was between 1% and 1.5% of the cost of the building (Home Fire Sprinkler Coalition, 2011).

Today, there are many sources of information and statistics to help justify this type of code change. For example, in 2006, the NFPA approved a provision supporting fire sprinklers in its Life Safety Code. Additionally, in 2008, the International Code Council began requiring fire sprinklers in new residential construction. Although national building and fire codes now require installation of residential sprinklers in all dwellings, including one- and two-family houses, the next step is for states and local governments to also adopt this requirement. Even though these systems are primarily designed to allow the occupants enough time to escape, their record of reducing property loss is significant.

Fire Safety Inspections

Many FES departments use line personnel to provide fire inspections of existing buildings. Although this is a good use of standby time, scheduling conflicts seem to be more and more common for station personnel. Such duties as fire calls, EMS responses, training, physical fitness, station and apparatus maintenance, and fire prevention can result in some critical non-emergency functions not being done or being done marginally.

Even though all of these objectives are very important and more than justified, there is not time for everything in a busy fire company. Therefore, tasks must be prioritized by the administration by directives to first line supervisors. The number one priority has to be response to emergency calls and the training (including physical fitness) that prepares these companies to do an efficient and effective job. Along with the skills training, prefire plans and familiarization inspections are very useful and can save firefighters' lives. All of these functions are necessary if the FES are going to be able to give 100% effort when responding to a structural fire.

There is no better time to improve (or damage) the professional reputation of an agency than during fire inspections or prefire planning visits. Therefore, inspectors and firefighters should be trained and educated in public relations. Inspectors should maintain an amicable, yet professional, demeanor when presenting the findings of the inspection to the owner or occupant **FIGURE 7-5**. A professional fire safety program has a favorable effect on the

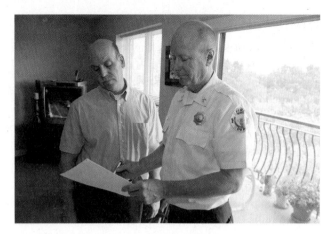

FIGURE 7-5 Presenting inspection results to homeowners.
© Jones & Bartlett Learning. Photographed by Glen E. Ellman.

ability of the fire department to obtain funding for needed equipment, personnel, salary raises, and new stations.

If a correction order must be issued, it is important that it is issued professionally, because not doing so can leave a bad impression of the agency. Inspectors should ensure that the violations are documented and validated. It is vital that inspectors cite the correct code while issuing violations to maintain public trust and respect. For example, a company fire inspection of a new car dealership observed that a fire door had been removed between the show room and the service area. Because the service area was protected with automatic fire sprinklers, this was a violation of the fire code because the show room was not. The company officer made an appointment with the owner and delivered the correction order. At the same time, the owner was advised to contact the fire insurance company to identify any rate penalties that had been applied to the fire insurance premium. It is typical for insurance companies to not notify and penalize business owners for fire code violations rather than force them to make corrections. Within a short period of time, the owner made arrangements to have the fire sprinkler system extended into the show room area. It can be assumed that the reduction in fire insurance was substantial for correcting this violation.

Fire inspections also provide knowledge that is helpful in firefighting. These fire inspections and prefire planning surveys present an opportunity to learn about conditions and construction features of buildings in the community before a fire occurs.

Home Inspection Programs

Although some fire codes contain fire safety requirements for existing homes, most inspections are done on a voluntary basis because of political realities and expectations of privacy. For example, NFPA 1, *Uniform Fire Code*, contains a requirement for existing homes to have a smoke alarm outside of sleeping areas and on each level of the house. For those jurisdictions that have adopted this fire code or another with a similar requirement, the code is more encouraged than strictly enforced.

Most home inspection programs are primarily devoted to fire safety education. As noted by the NFPA, 80% of fire deaths occur in the home (Home Fire Sprinkler Coalition, 2011). Therefore, it is extremely beneficial to the customer if smoke alarms can be installed at no cost by fire inspectors or firefighters during these inspections. One idea is to contact local civic organizations and request donations to cover the cost of a give-away smoke alarm program. If necessary, some can be bought with public funds. Programs such as these are great public service opportunities and demonstrate the concept of leading by example.

There are two ways to schedule home inspections. One is to advertise the service by cable and regular television, social media, local newspapers, and mailings. The advertisement would ask occupants to call a central number to schedule the inspection. The second, which can be in addition to notifications, is to have fire companies go into the field and knock on doors for entry. Home inspections should be conducted in pairs because this helps protect against any potential accusations of misconduct. A four-person engine can split up into two separate teams. If a paramedic unit is also in the field, the two-person unit is the right crew size for these inspections. It is a good idea to place signs on the fire and rescue apparatus indicating that the firefighters are performing home safety inspections so the local community knows correctly why the fire company is in their neighborhood.

Timing is very important in these visits. Many more people are home in the evening than during the day. Therefore, the best time to do home inspections is between 7 and 9 PM. It is not a good idea to go too late in the evening, such as at dinnertime or when people may be sleeping. Be prepared to return on a different day or time. If a resident is not home, leave a brochure on the door.

For serious violations of the fire safety code, entry can be gained either voluntarily or with a search warrant. For example, after interviewing several teenagers who were found with illegal fireworks, a fire department discovers that an individual is selling fireworks out of a home. With this information, the department's investigator can go to the state's attorney and ask for a search warrant issued by a judge. In situations such as this, depending on local or state laws, police should assist in gaining access and enforcement.

Enforcement of Fire Safety Codes

An effective enforcement program is made up of the following steps:

1. **Code adoption.** New or updated fire safety code ordinances need to be publicized when enacted. This should include an effort to highlight any potential controversial issues. Departments may consider giving a presentation outlining the major proposed changes and invite architects, engineers, developers, owners, and the media. It is counterproductive for a department to attempt to hide something in code language that may only become obvious after adoption.

2. **Plan review.** For new construction, it is imperative that a building be designed to comply with all fire safety codes and standards. In the typical building code, about one-half of all requirements are fire safety–based. Because the FES agency has to respond to any fire emergency and put their firefighters in harm's way, it has a greater motivation to make sure the building is constructed in accordance with the fire safety aspects of the building and fire codes. If the jurisdiction uses the NFPA Life Safety Code, it is good practice to have the FES enforce this code. In other jurisdictions that use only a building code, the FES plans examiner can be tasked with reviewing fire hydrants, access to the building for fire apparatus, fire protection systems, means of egress, fire-resistive construction, and smoke control systems. Beware, sometimes these situations can lead to territorial differences between the building and fire prevention departments. Local and state laws usually stipulate who has jurisdiction.

3. **Fire inspections.** The inspector should be tasked with enforcing those items in the building code that are fire safety–related. If some code items are missed during construction, it may be impossible to gain compliance at a later date. The success of these inspections is directly related to the correctness and thoroughness of the plan review.

4. **Maintenance fire inspections.** These are needed to ensure that fire protection systems, means of egress, and other fire safety items are properly maintained. To do this task properly, it is mandatory that inspectors know the fire code requirements and the fire safety aspects of the building code. In addition, housekeeping

items that, for example, could mean the accumulation of excessive amounts of combustible trash in the structure, must be enforced. These inspections can be combined with prefire plan visits.

Emergency Medical Services

The provision of EMS is held in high regard by the public and elected officials. At the most basic level, the fire engine responds to 911 calls and renders care and support to patients and their families. The engine may be staffed with emergency medical technicians and provide initial care until advanced life support (ALS) arrives; they may have paramedics that render ALS-level care, then turn over the patient to a private ambulance company for transport; or they could be an ALS fire department that also provides patient transport.

As reported by NFPA, "Nationwide 39.1% of fire departments provided no emergency medical services (primarily small volunteer fire companies), 45.4% provided basic life support (BLS) and 15.5% provided advanced life support (ALS)" (Hayes and Stein, 2017). It is alarming that these percentages are not 100% given the good public relations EMS can bring to the fire department.

However, fire departments that provide EMS transport have an additional challenge over first responder–only or non-EMS fire departments: potential competition with other agencies for patient transport. In many cases, transport fees are paid by medical health insurance or through state or federal funds, such as Medicaid or Medicare. Recently, private ambulance companies have proposed taking over patient transport services in several large cities. The main argument for this change is that the private companies can provide service more cheaply than a municipal fire service.

Disparity of Duties

In departments that respond to every call for medical help, the percentage of a fire unit's calls for EMS can approach 80%. For departments that respond only to critical emergency medical calls, the EMS fire responses hover at 45% to 50%. More and more often, firefighters are serving as primary care providers.

This issue is essentially a political decision: whether to send the closest emergency unit (normally a fire company) to all medical calls, or just true emergencies. Many elected and appointed officials prefer to be on the safe side by responding to all medical calls, despite the additional costs and increased risk of mental or physical harm that may be placed on the fire company.

The dilemma is best handled through **tiered response** because it allows the fire service to make the most efficient use of their human and material resources. Rather than immediately sending an engine for every call, response decisions are made after dispatch call screening and medical evaluation have occurred. This is also known as "priority dispatching." Questions that must be considered include the following: Does every call warrant an emergency response much less a large fire engine and staff? Will a basic life support unit suffice rather than an ALS unit? Is an engine necessary at all when the call can be managed by an ALS ambulance?

The only real advantage to running every call as an emergency is that it raises call volume and awareness, giving the community a sense that the fire department is serving their needs rather than sitting around the station waiting for the next fire call. This awareness might also increase the likelihood of justifying funds for additional personnel and equipment. Visibility to the public equates to survivability for many fire officials, although this is debatable.

However, increased call volume taxes resources, wears out expensive fire apparatus, and can take a physical and psychological toll on crews. If a department is forced to cut resources but continues to receive the same call volume, time to the scene will be impacted or mutual aid resources will need to be called in. More importantly, scarce EMS resources may be tied up on a minor medical incident while the patient with cardiac chest pain is made to wait. If a fire department cannot handle the load, a private ambulance company might convince local government that it can.

Another issue involves the risk of hazards present to the public and the fire department when making emergency responses such as when a fire truck crashes into a vehicle or a vehicle driver thinks it is necessary to swerve out into the fire trucks' path without looking for other vehicles or objects. If only 20% or less of the calls require ALS, the risk seems unjustified. Studies have shown that time saved by emergency responses is minimal and would not have much effect on a non-emergent patient.

Controversy over the issue of firefighters as paramedics also rages within the fire service itself. Norris W. Croom, Division Chief of Operations for Colorado's Castle Rock Fire and Rescue Department, notes his experience with members of the fire service who believe that firefighters should not provide (advanced) EMS. Croom goes on to state, "A more disturbing trend, though, is that some new firefighters are adopting the attitude that 'I joined to fight fires, not be a paramedic'" (data from Croom, 2010, p. 1).

Studies have conclusively shown that when fire engines with EMTs are dispatched with separate EMS, heart attack patients have a better chance of survival: "The American Heart Association estimates that there are more than 356,000 out-of-hospital cardiac arrests (OHCA) annually and that about 90% are fatal. Reducing response times to OHCAs and ensuring quicker initiation of CPR and defibrillation is critical to saving lives." Further, "Dual dispatch of EMT-trained firefighters shortens response time and leads to a significant increase in the chance for survival" (Raun, Pederson, and Campos, 2018). In addition, there is a growing trend in using paramedics to carry out non-emergency medical house calls. This should be approached with a lot of caution on many levels and is discussed in the following section "EMS Revolution."

The International Association of Fire Chiefs (IAFC) and the International Association of Firefighters support fire service–based EMS and urge public officials to support all efforts to broaden emergency medical care and ambulance transport by the fire service. In a 2009 position paper, the IAFC states, "The American fire service is strategically and geographically well-positioned to deliver time-critical response and effective patient care rapidly" (data from IAFC, 2009). Keeping in mind that a goal of all bureaucrats is to increase the number of employees and a goal of unions is more dues from total paying members, the justification for taking over EMS, especially transport, may have other motives. A thorough check of all financial justifications would be a good policy in these situations.

EMS Revolution?

A new concept is evolving and claims to help with the spiraling costs of primary emergency medical service, including reducing non-emergency demand for emergency room visits. In its simplest form, a paramedic unit would respond to a medical call and provide some level of triage and medical treatment of the patient at the scene. This new type of service is sometimes called either Mobile Integrated Health Care or Community Health Paramedicine. Training programs including textbooks to guide this expanded medical service have been created (Pollak, 2018) and a new NFPA 451 *Guide to Community Health Care Programs* (NFPA, 2020) has been created to help design and manage these new medical services. At one time it was a common practice for doctors to make house calls but this type of service disappeared over 50 years ago. Even some ambulances were staffed with doctors.

There is a growing number of these programs that are essentially "medic house calls." These programs may more recently be augmented by the "telemedicine" practice being recommended to protect doctors and medical providers from the contagious coronavirus. Because this saves the patient from visiting the local emergency room, it also protects hospitals, urgent care facilities, and other patients in the emergency rooms and doctors' offices from being exposed or exposing others to this highly infectious virus.

These programs have two distinct goals. The first is to provide a cost-effective way to treat patients that are not true emergencies. They also are effective at reducing 911 calls and visits to the hospital emergency room. Only recently have insurance companies, including Medicaid and Medicare, provided payment for these services. At the basic level wellness checks such as evaluating recently discharged patients, blood pressure, weight (BMI), and temperature are routinely offered. After gathering basic medical information, appropriate preventive health recommendations could be offered to the patient. These systems need the consent of medical professionals and would probably start with the local paramedic physician. This is a time-consuming and complex new customer service that is worth considering. It will also result in a new revenue source for the department.

EMS Challenges

EMS organizations continue to face challenges of their own. The following issues need to be addressed if an EMS organization is to provide the public with the best EMS. This summary list was compiled from an article in the *Journal of Emergency Medicine* addressing, "10 Harmful Realities of Modern EMS" (2016) (https://www.jems.com/2016/05/31/addressing-10-harmful-realities-of-modern-ems/):

- Increased volume of medical calls
- For community paramedicine, going from a skills-oriented trade to an education-oriented profession will be challenging.
- One measure of satisfactory consumer service is a response time of 8 minutes; 90% of time that must be monitored in real time.
- Integration with the healthcare system
- Disturbing trend to longer shifts of 48–72 hours, causing dangerous fatigue
- Leadership ethics resulting in fraud and other fiduciary troubles
- Patient abuse as a result of burnout, domestic issues, bigotry, fatigue, etc.

Increased Volume of Medical Calls

Public education programs to help reduce the number of medical emergencies in the United States can be very effective in reducing unnecessary 911 calls. Increasing volume of medical calls cannot be tackled simply by continually adding paramedics and ambulance units. Also, the model of fire prevention programs could be replicated to address non–fire-related injury and death in communities. The new programs using Mobile Integrated Health Care or Community Health Paramedicine would have a big impact on this issue as discussed previously.

Issues Surrounding Healthcare Reform

In 2010, a healthcare reform bill called Affordable Care Act (ACA) was adopted to improve access to and the quality of medical insurance including reduced costs in

CASE STUDY

Some Thoughts from the Field

In some agencies, firefighters can be heard saying "I'm here to fight fire, not take care of sick people." This attitude is dependent on the culture of the department and may be more common in older departments that have not traditionally had EMS responsibility. Although firefighting and EMS duties both require time-sensitive lifesaving skills, the skills themselves are very different. Some members may find they have a strong preference for one of the two; still, others enjoy both.

Jane, a paramedic and assistant fire chief said, "I have always felt that I am both equally. That is not true for everyone, but it is for me. I love the medical part, and love fighting fires. I have always thought that it's the best job in the world. Also, I was never very good at working in the for-profit environment. I was always drawn to public service, so that was another reason to move away from private ambulance service."

Michael, firefighter and paramedic: "I wanted to do both. The meat and potatoes, or what makes me feel good inside, is helping people (the paramedic stuff). The dessert, the fun adrenaline rush stuff is the fire service areas of expertise. So by far and away being a firefighter medic is much, much more rewarding both personally and professionally than being either a medic or a firefighter."

the United States. Fire and EMS expert Chief Gary Ludwig cites a number of issues that failed to be addressed with this reform, including the following (Ludwig, 2010):

- The failure to depart from the current model of treating and transporting patients to an emergency room.
- The failure to create injury and illness prevention programs that would change the current model of EMS delivery.
- The failure to fund "treat and release" programs that would save money by not transporting patients who do not need it.
- The failure to implement programs that would reimburse EMS for evaluating individuals on the scene and transporting them to clinics and urgent care centers when more intensive care is not warranted.

A case in front of the U.S. Supreme Court may result in this whole act being found unconstitutional. This entire law may be repealed, modified, or found unconstitutional but is not expected to be decided until after the publishing of this text.

In February 2019, the U.S. Department of Health and Human Services (the department that administers Medicare and Medicaid) announced a pilot program to pay for medical treatment outside of emergency rooms. These are called community paramedicine programs and are designed to reduce the number of patients using emergency rooms. Today many EMS providers are only paid for transport to an emergency room. Changing this paradigm would open up options, including treatment on scene and transport to doctors' offices or to clinics, including telehealth-type portals.

Overloading of System

Nationwide, EMS units have become tasked with far more duties than they have the resources to handle—far more, in fact, than they were ever intended to handle. This has given rise to a whole host of risks, including sleep-deprived EMS crews; long patient wait times; and emergency workers who lack the training to deal with large-scale catastrophes, such as chemical attacks or natural disasters.

The number of ER visits that was reported in a 2017 study is steadily going up as a result of the ACA and expansion of Medicaid in some states (https://www.healthaffairs.org/doi/full/10.1377/hlthaff.2017.0218?url_ver=Z39.88-2003&rfr_id=ori%3Arid%3Acrossref.org&rfr_dat=cr_pub%3Dpubmed&). It is expected that the increase in illegal immigrants may also result in increases in ER visits because they are less likely to have health insurance and thus use free medical services. Remember, since 1986, federal law requires all emergency rooms to provide emergency care even if the patient cannot pay the fees.

The exact depth of America's EMS problem is unknowable: federal patient confidentiality laws limit the amount of information made public, and immunity statutes minimize the number of lawsuits initiated by families. Several years ago, however, a study commissioned by *USA Today* reached a startling conclusion: an estimated 1,000 cardiac-arrest victims die in the United States each year, in part because of slow EMS response (*USA Today,* 2005).

EMS members also leave the profession because of the long hours—another result of budget shortfalls. A survey of EMS providers published in the *Journal of Emergency Medical Services* found that 24-hour shifts remain normal, raising concerns about crew and patient safety. A 2008 study in *Sleep Medicine* found that moderate, repeated sleep deprivation is comparable to a blood alcohol concentration of 0.08% (Caruso, 2012). "That's the legal limit for noncommercial drivers," says Brian Maguire, a clinical associate professor of EMS at the University of Maryland at Baltimore County. "That means if the man or woman who comes to treat you is nearing the end of a shift, he or she might as well be drunk" (data from National Association of EMS Officers, 2011).

Integration with the Healthcare System

In 1996, the 30th anniversary of the EMS industry, the National Highway Traffic Safety Administration (NHTSA) looked at the status of EMS nationwide, developed a vision for the future of EMS, and published the findings in the paper "EMS Agenda for the Future" (1996). NHTSA predicted that EMS systems of the future will be community-based and fully integrated with other healthcare providers and public health and public safety agencies. NHTSA also predicted that future systems will "reduce illness and injury risks through prevention, provide acute illness and injury care and follow-up, assist in the treatment of chronic conditions outside hospitals, and provide community health monitoring" (data from NHTSA, 1996). However, even in light of predicted changes, NHTSA's "EMS Agenda for the Future" recognizes that EMS providers will remain the public's emergency medical safety net.

Preventive Medicine

Similar to fire prevention, emergency medical problems can also be prevented in many cases; this has spawned a whole field of medicine called preventive medicine. It is always preferable for a person to lose

FIGURE 7-6 EMT taking blood pressure.
© Jones & Bartlett Learning.

weight, quit smoking cigarettes, and start exercising before a serious illness occurs. Although fire safety education is used to help with the fire problem in existing and nonsprinklered structures, preventive medicine education and treatment are the method of choice to prevent diseases and injuries.

The FES organization can provide the customer with some preventive medical services, such as high blood pressure and cholesterol screenings **FIGURE 7-6**. These types of customer services can be effective at identifying symptoms that may lead to serious illnesses, thereby improving the health of the community one individual at a time.

Customer Service at Emergencies: The Extra Mile

Many experienced fire protection experts estimate that each person will experience a hostile fire in his or her home once in a lifetime. These relatively rare occurrences can earn the department a lifelong friend. The same can be said for the more common, but equally traumatic, medical emergencies. If a death or serious medical emergency has occurred, these painful events become even more distressing for those close to the victims. Department members can be very helpful as a source of comfort on scene and in locating resources to help a grieving family.

FES personnel should be provided with information and techniques for handling the grief process. Always contact professional psychiatrists, psychologists, or therapists in your city/county administration who would be helpful at setting up a training program. There are many YouTube short presentations that can be useful also. Thoughtful communication must be taught; careless comments such as "I understand exactly how you feel" or "Calm down; everything is all right" can be hurtful to the bereaved. Short seminar courses can provide members with the knowledge, sensitivity, and awareness they need to provide compassionate support in distressing situations.

The following is a list of services or information that may be provided to the victims of fire or other emergencies:

- Explain all FES operations
- Separately talk to owners and occupants
- Salvage furnishings, especially personal items
- Provide temporary lodging

- Provide meals and clothing
- Provide transportation
- Notify relatives and friends
- Notify religious services, if appropriate
- Contact the Red Cross and Salvation Army
- Provide temporary pet services
- Contact the insurance company
- Contact the police for security patrol
- Secure the building
- Install smoke alarms, if needed

Wrap-Up

CHAPTER SUMMARY

- What constitutes "customer service" in the FES occupation?
- Traditional FES department services and other more contemporary services that could be added are discussed.
- Special issues of EMS are examined.
- Major service areas are discussed and improvements are offered for consideration.
- Extra nontraditional customer services are presented for possiblity of providing.

KEY TERMS

Customer service More commonly found in for-profit sales, it is the person-to-person interaction to assure the customer is completely satisfied with the purchase and company.

Home (fire safety) inspection programs A voluntary service offer to the homeowner for fire department personnel to visit and inspect the home for fire prevention problems. Fire safety education tips are also typically offered to the homeowner.

Preventive medicine A policy or program whose goal is the infrequency of a disease, either by preventing the occurrence of a disease or by halting a disease and averting resulting complications after its onset. Preventive medicine can be practiced by governmental

agencies, primary care physicians, and individuals themselves.

Public relations An agenda of persuading the public to have understanding for and support toward a person, idea(s), philosophy, or organization.

Public survey Used to recognize how the public perceives the department's present service, identify areas in which they would like improvement or expansion of service, and determine how much the public is willing to pay for improvements in service.

Tiered response Also known as "priority dispatching," this practice allows for efficient use of resources by making response decisions after dispatch call screening and medical evaluation.

REVIEW QUESTIONS

1. What do you think is the most important form of customer service discussed in this chapter? Write a brief essay (~300 words) outlining the pros and cons of this practice for FES.

2. Is there any way for the FES to provide good customer service without simply adding more

tasks to their already lengthy job descriptions? Discuss.

3. Does your locality require fire sprinklers in new residential construction? Briefly research the subject and ascertain the history of this issue in your community.

4. Does your community participate in a para-medicine program? How is it organized and what services are provided?

5. When considering a customer service goal in your community, are there adequate resources and training? If not, what needs to be added to effectively complete this goal?

CHAPTER ACTIVITY #1: TRAGEDY STRIKES

At 2 AM on a cold winter night, a department is dispatched to an apartment fire. The caller, a resident who lives on the third floor, said she can smell a strong odor of smoke and the fire alarm is ringing.

When the fire company arrives, they find a large fire on the second and third floors of a four-story garden-style apartment building. The initial dispatch is three engines, two ladders, and a battalion chief. The first engine to arrive reports a working fire and requests the assignment be upgraded. This calls for the dispatch of two more engines, one ladder, a heavy rescue, a second battalion chief, the safety chief, and a deputy chief.

After the fire is extinguished, all of the occupants escape without any deaths. However, several occupants are burned and require hospitalization. In addition, 10 apartment units are so heavily damaged by fire that the occupants are not allowed to return, rendering them homeless.

Discussion Questions

1. Make a list of the services that the FES personnel could either provide or direct these victims to for help and identify a contact person and telephone number (24-hour access) for each.

2. If an elected official or the press shows up as part of a customer service plan, describe how they should be treated and what information should be provided to them.

3. Is this a good opportunity to get a fire safety message to the public?

CHAPTER ACTIVITY #2: A NEW MODEL OF CUSTOMER SERVICE

The emerging trend known as "Community Paramedicine" began with EMS providers in Colorado, Texas, and Minnesota providing primary and preventive healthcare through house calls. The program addresses the issue of shortages in primary care, especially in rural communities, but it is also being considered in urban communities to reduce the dependence on emergency rooms for primary care.

Through this program, paramedics are given specialized training and use down time between emergency calls to visit patients. They check blood sugar and blood pressures; confirm adherence to medications; change bandages; and occasionally run diagnostic checks, such as an electrocardiogram. They can also evaluate an older person's home for fall hazards and advise on other safety issues.

Community paramedicine allows many patients to obtain the routine care that would otherwise not be available. It also gives the paramedic the opportunity to recognize a more serious problem before it becomes an emergency. The paramedic's report is evaluated by a physician, who can recommend further care options for that patient.

The Western Eagle County Health Services District has created the "Community Paramedic Program Handbook" to explain its program in detail. The handbook is available through the following link: http://www.communityparamedic.org

Discussion Questions

1. Do you consider community paramedicine to be a viable option in your community? Why or why not?

2. In considering existing staff, would you expect any issues with morale, motivation, competencies, etc.? Why or why not?

3. Do you consider this to be a valuable asset to FES or EMS in terms of garnering support or funding? Why or why not?

REFERENCES

Caruso, C. C. 2012. *Running on empty: fatigue and healthcare professionals.* Retrieved from http://www.medscape.com /viewarticle/768414_2

Croom, N., III. 2010. "I didn't join to be a paramedic: Challenging bad attitudes toward fire service-based EMS." *IAFC On-Scene,* 2 (May): 8.

Drennan, V. 1998. Forty more are alive today. *The Voice, International Society of Fire Service Instructors* (November /December).

Hayes, Hyulton, and Gary Stein. 2017. *U.S. Fire Department Profile Through 2015.* Quincy, MA: National Fire Protection Association.

Home Fire Sprinkler Coalition. 2011. *Scottsdale report 15-year data.* Retrieved from http://www.homefiresprinkler.org/ fire-department-15-year-data

Insurance Service Office. 2012. *Fire Suppression Rating Schedule.* Jersey City, NJ: Copyright Insurance Service Office Properties, Inc. Retrieved from http://www.ciclt.net/ul/gsfa /ISOFireSupressionRatingSchedule.pdf

International Association of Fire Chiefs. 2009. *Position statement: Fire-based emergency medical services.* Retrieved from https://www.iafc.org/about-iafc/positions/position /iafc-position-fire-based-emergency-medical-services

Laviue, D. 2006. R.I. attorney general says he opposed deal, sentences in nightclub fire case. The Associated Press. Retrieved from https://www.denverpost.com/2006/09/21 /judge-rebukes-ag-in-nightclub-fire-case-4/

Ludwig, G. 2010. EMS misses chance for health-care. *FireRescue Magazine.* Retrieved from https://www.firehouse.com /leadership/article/10467268/a-missed-opportunity -in-health-care-reform)

National Association of EMS Officers. 2011. *National EMS assessment.* Retrieved from http://ems.gov/pdf/2011/National _EMS_Assessment_Final_Draft_12202011.pdf

National Fire Protection Association. 1996. Americans underestimate fire risk. *NFPA Fire* 77 (May/June).

National Fire Protection Association. 2020. *Guide for Community Health Care Programs.* Quincy, MA: National Fire Protection Association.

National Fire Protection Association. 2008. *Fire Protection Handbook* (20th ed.). Quincy, MA: National Fire Protection Association.

National Fire Protection Association. 2009. Comparative analysis of housing cost and supply impacts of sprinkler ordinances at the community level. Retrieved from http://www .firesprinklerinitiative.org/~/media /Fire%20Sprinkler%20 Initiative/Files/Reports/FSI _Comparison_Analysis_Final _Report.pdf

National Highway Traffic Safety Association. 1996. *EMS agenda for the future.* Retrieved from https://www.ems.gov/pdf /advancing-ems-systems/Provider-Resources/EMS _Agenda_Imp_Guide.pdf

Phoenix Fire Department. 1999. Phoenix Fire Department national survey on fire department operations.

Pollak, Andrew. 2018. *Community Health Paramedicine.* Burlington, MA: Jones & Bartlett Learning.

Raun, L., J. Pederson, L. Campos, et al. 2018. Effectiveness of the dual dispatch to cardiac arrest policy in Houston, Texas. *Journal of Public Health Management and Practice* 25(5): E13–E21.

USA Today. 2005. *The price of just a few seconds lost: People die.* Retrieved from https://clkrep.lacity.org/onlinedocs /2015/15-0719_pc_2_8-10-15.pdf

Weatherby, S. 2009. *Benefits of residential fire sprinklers: Prince George's County 15-year history.* Retrieved from http://www .homefiresprinkler.org

CHAPTER

Training and Education

LEARNING OBJECTIVES

After studying this chapter, the student will be able to:

- Understand the necessity for continuous reinforcement of basic emergency services skills and knowledge (pp 150–151).
- Identify training needs for fire and emergency services (FES) personnel (pp 151–155).
- Recognize the best training practices that lead to safe, professional emergency operations (pp 155–156).
- Understand the requirements for emergency medical technician (EMT) and paramedic certification (pp 153–154).
- Examine local and national developments that require new training and education efforts (pp 157–158).
- Recognize ways to determine acceptable competencies for training (pp 157–158).
- Identify advanced educational opportunities for FES professionals and chief officers (pp 159–162).
- Understand the National Professional Development Model (pp 162–163).

Case Study

Collaborative Training

In 2005, the County of Contra Costa, California's emergency medical services (EMS) Agency piloted a unique collaborative training model called the Contra Costa County Fire-EMS Training Consortium. This model pooled the resources of nine fire departments, a large ambulance company, a trauma center, and two air ambulance providers. The curriculum for this training is created quarterly by different system partners. Time, energy, and money are saved by eliminating redundancy in curriculum creation and sharing resources and materials. Relevancy is ensured by using quality assurance and quality improvement data. Easy access to standardized curriculum provides a forum for FES providers to train together in a real world setting, thus creating better cooperation, understanding, and improved patient care in the field.

The Contra Costa County Fire-EMS Training Consortium Advisory Board uses quality improvement data collected by the ambulance provider, county fire departments, and the EMS agency to determine the subjects that are targeted for training. This not only ensures that training is pertinent but also addresses identified system deficiencies. It also provides the opportunity for the EMS agency to provide input and ensure standardization of the message and the materials.

The program has its own quality improvement process, which includes instructor education and development, an annual needs analysis from field providers, feedback and evaluation on each training module, and regular updates on project design and implementation.

Experienced instructors from multiple departments provide variety and expertise. Educators are respected people to whom the students can relate. Training needs are easily and immediately adapted to actual practice. This program has proved it can provide challenging, satisfying training with relevant curricula and realistic skills training. The end result is improved motivation and self-esteem on the part of Contra Costa County's FES providers.

Discussion Questions

1. Does your department have any cooperative training programs with other FES departments?

2. Is in-service training required for fire or EMS providers? If no, explain.

3. Do you see a need for recruit or in-service training to be coordinated with other departments?

4. Would there be a cost savings?

Introduction to Training and Education

Administrators are responsible for the efficient and effective emergency operations of the department. They should be aware that there are no shortcuts to having members who are well trained and equipped, including using standard operating procedures (SOPs) that are followed for emergency operations. Success at the scene of an emergency is the result of a well-trained team, meaning the providers have the ability to complete manual skills with little or no conscious effort. This level of ability is achieved only after practicing the skills many times, beginning with an intensive recruit school.

In 1873 (interestingly, the same year that the International Association of Fire Chiefs was founded),

Captain Eyre Shaw, a chief officer in the London Metropolitan Fire Brigade, made the following comment after visiting several fire departments in the United States: "The day will come when your fellow countrymen will be obliged to open their eyes to the fact if a man learns the business of a fireman only by attending fires, he must of necessity learn it badly....I am convinced that where study and training are omitted, the fire department will never be capable of dealing satisfactorily with great emergencies" (Parow, 2011). Jack Parow, former president of the International Association of Fire Chiefs, poses the following question regarding this account: "Did we listen to his words of wisdom or did they fall on deaf ears?" (Parow, 2011). Parow reminds us that there are still issues regarding contemporary FES training and education that need to be resolved.

Firefighting Training and Certification

To be proficient at emergency operations, members have to be thoroughly trained before participating, and must receive updated training regularly. Because some essential skills, such as those required for structural fires, are used infrequently, some officials make the argument that there is less value to training for firefighting. Actually, the opposite is true. This profession requires a higher level of training because of the nature of emergencies and the infrequent use of certain essential skills. In emergency operations the job must be timely and done right the first time; otherwise, the result is the potential for increased property loss, deaths, and injuries. For this reason, administrators must assure that training is an ongoing and never-ending process and is especially needed for scarcely used knowledge and physical based skills.

Take the example of the New York City Fire Department. This department is the largest and busiest fire department in the United States. In 1995 the department's administration instituted a back-to-basics training program designed to ensure that firefighters maintain their skill level in the most important aspects of fire suppression operations while also receiving training in new areas **FIGURE 8-1**. Even with a large number of experienced firefighters (as a result of the high number of working structural fires in their jurisdiction), this department saw a need to reinforce basic skills by using a structured mandatory **in-service training program** (U.S. Fire Administration 1997, p. 13). If the largest, busiest, and most experienced fire department in the country sees a need for in-service basic training, all fire departments could benefit from this type of training.

Although training can sometimes present a problem to smaller departments in which finding the time and resources for exercises, especially multi-company,

can be difficult, training (including live fire exercises) is essential to retain skills and ensure safety. Smaller departments will find that partnering with other departments can often provide the necessary resources. Larger departments might also need outside sources to provide training in specialized areas such as hazardous materials incidents, terrorism, technical rescues, wildfires, catastrophic disasters, high-rise fires, multialarm mutual-aid incidents, large-scale emergency medical incidents, and other infrequent events. An instructor experienced in these types of incidents can enhance the training program and the retention of information by the students, so it is critical to find someone with the correct expertise.

The fewer working fires a department experiences or the less frequently a skill is used, the greater the need for training. It is common for members to have a better memory for a given skill than their ability to actually perform the skill. Skills that require muscle coordination can degrade rather quickly. This varies for each individual, but because firefighters work in teams, training must be directed to the least able member. All members, including those assigned to staff positions, should receive **refresher training**. It is common for firefighters working in staff functions, such as fire inspectors, to not experience actual firefighting for years at a time.

At a minimum, the FES administrator should schedule drills for incumbents that cover all the basic essential job functions throughout each year. Technical competencies are most successfully taught to the adult learner through hands-on training. Many higher-level skills, such as emergency command, hazardous materials evaluation and operations, fire prevention, and EMS, can be learned through **simulation drills**. Although many of these skills require massive amounts of special knowledge, the highest level of knowledge retention comes from applying the knowledge to an actual incident or training exercise and practicing the skill or administrative task. This type of learning exercise also tends to keep the interest of the adult learner better than just memorization of facts and regurgitation on an examination. Although some multimedia presentations and reading assignments are necessary for the familiarization process, an exercise should be included in the program of instruction to ensure long-term retention and immediate recall at the emergency incident.

Most of the in-service training drills can take place at the fire station, ensuring a convenient location without requiring the company to go to a training facility. Still, there are also times, such as for live-fire training, when fire companies must go to a central training facility and the department may have to make provisions for a stand-by crew to be available at the station.

FIGURE 8-1 Firefighters stretching a hose line.
© Glen E. Ellman/Jones & Bartlett Learning.

For example, a metropolitan department may choose to provide a special shift consisting of four 10-hour days per week that can fill in for companies or members who need to attend multicompany drills or paramedic recertification classes.

For station drills, one effective technique is to have a rookie administer basic skill drills. This new firefighter will have recently learned the most current methods at the training academy, because these agencies are often the first to adopt new skills and equipment. In addition, having a rookie administer the drills also helps reinforce the skills in new members. Often, the best way to learn something is to teach it.

A firefighter's job is complex; there is a need to provide challenging, relevant training. A department with a negative training culture will find lack of motivation, inattention, poor retention, and nonadherence to expectations in the field. The culture of the department needs to set the tone that training is taken seriously and is demonstrated in practice. Unfortunately, training programs are often the first thing to be cut during a budget crisis.

Therefore, the allotted time, opportunity, and budget for firefighter training have to be carefully considered. For training to be effective, it needs to be timely and relevant. Curriculum needs to be meaningful, and easily and immediately adapted to actual practice. The instructor not only needs to be knowledgeable but also relatable and someone upon whom the employees can confer respect.

Motivation is one factor that determines successful training, but another important consideration is the delivery method. Training programs need to have a variety of delivery methods, a challenging and relevant curriculum, and a direct application to the field. There is added value when firefighters from different departments, with a different set of experiences to share, can train together. Further, integrating multiple companies in the training exercises, exchanging instructors, and promoting other regional efforts to share training allows the departments to more easily be able to cooperate when a large-scale emergency calls for them to work together.

The Adult Learner

The FES administrator when designing an in-service training program needs to keep the following items in consideration when using adult education techniques:

- In all cases, the adult learners have specific motivation norms that need to be considered such as:
  - The reasoning behind the training/education must be clearly understood by the learner.
  - The benefits to the learner must be unmistakably recognized.
  - Must be goal oriented, such as safety at emergency incidents.
- Must decide if the training or education is better presented at the fire station or a more formal setting in a classroom.
- Can it be presented in a distance learning format with follow-up testing? In many cases, self-paced learning courses can be very effective for these motivated learners.
- Even the most unexciting physical skills practicing may benefit from the example of their favorite sport where players never stop learning or repeating essential skills. Finding common analogies to motivate the adult learner can be very effective in providing enthusiasm.

Basic Firefighter Training

As a guideline, most firefighter basic training uses NFPA professional qualifications standards, such as NFPA 1001, *Standard for Fire Fighter Professional Qualifications*. There are a few states that have a mandatory statewide basic fire training or certification system; generally, however, each state and even larger city fire departments have their own system. In addition, many states and cities have different certification requirements for volunteers. Therefore, although they follow the same training and education outlines from the NFPA, no two systems are identical and many often vary in length of instruction and hands on practice. This variance is addressed by the fifth future goal of the National Fallen Firefighters Foundation's Everyone Goes Home Firefighter Safety Initiatives: "Develop and implement national standards for training, qualifications, and certification (including regular recertification) that are equally applicable to all firefighters based on the duties they are expected to perform" (National Fallen Firefighters Foundation, 2013).

As noted by the National Fallen Firefighters Foundation, firefighting training and education should be standardized nationally. Because of an increasing amount of manmade and natural large-scale incidents, there are going to be far more situations in the future that require responses that cross local, county, and state lines.

This kind of cooperation might also be precipitated by reduced resources from a struggling economy, prompting more departments to ask for help from their neighbors. FES agencies have to continue to figure out how to work together seamlessly and be confident in their neighbors' abilities **FIGURE 8-2**.

FIGURE 8-2 Staging area for a major catastrophic emergency.
© Glen E. Ellman/Jones & Bartlett Learning.

Standards and credentialing for firefighters and officers need to be national and portable.

Traditional in-service training programs rely on company officers and their knowledge, experience, and individual approach to training. Even with the most competent officers, many of these programs lack consistency because of the numerous fire station locations and different shifts. The training given in one station on a particular day is not the same as that given in another. Therefore, any in-service training program should start with department-wide planning and implementation.

For consistency, NFPA 1001, *Standard for Fire Fighter Professional Qualifications*, can be used as a basis for a departmental in-service training program. This program should be in addition to a new member's basic skills training. The key to providing the best fire and rescue service is a comprehensive, exhaustive recruit school for new firefighters and an in-service training schedule that stresses and reinforces basic skills. NFPA 1500, *Standard on Fire Department Occupational Safety and Health Program*, calls for monthly structural firefighting training, totaling a minimum of 24 hours per year. This is a minimum number of hours and is only appropriate for departments that protect a small number of single-family dwellings.

Live-Fire Training

One of the best training drills that results in higher levels of competency and morale is the live-fire drill. It can also be one of the most dangerous and should not be done without extensive planning and research, and the supervision of certified, experienced fire instructors. Because this drill is very dangerous and safety is vital, some states have a requirement for live-fire instructors to successfully complete an additional training course. A great place to start in planning a live-fire drill is NFPA 1403, *Standard on Live Fire Training Evolutions*.

This document has a thorough process that, if followed, provides a safe but valuable experience for the firefighter. Two categories of structures are defined for use in a live-fire drill: acquired and live-fire training structures. Each has its own pros and cons; however, the acquired structures are more realistic. The biggest hurdle is in obtaining permission to burn a structure, especially in urban areas, because this often generates large quantities of smoke.

Because NFPA 1403 is heavy on safety aspects, some skills are not tested. For example, in a real fire it is not possible to perform a walk through before arriving on the scene. Therefore, it is valuable to also include some other confidence-building drills, such as navigating through a maze with a self-contained breathing apparatus with the face mask blacked out to simulate heavy smoke conditions.

This training is also a good time to put members through other seldom-used but critical skills, such as climbing to the top of the department's ladder truck, forcible entry, and ventilation skills. Generally, a whole day is set aside to complete this training and several companies are required to participate. Because many of these drills are physically demanding, members should be checked for minimum medical and physical benchmarks at the start of training.

EMS Training and Certification

Departments that have EMS-trained firefighters have to accommodate both EMS and firefighter training. EMS has a national standard for training and testing. However, some minor differences can be found in a few states. Formal training and certification are required to become an EMT or paramedic in all states. Training requirements and minimum hours are recommended in a national curriculum created by the U.S. Department of Transportation and the National Highway Traffic Safety Administration.

A certified EMT needs to accumulate 24 continuing education hours in each 2-year certification cycle, or he or she can take a 24-hour refresher course for recertification. A paramedic, however, needs to accumulate 48 continuing education hours. These hours are generally acquired through merit badge refreshers, such as advanced trauma or pediatric courses.

In addition, many states also require testing through the National Registry of EMTs before final certification or licensure. In some states it is a first-time certification requirement; in other states it is required as a basic part of continued certification or licensure.

It does not make a difference if an EMT is in a city, county, or state agency, or if the area served is urban or rural; each person has essentially the same training and certification testing. The only variable in practice may be optional skills as appropriate to an EMT's or paramedic's scope of practice, but those skills must be requested and approved by a local medical director. These local options can vary by state, county, special district, or city. One common optional treatment is the medication naloxone (Narcan) that reverses opiate overdoses.

Technology-Based Training and Certifications

Interactive multimedia training is a very useful technology to FES. These are personal computer-based training and education courses that include some combination of texts, graphics, still images, animation, video, and audio. This means of training is especially useful for required yearly recertification training, such as that needed for hazardous materials operations. It is even being used in basic training in lieu of classroom lectures and to reinforce self-study material where actual hands-on physical skills are not required. The big advantage to this technology is the ability to obtain the certified training at the local fire station or at a member's own home. These are self-controlled modules and can be taken any time/any place, resulting in being very adaptable to the employed or volunteer learner.

The National Fire Academy and the Emergency Management Institute of the Federal Emergency Management Agency (FEMA) offer numerous online courses that when completed successfully will result in the student receiving a FEMA certificate. Some of these courses can even be applied to college credit and may help document the student's knowledge for a promotion.

Podcasts are another technique for distributing training to many FES members separated by distance. A 2009 study of podcast-based learning conducted by the State University of New York at Fredonia discovered that students who listened to a podcast lecture outperformed those who attended in-person lectures. Furthermore, these same students performed better when they took notes, which was indicated by their use of the pause and rewind features (SUNY Fredonia News Services, 2009).

Although written, supervised examinations may still have to be administered in person, many certification tests can also be taken over the Internet. Web technology has the greatest potential for reducing the off-duty time needed for recertification. In addition, some computer-based training packages have the ability to administer a test and record the results in a tamper-proof computer file. Although these training packages may seem pricey, if costs are compared with classes that require a certified instructor to provide the same training, the interactive program may be more cost effective. Plus, costs can be divided if several small departments share the use of the program. As with any new trend, changing the institutionalized certifying agencies is difficult but not impossible.

Seminars and Training Sessions

Many FES organizations send their members to seminars or training sessions. Because there are more than 30,000 fire departments in the United States, it is impossible for one organization to have all the knowledge and experience needed for every potential emergency situation. Seminars are a great method for increasing knowledge and gaining valuable experience from national experts on advanced or specialized topics.

There are numerous state, federal, and national training programs and conferences that have excellent content. These training sessions are sometimes provided at a local training academy by outside experts. In some cases, a local fire department may be fortunate enough to have a nationally recognized expert to use as a resource. The training bureau may choose to record a webcast of the session for those who cannot physically attend, for circulation, or for future use. Another alternative may be to have the attendees teach their new knowledge to all members when they return home.

Recertification

A weakness in many certification systems and in the NFPA fire service professional standards is that there are no recertification or continuing education requirements. In the future this may change and existing members may be required to update their knowledge and skills on a yearly basis. For example, the State of Florida already requires people holding fire officer, fire inspector, and fire instructor certifications to complete 40 hours of continuing education every 3 years. For departments located in areas where there is no requirement for recertification, however, administrators can include items in the yearly performance appraisals that require self-study or the attendance of seminars and formal courses to update knowledge and skills.

Over a 20- to 30-year career, many things change and it is very difficult to retain information that is used infrequently. Therefore, each department should create a standardized in-service training program and schedule. This approach helps to ensure that all members have equal abilities, knowledge, and skills, and can easily move from assignment to assignment. It also ensures that an engine company that services a rural farm district can operate with efficiency and skill when dispatched or transferred to an area with high-rise buildings, and vice versa.

Training to Fit a Need: Standard Operating Procedures

SOPs are rules that clarify consistent actions that result in proven outcomes and safe practices. There is a reliance on SOPs in FES organizations because of the nature of emergency incidents. In many cases, split-second decisions must be made by the first arriving company officer, and these decisions can have a huge effect on the outcome of the incident. The likelihood of deaths and property damage increases when there are no SOPs in place; therefore, SOPs should be followed with few exceptions.

To illustrate the complexity of the initial actions at an emergency, the following list includes major questions that could confront the officer of a first-arriving engine company at a single-family dwelling fire:

- Should a supply line be laid for water, or should tank water on the engine be used?
- From the first observations, can it be determined where the fire might be in the structure and if there are any potential victims inside?
- What size hose line should be used?
- What type of nozzle is appropriate?
- Are there enough firefighters to attack the fire safely?
- If the building is locked, how do the firefighters gain access?
- Is there any information or observations that need to be communicated by radio to other responding companies or the incident commander?
- Which door should be used to enter the structure for fire attack?
- Does the building show any indications of a backdraft or flashover condition?
- Are there signs of structural weaknesses?
- What method will be used to account for the progress and location of all firefighters?

- Can it be determined if more resources will be needed than are presently on the scene or responding?

Answering questions such as these can be challenging because most of the variables are typically unknown when the first responding company arrives. Most of these questions can be resolved by SOPs. Many of these SOPs are safety oriented and a standing safety committee is the ideal way to find a solution. Therefore, after SOPs are adopted, training sessions must be required to ensure that all members are proficient in following procedures. Any actions or situations that might require variance need to be addressed and incorporated into the SOPs and training sessions. For example, even though many structural fires can be extinguished with a small amount of water, most departments require the use of a minimum-size hose line because the size of the fire might not be known until the company has spent a few minutes on the scene and has stretched its initial handline. SOPs such as these should be reviewed frequently and updated for their technical correctness and relevance.

Consistency and Reliability

SOPs are the basis for consistent and reliable levels of service. They should be followed with few exceptions, and must be studied, reviewed, and practiced frequently. The public expects a high degree of reliability in the service provided by the FES agency. SOPs that are rigorously enforced—similar to the military's commitment to obey orders—will provide the consistency the public expects.

Variances in SOPs

There have been numerous arguments and discussions over the years involving the use of different fire tactics and equipment, such as the type of nozzle to be used (solid stream or fog). For controversial subjects, the chief officer should create an advisory committee that can analyze and make recommendations on SOPs. In addition, the chief officer should contact other FES agencies and request copies of their SOPs, as well as find out if they are actually followed or if there are enforcement issues.

The advisory committee would then be instructed to find the best solution to the issue, with the understanding that exceptions may be handled in the field. For example, a department has a fireground SOP that states that the first due engine take action to extinguish the fire. On one rare occasion, however, when approaching an apartment building, an officer of the first due engine spots a woman holding a baby at a third-story window. The officer orders the crew to remove the ground ladder from the engine and use it to rescue the victims. The officer should immediately radio to other companies and any responding chief officers that his or her company varied from the SOP and another company must accomplish the original assignment. Any time a variance such as this occurs, it needs to be communicated immediately to the command officer and the other companies that are responding to the same incident.

Critique of SOPs

In the FES business, a critique of operations can point out deficiencies that, if not corrected, may cause poor firefighting practices, civilian or firefighter deaths, or other negative outcomes. An example of an honest and comprehensive critique of a firefighting operation was performed by the Prince George's County, Maryland, Fire Department after a single-family dwelling fire response required the rescue of a downed and unconscious firefighter. The critique by an ad hoc committee of experienced firefighters and officers required many hours of investigation, interviews, and research to produce. It pointed out actions that did not conform with generally accepted safety practices and the department's own SOPs. The department then made these discrepancies known to all members.

Although only a larger department with extensive resources is capable of a comparable effort, small departments also benefit from a brief critique. Comparing the actions taken with the department's SOPs is very valuable. A checklist of actions required by each company with an evaluation of yes/no is helpful. Even if this critique indicates that some companies and their officers do not comply with departmental SOPs, the critique can reduce the chance of a reoccurrence by the officer or other departmental officers. A critique should be primarily a learning experience, although its results can also be useful during the annual performance evaluation of members, especially officers.

Even the simplest SOPs for emergency operations are very valuable. For example:

SOP for Structural Fire

1. First due engine: Size up and take action for any obvious rescues; gain entry and extend a hose line to the fire area.
2. Second due engine: Gain access to the rear, completing a 360° view of the structure; extend a second hose line to back up the first due engine.
3. Third due engine: Assure a continuous water supply for the first due engine; provide a rapid intervention team for rescue of any firefighter who is lost, trapped, or disoriented.

Training to Fit a Need: Hazardous Materials

In the past two decades, there has been an increase in the number of hazardous materials teams created in the public sector. For example, because of the high density of oil and chemical industries in the area, the Houston (Texas) Fire Department has one of the busiest hazardous materials units in the world, with 813 runs in 2017 (*Firehouse*, 2018).

Emergencies involving hazardous materials require a rethinking of many of the traditional approaches the fire service has relied on in past firefighting and rescue work. The speed with which an incident is approached, the level of preincident planning required, and the dependence on specialists from other fields, including the private sector, are all radically different from traditional firefighting. Training with the facility emergency response team is vital to understand the hazards and how to mitigate them. This would include familiarization with the facility, including fire protection features and physical structure. Although the experience obtained from an actual incident is a team's most valuable resource, it is still very important to practice hands-on drills and field training with team members and teams likely to be called in for mutual aid.

Still, many communities have found that despite formation, these specialized teams—which are generally under the jurisdiction and command of a local fire department—have rarely been used. According to incident records collected from state and local jurisdictions, and discussions with members of various operational teams, hazardous materials incidents account for only 1% of the fire calls made by local departments (NFPA, 2018a) (**FIGURE 8-3**). Approximately 50% of these incidents involve very small quantities (less than 55 gallons) of the hazardous materials, and 75% of these incidents occur at fixed facilities where the material is routinely stored and used.

FIGURE 8-3 Chemical fire.
© ASSOCIATED PRESS.

In the initial rush to create hazardous materials teams everywhere, many experts originally supported the creation of teams. However, in the face of infrequent calls, limited chances to test their training, and an increased need to broaden their experience and knowledge in the field, hazardous materials team members often become discouraged and burned out. In addition, because a hazardous materials response team should consist of approximately 20 members to provide coverage on a 24-hour basis (including holidays and weekends), costs of staffing, training, and equipment are a major expense.

The increase in demand and costs for this type of training often reduces the team head count to less than optimum levels. Department chiefs who experience few hazardous materials–related runs have difficulty justifying the continued cost of training and investment in new technologies and equipment that keep the team at the highest state of readiness. This is one of the primary reasons that many smaller communities have consolidated their teams into regional or district hazardous materials response units. A great example of a regionalized approach (several counties in each region) was created in Massachusetts **FIGURE 8-4**.

This same type of analysis can be used for any special service that an FES agency considers offering or presently provides. Many special services can be better provided through the cooperation of several departments to share the costs and staffing. Such services might include ladder or rescue services that are difficult to staff and whose personnel are difficult to train adequately in a small department with few calls.

Training to Fit a Need: National Incident Management System

After several national emergency incidents resulted in poor coordination of resources, the federal government created the National Incident Management System (NIMS). NIMS establishes a uniform set of processes, protocols, and procedures with which all responders, including EMS personnel, must be compliant. Primarily, the function of this new management system is to command and control operations at catastrophic emergency incidents.

NIMS works with the National Response Framework (NRF). NIMS provides the template for incident management and is the command system used to control and allocate the resources provided by the NRF. The NRF provides structure and mechanisms for national-level policies, allowing emergency response partners to prepare for and provide a unified national response to disasters.

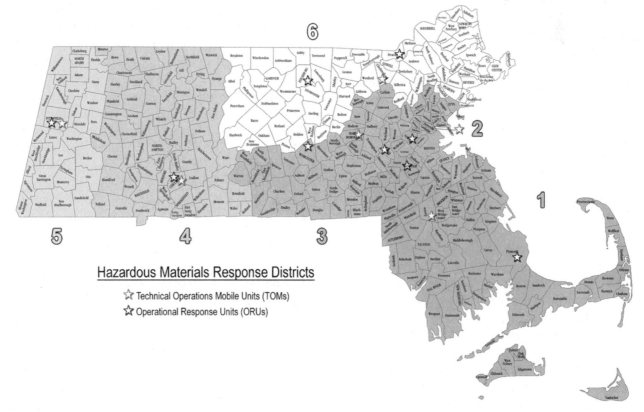

Hazardous Materials Response Districts

☆ Technical Operations Mobile Units (TOMs)
☆ Operational Response Units (ORUs)

FIGURE 8-4 Hazardous materials districts map of Massachusetts.
Courtesy of Massachusetts Executive Office of Public Safety and Security.

Incident Command System (ICS) is a subcomponent of NIMS and the initial model of command for local incidents. With the exception of large disasters, all emergencies start as local emergencies. When the situation gets more complex and starts to involve multiple local, state, or federal jurisdictions, then an expanded model of management is necessary. Fire departments are encouraged to use this system for everyday incidents so that when large incidents occur, the use of NIMS is second nature. As is true for all emergency operations, practice ensures emergency skills and procedures are accomplished successfully and safely. Making anything routine more likely results in it being done correctly.

Key incident command positions at an emergency medical incident may include the following:

- Triage unit leader
- Treatment unit leader
- Medical transportation officer
- Staging supervisor
- Safety officer

All public safety and EMS agencies must preplan and become competent in the use of tools, techniques, practices, and the vocabulary of the ICS. According to FEMA, each agency should do the following (Kirkwood, 2008):

1. Develop policies and procedures for training, implementation, and evaluation of NIMS.
2. Have a cache of tools necessary to function in an ICS/NIMS environment. These tools include radios programmed with common channels, reflective vests, and other incident management tools, such as ICS forms and tracking boards.
3. Develop ICS/NIMS competence by getting involved with local planning, exercises, and after-action reports.
4. Be prepared to act as the incident commander in cases where they may be the first arriving unit on an incident.

NIMS-compliant ICS should be used for every incident involving more than one resource. The first arriving unit—whether a police vehicle, private ambulance, or fire engine—initiates initial command. As other agencies appear on scene, command should be passed to the most appropriate and qualified agency. Which agency takes that role depends on the type of incident, the local structure, and the resources available.

For example, in situations involving multiple-patient injury or illness scenarios, the initial command or operations section chief would most appropriately be an EMS expert. Another option in these scenarios might be to have an experienced and skilled EMS provider serve as a medical group supervisor to help with patient care–related decisions. How these roles are determined depends on the level of the agencies' ICS expertise.

In some cases, a private or nonfire EMS unit may arrive on the scene and set up a parallel command, which may or may not coordinate with the fire service. This could result in redundancy of requested resources, incomplete or inaccurate information, and a number of other issues. These issues need to be addressed and formally agreed to with the management of the private ambulance company. For reasons such as these, unity of command is very important and must be included in pre-incident discussion and arrangements.

Facts and Figures

As a result of the mutual-aid agreements created by the Metropolitan Washington Council of Governments in the Washington Metropolitan Area, a standard numbering system was put into place for area fire departments. All units have a three-digit designator, with the first digit denoting what jurisdiction the unit is from. The following are examples:

- 000 series: Washington, DC
- 100 series: Arlington County, Virginia (includes City of Falls Church)
- 800 series: Prince George's County, Maryland

In this system, Engine 101 would be an engine from Station 1 in Arlington County, Virginia. This number-identifier system is a great help for mutual aid calls across city, county, or state jurisdictions.

EMS supervisors and senior staff should be encouraged to get higher-level ICS training as provided through FEMA (https://training.fema.gov/emiweb/is/icsresource/) and experience, and to participate in state and regional Incident Management Teams. Not only will this teach the EMS agencies practical application of the system, but it will also show other public safety agencies that they are experts and can function in a NIMS/ICS incident with a unified command.

Training Goals: Initial Fire Attack

This section provides an example of one portion of the successful extinguishment of a structural fire and is a good example of how to determine acceptable competencies for firefighting. Similar approaches can be derived for other emergency services operations. Many other skills and knowledge are needed to manage the varied emergencies a department is called to handle.

To determine a department's capabilities for initial fire attack performance at structural fires, NFPA 1410, *Standard on Training for Initial Emergency Scene Operations*, can be used. This standard describes methods for evaluating the adequacy of a department's training program, including typical fireground evolutions broken down into three separate items (handlines, master streams, and automatic sprinklers) and recommended completion times for each. The fire company or department can evaluate the effectiveness of its training program by comparing the times that its crews attain performing the evolutions with those of other jurisdictions, historical times, and those contained in NFPA 1410.

Experience has shown that with appropriate training, SOPs, and adequate staffing, most fire departments can complete these evolutions within the suggested times. If the times are slower than those recommended in the standard, administrators may want to evaluate the following:

- Frequency of training sessions or actual fireground experience. If firefighters perform these basic evolutions frequently at actual fires, their skill level is higher. If the fire experience is low, the frequency of in-service training needs to be increased.

- SOPs. These may contain evolutions that slow the emergency operations. The critical importance of SOPs has been discussed previously.

- Staffing. Staffing can affect the times of each evolution. Staffing is described in NFPA 1410 as "the average number of personnel that ordinarily respond" (NFPA, 2015). However, the staffing provisions listed in NFPA 1500, *Standard on Fire Department Occupational Safety and Health Program*, must be followed to ensure the safety of firefighters participating in the evolutions and on the fireground. For handline evolutions, NFPA 1410 explains, "A minimum of two firefighters shall be used on each hose line to keep interior attack lines under control" (NFPA, 2015). Similar wording is found in NFPA 1500, which states, in part: "Members operating in hazardous areas at emergency incidents shall operate in teams of two or more" (NFPA, 2018b).

All three of these areas (training frequency, SOPs, and staffing) should be considered individually and in combination when analyzing a department's problems with meeting competencies.

Other training goals can be formulated from standards, such as NFPA 1001, *Standard for Firefighter Professional Qualifications*, which specifies the minimum competencies for a firefighter, and NFPA 1500, *Standard on Fire Department Occupational Safety and Health Program*, which provides the framework for a safety and health program.

Finding Time for In-Service Training

In most departments, there is not enough uninterrupted time to realistically complete In-Service Training, such as classroom attendance or physical fitness activities. A few reports of creative policies for providing on duty time for EMS recertification, in-service firefighter training, physical fitness, etc. are listed below:

- Allowing companies' first due areas to be covered by adjoining companies for non-emergency calls, except true medical emergencies and first due structural fires.

- Creating a day shift that fills in for companies to participate without interruption.

- Using volunteers or overtime to fill in for firefighters during daytime training sessions.

Higher Education

There are many incentives administrators can use to encourage members to take advantage of a formal college-level education. For example, some departments use educational incentive pay or policies regarding credit or minimum educational requirements for promotional opportunities.

Most higher education programs in FES are offered at community colleges, with a few universities offering related bachelor degrees. Many community colleges have a committee made up of representatives of FES organizations to provide curriculum advice. In addition, the National Fire Academy's Higher Education program has developed a recommended core curriculum for fire science programs. The model curriculum can be found at the U.S. Fire Administration Web site (https://www.usfa.fema.gov/training/prodev/model_courses.html).

When considering a college-level education, it is important that individuals are sure a particular program meets their needs. To check whether the college is **accredited**, visit the websites of the U.S. Department of Education (https://ope.ed.gov/dapip/#/home/) or the Council for Higher Education Accreditation (https://www.chea.org/). Each contains a list of accredited colleges. In addition, individual courses that are recommended by The American Council on

Education may not be accepted at many of the regionally accredited schools. Individuals should ensure that course work will transfer if there are potential plans to go to another institution of higher learning.

Every college and employer has the right to set standards and refuse to accept transfer credits and college degrees from institutions they deem substandard. Students who have gone to a nationally accredited school may find it especially difficult to transfer credits. In a 2004 to 2009 study, the U.S. Government Accountability Office (GAO) found that "...students who transferred from private for-profit schools to public schools accounted for 4 percent of all transfer students but lost an estimated 94 percent of their credits" (Ref: https://www.gao.gov/products/gao-17-574). Previously, in a 2005 study by GAO found that, in making decisions on college credit and degree acceptance, about 84% of the country's higher education institutions considered the accreditation of the institution; many had policies stating that they would accept college credits and degrees only from "regionally" accredited institutions. One reason given for "regional" institutions' reluctance to accept credits from nationally accredited institutions is that national accreditors have less stringent academic standards, such as faculty qualifications and library resources (U.S. Government Accountability Office, 2005, p. 3).

In addition, examine the courses offered by the institution. Many institutions offer only fire science courses but do not offer courses on EMS or emergency preparedness management and operations, which may be of interest to those who would like a more inclusive education in FES.

Finally, look at the official degree title. Some institutions add a small number of courses to an existing program, such as business or public management. On a transcript, a degree with an official title of Bachelor of Science in FES means more to employers than a Bachelor of Science in Public Administration with a concentration in fire science.

Many bachelor degrees and some associate degrees can be achieved by distance learning over the Internet **FIGURE 8-5**. This is the most common method of instruction for FES education at the bachelor degree level because most students are physically spread throughout a state or the country and, in addition, it fits in with the FES member's work schedule.

There are some drawbacks to Internet-based programs. They require a great deal of self-discipline in setting aside the time for studying and completing homework. However, those who have adapted to the rigid structure of firefighting and EMS generally do well. In fact, many of them complete entire degree programs online.

FIGURE 8-5 A distance learning student.
Courtesy of FEMA.

Officer Education and Training

The methods for selecting FES officers range from popular election to sophisticated testing, education, and experience criteria. Currently, there is no national consensus for the minimum training, education, and experience for a fire officer. The closest thing that FES has for professional qualifications is the comprehensive guide to education, training, and self-development discussed in the International Association of Fire Chiefs' *Officer Development Handbook,* second edition (2010). This book is an excellent resource for a career fire officer, because it outlines the training, experience, and education needed to successfully gain a promotion and be effective in an officer position.

The following is a checklist of training and characteristics that are helpful in preparing to function as a professional fire officer:

- NFPA professional fire officer qualification certifications
- College degrees in fire science or administration
- Promotional examinations
- Officer candidate schools
- In-service training
- National Fire Academy courses
- FES textbooks (self-study)
- Professional periodicals
- Conference and seminar attendance
- Experience and common sense
- Knowledge from the experiences of other departments of all sizes and types, including respected officers from these departments (there are good ideas everywhere)

Chief officers need to build on the experience and training they gained during recruit and in-service

FIGURE 8-6 Fire Administration class.
© Glen E. Ellman/Jones & Bartlett Learning.

training, company officer training and education, and formal higher education (bachelor level is recommended) to be effective **FIGURE 8-6**. As a prerequisite, the general education (i.e., reading, writing, math, history, humanities, and so forth) for the company officer is relevant to the chief officer and is offered at many community colleges. A chief officer should look at formal education as another tool to help them be effective at their job. Many of the other agency heads, appointed officials, and elected representatives will have at least a bachelor degree. A good educational background puts the officer and any staff on equal ground with contemporaries, and that equality gives the chief many opportunities, such as to acquire a fair share of the tax revenues. In most cases, education gives the chief the tools needed to be a change agent.

Conferences and seminars have presentations for advanced and infrequently used skills as well as sessions for innovative management and leadership practices. These are great opportunities to gain contemporary knowledge and education, and to reinforce basic management and leadership skills. However, there is no peer review of the content contained in these educational and training sessions. Therefore, it is good to confirm what is heard with others' knowledge and one's own experience. Have an open mind and be prepared to accept new ideas that may be different than what you previously accepted as the truth. Many administrative procedures and techniques evolve and change over time. A great example can be seen on the use of personal protective equipment (PPE) for EMS calls, especially potential virus-contaminated patients.

Another great opportunity for an administrator is to be a presenter at one of these sessions. With the appropriate knowledge and experience, the seasoned administrator has a lot of wisdom that can be passed on to others. Ideas become clearer as they are justified and explained to an audience. During the presentation,

CASE STUDY

Studying for a Promotional Exam: An Example

It is very difficult to determine how much studying a person must do to be successful on exams. One firefighter was faced with the fact that he was slightly dyslexic and was not a good test taker. With these handicaps, he knew he would have to study harder than someone who had a photographic memory, for example. Also, promotional exams are very competitive and a better score relates to a better chance of being promoted. The study material for the Lieutenant's exam included six fire science college textbooks, operational procedures, general orders, a manual of uniform training procedures, and several career services directives. This would be equivalent to taking a college level final exam of approximately 10 courses at one time.

The firefighter went through each reference source and highlighted those items that would be helpful when promoted or seemed like good test questions. After the first comprehensive review and highlighting of the reference material, he went back through the highlighted materials 12 more times. Although this seems excessive, he ended up first on the promotional list.

Most people will find that they are the kind of student who reviews materials somewhere between 1 and 13 times. Each person is different. Once the right number of reviews is determined, that is the studying effort that is necessary to be successful. This also applies to any college course exams. Hard work does pay dividends.

there may be questions or feedback from the audience. A good lecturer tries to learn something from every audience. For example, one place to start looking for conferences is in a list at https://stationalerting.com/top-fire-conferences-to-attend-in-2020/. Each has a web site and information on dates, location, and how to become a presenter.

For emergency management competency, the Emergency Management Institute of FEMA provides a nationwide training program of resident courses, nonresident courses, and interactive Internet courses to enhance U.S. emergency management practices (https://training.fema.gov/emi.aspx). Also, several colleges offer degrees and courses in Emergency Management (https://training.fema.gov/hiedu/collegelist/).

In addition, officer training and continuing education are crucial for quality customer service (see the chapter on *Customer Service*). It takes a strong commitment to train at the level needed to keep all basic skills sharp over 20–30 years of service alongside more advanced officer and administrative skills. It also takes a disciplined and rigorous promotional and in-service training program to keep officers at their peak of effectiveness. It is more common to replace an inefficient fire engine than retrain a seasoned firefighter to the skill level achieved earlier in his or her career. Just as water pumps are tested once each year, testing FES officers (and firefighters) for basic skills once each year is good practice.

National Professional Development Model

High-quality, professional, public FES is primarily the result of two basic factors. The first of these, physical resources (e.g., equipment and facilities), is the most obvious and typically gets the most attention. The other factor is human resources: members who must be in good physical condition, well trained, and educated. If either of these elements is weak, critical effective emergency service suffers.

FIGURE 8-7 is a representation of the balance of training and education necessary in the FES. The model was developed by the U.S. Fire Administration/ National Fire Academy's Fire and Emergency Services Higher Education program. The right side of the graphic represents training, the middle represents the experience and self-development attained over many years, and the left side represents formal education. Training is typically attained in firefighter basic training, fire officer courses, seminars, and company drills. The basis for most of this training is contained in the NFPA fire-service professional qualifications standards, a series of distinct criteria, starting with those that pertain to entry-level firefighters and going up through four levels of officers. For instance, Fire Officer I and II are generally company officers, or officers who may command up to three companies. These positions normally carry the job title of lieutenant or captain, although in smaller one- or two-station departments, the position title may be chief. On the education side of the pyramid, the banner notes that an associate degree should parallel the positions of Fire Officer I and II.

The best practice to ensure effective human resources is to offer combined fire officer training and education. Many community colleges offer fire officer courses that also meet the requirements of the state certification system. This is beneficial to the students because they do not have to take two identical courses: one for college credit and the other towards certification.

The most common associate degree is in fire science. Fire science degrees primarily contain coursework that focuses on fire prevention, management, and incident command skills. The modern, progressive fire department realizes it can play a leading role

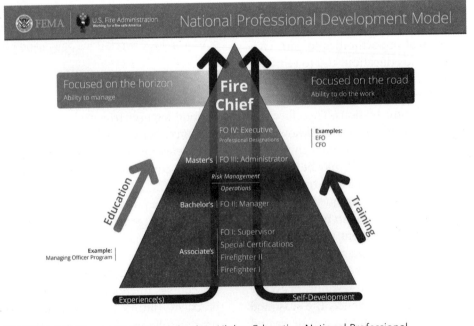

FIGURE 8-7 Fire and Emergency Services Higher Education National Professional Development Model.

in reducing loss of life and property by preventing fires and lowering their risk. Achieving this goal requires officers who are professionally educated to offer fire safety education and code enforcement as well as modern emergency incident skills. As is true in all professions, it is not realistic to expect fire officers, especially newly promoted ones, to be proficient in all situations as a result of only their experience. In addition, it is always better to learn from the mistakes of others, because these mistakes have the potential to result in death and property loss. For these reasons—along with the new likelihood of being first to arrive at a terrorist incident—the knowledge and skills needed to respond successfully to an emergency can be gained primarily through a formal education.

As the complexity of administrative roles increases, so does the need for officers within the roles of battalion, division, deputy, or assistant chief. The National Professional Development Model (Figure 8-7) recommends that officers with these types of titles have a bachelor degree. This model was created by "…the U.S. Fire Administration establishing the Fire and Emergency Services Higher Education (FESHE) network of emergency services-related education and training providers. The FESHE mission is to establish an organization of post-secondary institutions to promote higher education and to enhance the recognition of the fire and emergency services as a profession to reduce loss of life and property from fire and other hazards" (https://www.usfa.fema.gov/training/prodev/about_feshe.html). Although implementing educational requirements can be challenging, discussions with department members and their representative employee organization are always helpful. Such incentives as educational pay and tuition reimbursement may be used. Two additional techniques that follow may also be helpful.

The use of the first technique was exemplified by the New York City Fire Department. In 2001, days before the World Trade Center terrorist attack, the department announced that for an officer to be eligible to take the battalion chief and deputy chief examinations, the officer needed to have a bachelor degree. However, it was not until 2008 that the requirement took effect. This type of phase-in is common for education requirements to allow existing department members time to acquire the education.

Another method used to implement educational requirements is to award bonus points to written promotional test scores depending on different levels of higher education. These bonuses may be modest at the beginning of the policy's implementation and become substantial after several years. An appropriate time frame would allow a member the opportunity to gain the necessary education. In addition, as the need for higher education increases with higher officer ranks, point level can be increased accordingly.

Awarding additional points on the promotional list score offers an advantage, especially in smaller departments. This practice is more flexible than the New York City Fire Department plan during the implementation phase, but it does not guarantee that the people promoted have a minimum level of education. For example, a department that has only a few individuals with the minimum educational requirements might still find that no one on the promotion list is suitable. This department's only option might be to hire from outside. This option usually has merit only at the fire chief's level because junior officers have a need to be familiar with local conditions.

It is also helpful and justified to pay for tuition for courses leading to a degree in FES administration or comparable degrees. Many veterans have education benefits to cover these costs, but not all new members have this benefit. These should be payments once the FES member student has successfully completed the course in an approved curriculum.

The National Professional Development Model effectively takes the fire service to the final level of professionalism. Obtaining the knowledge and skills associated with this model results in a highly professional fire service that delivers better service to its customers. Still, training and education are only one part of the effort to become a professional; experience is also needed. Gaining knowledge from past experience and seeking to understand new ideas should be a never-ending effort for the FES officer.

To make this professional model adopted throughout the FES community, it must be accepted for promotions and hiring from outside. Nevertheless, a major Florida county just promoted a fire officer to chief who did not meet the minimum educational requirements as contained in the job announcement. Although this is an outlier, some political officials do not see the need for formal education in the FES. At this time, at a minimum, the FES National Professional Development Model should be used as a personal goal.

Road Blocks to Professional Recognition

What would you say to your children if they asked, "Mom/Dad, I want to be the fire chief, just like you. What do I need to do?" Most of us could answer this question about being a doctor, lawyer, engineer, etc. There is an accepted combination of education, experience, and formal testing that leads to being recognized as a professional in these disciplines. However, it is a little more complicated for fire chiefs.

Becoming a fire chief is a complicated process and varies from place to place depending on the state, governing authorities, and the organization. As discussed in a 2005 paper by Dr. Denis Onieal, Superintendent of the National Fire Academy: "When physicians or lawyers or nurses move from state to state, they may have to present their credentials to the professional board in their new state. They may have to take an exam, or perhaps take some refresher courses—but they don't have to go back to school to learn the basics all over again. You can learn surgery in Texas, and operate in Minnesota. You can attend law school in Massachusetts and appear in court in Oregon. You can learn electrical engineering in Montana and design computers in Silicon Valley in California" (https://usfa.kohalibrary.com/app/work/137492). So why is this not the case for FES?

For your information, President Abraham Lincoln did not attend any formal higher education institution to become a lawyer. These professions have slowly evolved over time. The fire service has most of the requirements for professional recognition in place, but there is little consensus on the process to achieve professional status. Again, Dr. Onieal notes that "One of the principal challenges we have is that aspiring fire service professionals are surprised by the number of independent systems of training and education. There is no 'one way' for the student to determine which is the most appropriate training and/or education. There's no 'one-way' to become the chief. Moving from fire department to fire department (or even more difficult—from a fire department in one state to a fire department in another), training or education already accomplished may not be recognized."

Therefore, individuals need to chart a course of formal education and experience that will be recognized by a potential employer as competent and professional. Normally, this always includes higher education.

Wrap-Up

CHAPTER SUMMARY

- Training and education in FES is discussed and explained in detail.
- Basic training and certification for firefighters, medics, and emergency services personnel are detailed.
- Training and education for special subjects are categorized and explained.
- Professional development model for FES administrators (officers) is discussed and validated.

KEY TERMS

Accreditation To certify (a school, college, or the like) as meeting all formal official requirements of academic excellence, curriculum, facilities, faculty, etc.

Critique of operations A post-incident evaluation of the actions taken after a large-scale emergency or death of a firefighter. In a committee format, participants involved in the incident and experienced FES members are gathered to investigate and discuss the incident, resulting in a formal report containing recommendations for a better outcome in the future. Details of a comprehensive policy and process can be found at *Special Report: The After-Action Critique: Training Through Lessons Learned* (2008) (https://www.usfa.fema.gov/downloads/pdf/publications/tr_159.pdf).

Curriculum A set of courses constituting an area of educational specialization or degree criteria.

In-service training An in-service program is a professional training or staff development effort where skills and knowledge to be proficient are discussed and demonstrated. In many cases, this includes new knowledge or newly generated SOPs.

Incident Command System A subcomponent of NIMS and the initial model of command for local incidents.

Live-fire drill Deliberately starting a structural fire for the purpose of safe successful extinguishment by firefighters. See NFPA 1403 *Standard on Live Fire Training Evolutions*, which provides minimum requirements for conducting live fire training to ensure they are conducted in safe facilities and a safe manner for participants.

National Incident Management System Establishes a uniform set of processes, protocols, and procedures with which all responders must be compliant to command and control operations at catastrophic emergency incidents.

National Professional Development Model Primarily a visual presentation of the process of formal education, self-education, and experience needed to be a "professional" fire officer.

Refresher training See in-service training.

Seminar A seminar is a form of academic instruction, either at an academic institution or offered by a commercial or professional organization.

Simulation drills Simulation drills are used to practice in a non-emergency setup the actions and process needed to manage a large-scale disaster situation. Used both as a training exercise and to identify areas needing improvement.

Standard operating procedures Rules that clarify consistent actions that result in proven outcomes and safe practices.

REVIEW QUESTIONS

1. Do some quick research on learning styles to determine your primary learning type- (i.e., are you a visual learner, an aural learner, a kinesthetic learner, etc.?). What are the advantages and disadvantages of this learning style? Which multimedia approaches work best for your style?

2. Discuss an SOP that was recently implemented at your work. How could or did training supplement the SOP's implementation? Was the implementation successful? How could this process have been improved?

3. Identify three lesser-used FES skills for your department (e.g., hazardous materials, vehicle rescue) and determine how often you think

members should be retrained on these skills. Defend your choices.

4. Define NIMS, NFS, and ICS and explain how all three are interrelated.

5. What level of higher education, if any, have you attained? What level would you like to achieve in the future? Identify any obstacles you have encountered to completing more formal education and how administrators can assist in overcoming these barriers.

CHAPTER ACTIVITY #1: WOULD TRAINING HAVE MADE A DIFFERENCE?

After a firefighter within a small community was killed fighting a house fire, the community was quick to blame the department's poorly organized training program. Many believed that the department ought to refocus on the basics of structural firefighting rather than spend time training for EMS, terrorist and hazardous materials incidents, and numerous

other services. Although these critiques may be valid, deaths such as this open the door to a very sensitive subject blaming the death of a firefighter on the FES department. Often, a person or organization cannot make needed changes without facing uncomfortable situations or controversy.

Discussion Questions

1. Would you support mandatory firefighter training, including on-the-job refresher training, by either the state or federal government? Why or why not?

2. Make a list of those training programs needed by every department and the frequency with which they should be held.

3. Review a local fire department's training program and analyze its training needs.

4. Would a reward after successfully passing a test on the knowledge content or demonstration of the skill be helpful?

CHAPTER ACTIVITY #2: INFREQUENTLY USED SKILLS

A suburban fire department gets a call for a trench rescue. The first due crew finds a contractor buried to his waist in an 8-ft deep trench. He is conscious and alert and explains that he was hired to replace some pipes. Trench rescues are not common for this area, and the battalion chief and firefighters feel unprepared.

A few firefighters jump into the trench to approach and assess the patient, but the battalion chief orders them to retreat until the trench can be stabilized. A paramedic then approaches the patient from the edge to assess and treat, but he does not remember specific treatments other than fluid infusion. It is quickly determined that the team does not have the equipment or expertise to mitigate the incident safely. A trench rescue team from a department across the county is requested, but it takes 2 hours for them to assemble and transport personnel and equipment.

Four hours later the patient is extricated and transported. Later discussion reveals that most members of the department had not had trench rescue training for 5–7 years. The program was considered less important than others and was cut because of budget woes.

Discussion Questions

1. What is the issue in terms of safety for the firefighters and the victim?

2. Could this incident have created political issues for the fire district?

3. What actions should be taken by the department after the fact to ensure the situation is not repeated in the future?

REFERENCES

Firehouse. 2018. 2017 national run survey. Part 3. *Firehouse* 42 (August).

International Association of Fire Chiefs. 2010. *Officer Development Handbook* (2nd ed.). Fairfax, VA: International Association of Fire Chiefs.

Kirkwood, S. 2008. *NIMS and ICS: From compliance to competence.* Retrieved from https://www.emsworld.com/article/10321278/nims-and-ics-compliance-competence

National Fallen Firefighters Foundation. 2013. *16 firefighter life safety initiatives.* Retrieved from https://www.everyonegoeshome.com/16-initiatives/

National Fire Protection Association. 2015. *NFPA 1410, Standard on Training for Initial Emergency Scene Operations.* Quincy, MA: National Fire Protection Association.

National Fire Protection Association. 2018a. *Fire Department Calls.* Quincy, MA: National Fire Protection Association. https://www.nfpa.org/News-and-Research/Data-research-and-tools/Emergency-Responders/Fire-department-calls

National Fire Protection Association. 2018b. NFPA 1500, *Standard on Fire Department Occupational Safety and Health Program.* Quincy, MA: National Fire Protection Association.

Parow, J. 2011. Higher education: The key to professional fire/EMS recognition. *IAFC On Scene* (April 15). Retrieved from https://www.dailydispatch.com/Columns/Leadership Messages/2011/April/Higher.Education.The.Key.to.Professional.FireEM.SRecognition.aspx

SUNY Fredonia News Services. 2009. *Study tests effectiveness of podcasts vs. lectures.* Retrieved from https://www.fredonia.edu/news/study-tests-effectiveness-podcasts-vs-lectures

U.S. Fire Administration. 1997. *The aftermath of firefighter fatality incidents, preparing for the worst.* Retrieved from https://www.usfa.fema.gov/downloads/pdf/publications/tr-089.pdf

U.S. Government Accountability Office. 2005. *Transfer students.* Retrieved from https://www.gao.gov/assets/250/248170.pdf

CHAPTER 9

Health and Safety

LEARNING OBJECTIVES

After studying this chapter, the student will be able to:

- Understand the importance of safety for fire and emergency services (FES) and its members (pp 168–169).
- Recognize the influence that the National Fire Protection Association (NFPA), U.S. Occupational Safety and Health Administration (OSHA), Everyone Goes Home, and other safety organizations have on FES (pp 169–172).
- Examine progressive health and safety trends in FES (pp 172–179).
- Realize the impact that work-related injuries and deaths have on an organization and its members (pp 179–181).

Case Study

A Recruit in Distress

In response to staffing issues, several recruit firefighters were assigned to fire stations before they completed the entire recruit training program, including the EMT portion, which was a separate block of instruction toward the end of the recruit school. Shortly after the new recruits arrived, the engine company was dispatched to a medical emergency of a patient who was not breathing. The four-person company arrived prior to the paramedics and went to work as one of the regular crew members started CPR. A few minutes later, the paramedics arrived on their ambulance and took over the treatment. They quickly determined that the patient had expired and called off CPR. Upon leaving the house, the officer found a new recruit standing by the street, staring down at the asphalt. The officer asked, "Is everything OK?" The obviously distraught recruit answered, "Was that person dead?" He was comforted by the officer and advised that was a standard emergency incident. Eventually, the recruit requested transfer to a position as a correctional officer.

This rookie had been recruited for the fire department by municipal recruiters during an affirmative action program and had no idea about the job and its demands. The recruiters never explained all the duties of firefighting or that it was common to be in contact with deceased people; this information could have helped the new recruit make a better choice for his career from the start. This case study illustrates why it is important to fully explain all the duties and expectations of a firefighter. Further, training programs should implement aptitude testing for potential members before acceptance, which could help ensure a good fit from the start.

Discussion Questions

1. Was this recruit's transfer appropriate or should he have received some emotional stress management counseling?

2. Are you comfortable that your department adequately explains all of the job duties prior to hiring or accepting new members?

3. Does your department have the ability to identify those behavioral characteristics in a potential member that are necessary for a good match with the duties of an FES member?

Introduction

Any organization's most valuable resource is the people that work for the organization, and the well-being of these people is vital to the organization's success. In a profession that is so dependent on physical skills—as is the case in FES—it is especially important for administrators to ensure the health and safety of its members. FES members' health and safety should be job one for the administrators.

Firefighter Safety

In the past, firefighter safety has been at times a controversial topic. When the first safety standard for the fire service was issued by the NFPA in 1987, many FES leaders, along with city and county managers, predicted that it would lead to departments going out of business, or that huge sums of money would be required to comply with the safety standard. Neither of these predictions has come true; instead, many FES departments have used safety standards to justify additional funding for important upgrades, such as more modern protective clothing or new hires for safe and effective staffing levels.

Nevertheless, there still seems to be selective compliance with safety standards by FES agencies. Recently, the NFPA reported, "While the number of structure fires and deaths at structure fires has dropped, the rate of firefighter deaths at structure fires has not followed the same pattern [2000–2009]" (data from Fahy, 2010, p. 7).

In addition, the results of a study conducted by the University of Minnesota's Carlson School of Management/City of Minneapolis indicated that firefighters who tended to ignore safety rules and regulations had accidents more frequently and suffered more severe injuries than firefighters who conscientiously followed safety rules (Liao, Arvey, and Butler, 2001). Although these results seem obvious in hindsight, the problem persists even today. It behooves FES agencies everywhere to review their compliance with standards and recommendations set out by the NFPA on a regular basis.

Emergency Medical Services Personnel Safety

Although there are fewer long-term sources of data identifying threats to and injuries of emergency medical services (EMS) personnel, one commonly used source is at the National Institute of Occupational Safety and Health (NIOSH), U.S. Center of Disease Control of injured EMS workers treated in U.S. hospitals (https://www.cdc.gov/niosh/topics/ems/data.html). These statistics list a number of reasons why EMS personnel face one of the highest risks of illness or injury compared with other professions, such as working outdoors in all types of weather; kneeling, bending, performing heavy lifting; and being exposed to communicable diseases or violent patients. In 2017, about 40% of these injuries involved the overexertion during lifting of patients, resulting in a back injury.

Long-term effects of multiple injuries to knees, backs, and shoulders could affect the quality of life, and certainly the job longevity, of EMS personnel. These types of non–vehicle-related injuries pose a challenge for researchers, because there are fewer established data definitions and repositories for collection and analysis.

Unfortunately, back injury is only one type of serious injury that can be career ending to personnel. FES administrators should also recommend changes to modifiable risk factors, such as smoking; encourage surveillance and management of weight, blood pressure, and cholesterol; and reward healthy behaviors, such as stretching muscles, staying hydrated, maintaining good bone health, getting enough sleep, and healthy eating habits.

NFPA Safety and Health Standard

NFPA 1500, *Standard on Fire Department Occupational Safety and Health Program*, was established in 1987 with the goal of making the FES profession less dangerous by reducing the risk of accidents, injuries, and fatalities in the course of conducting emergency operations and routine duties at the station. This standard outlines the minimum training, education, and physical fitness requirements needed to provide the skills and knowledge that are necessary to be an effective competent firefighter. NFPA 1500 is also effective at helping personnel become healthier and more physically capable. This combination directly benefits the individual, the organization, and the public they serve.

Although many FES organizations are in the process of adopting the requirements contained in this standard, very few actually comply with it in its

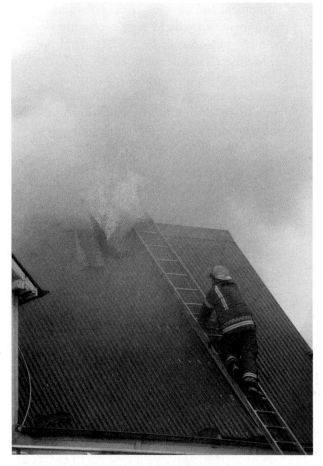

FIGURE 9-1 A hazardous profession.
© Arctic-Images/Stockbyte/Getty Images.

entirety. This partial adoption is not an unusual situation, and NFPA 1500 allows a phase-in schedule for compliance. Following the standards set in NFPA 1500 should be one of the chief officer's highest priorities for several reasons. The first is a moral obligation to provide a safe working environment for the members of this very hazardous profession **FIGURE 9-1**. Second, caring for personnel injured on the job is expensive. In the case of burns, the costs can be exorbitant and complete recovery can take years. Many FES personnel are better motivated if they know that their administration is looking out for their welfare.

NFPA 1500 Major Topics

The following is a list of major topics found in NFPA 1500, along with a brief explanation of each:

- Health and wellness. Health problems, including heart attacks, are consistently noted as the number one cause of firefighter deaths (USFA, 2019). FES has made progress in addressing this issue, even though the health and wellness requirements in NFPA 1500 are sometimes poorly defined. This topic is discussed in greater detail later in this chapter.

- Equipment and vehicles. Issues regarding equipment and vehicles have been addressed by many FES organizations. The federal government's Assistance to Firefighters Fire Grant program helps needy departments purchase newer and safer apparatus and equipment. Unfortunately, many rural and other small departments are still struggling to find enough money to buy safety equipment.

- National Incident Management System (NIMS). Progress in this area is ongoing but NIMS implementation is common in most larger fire departments. Online and in-person NIMS certification courses are offered throughout the United States.

- **Safety committee.** The safety committee provides input that can prevent deaths and injuries. These committees are a major part of implementing a comprehensive safety program and provide the avenue to plan and implement changes needed to comply with the standard. Safety committees also provide a real-time discussion of current, potentially unsafe equipment; employee behavior; and operations. Members of this committee should be selected from management, labor, and mid-level officers. Multilevel recruitment helps ensure that discussions focus on considering and assessing everyone's point of view. It is also useful to invite a representative from the municipality's risk management office. Besides providing the

opinion of a safety expert, the representative has extensive knowledge of the fiscal impact of proposed changes, including insurance costs and worker's compensation issues.

- Standard operating procedures (SOPs). These are sometimes called Standard Operating Guidelines (SOGs) and set out safe and efficient methods of managing specific emergency operations, such as attacking structural fires. Previously developed SOPs are available from multiple sources, but what works for one organization may not be effective for another. Therefore, blindly copying another organization's work is strongly discouraged. Instead, it may be useful to review the Federal Emergency Management Agency's (FEMA) publication *Guide to Developing Effective Standard Operating Procedures for Fire and EMS: Federal Emergency Management Agency United States Fire Administration* (1999). If the SOGs title is used, it is assumed by some that these are more like suggestions and not an actual rule for operations. Because these are for safe and efficient operations, this is not a good idea.

- Training and education. These issues are covered in more depth in Chapter 8, *Training and Education*.

Federal Safety Regulations

Because of language within the U.S. Constitution that prevents the federal government from

---

Facts and Figures

OSHA Regulated States

The following states (26) have OSHA-approved safety regulations that cover state and local employees:

- Alaska
- Arizona
- California
- Connecticut
- Hawaii
- Indiana
- Illinois
- Iowa
- Kentucky
- Maine
- Maryland
- Michigan
- Minnesota
- Nevada
- New Mexico
- New York
- New Jersey
- North Carolina
- Oregon
- South Carolina
- Tennessee
- Utah
- Vermont
- Virginia
- Washington
- Wyoming

In addition, Florida and Oklahoma have typically adopted OSHA regulations for use by fire departments.

Courtesy of National Fallen Firefighters Foundation's Everyone Goes Home® Program.

having too much power over the states, many federal regulations—including NFPA safety standards and OSHA regulations—are not enforceable for state or local employees. However, when a state adopts the federal OSHA regulations (and receives federal grants in return), the regulations then become enforceable for state and local government employers and employees. Many departments in nonparticipating states use the federal and NFPA safety standards, with the understanding that they are providing a safer working environment for their members and reducing the chance of liability.

Firefighter Injury and Fatality Prevention

Every year, approximately 100 firefighters in the United States lose their lives in the line of duty. Recent experience shows a slightly lower number of 87 for 2017. Over the past few years there has been a realization that, although firefighting is very dangerous, many of these deaths and injuries are preventable. Although various national leaders and organizations have attempted to convince the fire service to follow safety standards, it is clear that for some organizations, the culture of risk remains unchanged. For example, consider live fire-training exercises for recruit firefighters, in which instructors deliberately create unsafe, extremely high-temperature evolutions. Think about the firefighter with heart disease who continues performing regular firefighter duties, performing strenuous physical work even when it is medically unsafe.

Examples such as this point to an FES safety paradox. Every effort is made to recruit members who are courageous and take risks. When faced with a dangerous situation, this type of person looks for ways to rescue victims and save property from destruction, even at their own peril. Although these brave members are needed, the challenge—through discipline and training—is to make these members safety-conscious and have them practice safe behavior at emergency incidents, ensuring their own well-being and taking risks only when necessary.

The National Fallen Firefighters Foundation (NFFF) is working to do just that. In 2004, with the cooperation of the U.S. Fire Administration, the NFFF gathered about 200 prominent national fire service leaders together to focus on prevention of line-of-duty deaths and reduction of firefighter injuries. The NFFF described the meeting by stating the following: "The Summit marks a significant milestone, because it is the first time that a major gathering has been organized to unite all segments of the fire service behind the common goal of reducing firefighter deaths. It provided an opportunity for all of the participants to focus on the problems, jointly identify the most important issues, agree upon a set of key initiatives, and develop the commitments and coalitions that are essential to move forward with their implementation" (reproduced from National Fallen Firefighters Foundation, 2019).

The result of the meeting was the NFFF's Everyone Goes Home program, which cites the following 16 Firefighter Life Safety Initiatives:

1. Define and advocate the need for a cultural change within the fire service relating to safety; incorporate leadership, management, supervision, accountability, and personal responsibility.
2. Enhance the personal and organizational accountability for health and safety throughout the fire service.
3. Focus greater attention on the integration of risk management with incident management at all levels, including strategic, tactical, and planning responsibilities.
4. All firefighters must be empowered to stop unsafe practices.
5. Develop and implement national standards for training, qualifications, and certification (including regular recertification) that are equally applicable to all firefighters based on the duties they are expected to perform.
6. Develop and implement national medical and physical fitness standards that are equally applicable to all firefighters, based on the duties they are expected to perform.
7. Create a national research agenda and data collection system that relates to the initiatives.
8. Utilize available technology wherever it can produce higher levels of health and safety.
9. Thoroughly investigate all firefighter fatalities, injuries, and near misses.
10. Grant programs should support the implementation of safe practices and/or mandate safe practices as an eligibility requirement.
11. National standards for emergency response policies and procedures should be developed and championed.
12. National protocols for response to violent incidents should be developed and championed.
13. Firefighters and their families must have access to counseling and psychological support.
14. Public education must receive more resources and be championed as a critical fire and life safety program.

15. Advocacy must be strengthened for the enforcement of codes and the installation of home fire sprinklers.

16. Safety must be a primary consideration in the design of apparatus and equipment.

Initiatives 5 and 6 call for national standards for training and certification, and national standards for medical and physical fitness. Respectively, these specific standards have the potential to save the greatest number of lives and reduce injuries among firefighters. However, they both start with the words "develop and implement," indicating that a lot of work still needs to be done and action must be taken by a standards-making organization, such as the NFPA. Although the NFPA does already have standards that include these topics, they are more of an outline than a specific standard that would result in firefighters having equal abilities, health, and fitness (refer to the chapter *Training and Education*). At this time, there does not seem to be any movement for the appropriate NFPA committees to more specifically confront these issues, and there is always the potential that influential special interest groups might resist efforts to develop national criteria for training, education, and medical and physical fitness requirements for firefighters. In most cases, this leaves the responsibility of development and implementation to the local or state FES agencies.

EMS Injury and Fatality Prevention

The EMS Workforce Agenda for the Future, prepared by the University of California, San Francisco, with funding from the National Highway Traffic Safety Administration (NHTSA) and the Health Resources and Services Administration, investigates the issue of developing and retaining a professional, well-trained EMS workforce. This document, which was released in 2011, follows and compliments the 2008 *EMS Workforce for the 21st Century: A National Assessment*. It identifies four key research components that are considered critical to developing a thriving EMS workforce (NHTSA, 2011):

- Education and certification. Variations of EMS certification and licensing create a lack of reciprocity—an issue that is especially important for EMS personnel responding to disasters outside of their immediate area or state. The *Agenda* recommends a national EMS education process to eliminate this issue and establish a high standard of quality across the country.

- EMS data. It is necessary to research and evaluate workforce issues; however, until accurate data are available, attempts to improve issues for the EMS workforce are difficult to justify. The challenge seems to be centered on two issues: the lack of trained EMS researchers and the lack of reliable funding sources. One solution is for EMS agencies to contribute to the nationwide collection of EMS patient and system data, the National Emergency Medical Services Information Systems.

- Since this report was issued, NIOSH has been collecting injury data for EMS workers from participating hospital emergency rooms starting in 2008. However, the data is described as being the result of a probability-based result. This is a statistical technique to estimate what the real data numbers are from a small sample.

- Health, safety, and wellness. The *Agenda* recognizes that EMS can be dangerous and unhealthy because employees are vulnerable to infectious diseases (e.g., COVID-19), patient assaults, sprains and strains, and vehicle crashes. It recommends the establishment of a national EMS Workforce Injury and Illness Surveillance Program to study these issues and look for solutions.

- Planning and development. To predict future demands for workforce needs, planning and development of resources are essential. There are currently very few data sources on which to predict future needs. It is recommended that state agencies help collect information on what drives EMS requests, workforce supply and demand, and compensation issues. National EMS Information System (NEMSIS) has a pilot program to build a system to collect this data funded by NIOSH (https://www.cdc.gov/niosh/topics/ems/projects.html).

Emergency Vehicle Crash Prevention

States have provisions in their vehicle driving regulations that allow emergency vehicles to exceed posted speed limits and proceed through controlled intersections without coming to a complete stop. These emergency driving regulations contain some language that assumes the emergency vehicle driver will use all due caution in being granted this privilege. If the emergency vehicle hits another vehicle while exceeding the speed limit or proceeding through a stop signal or sign, the driver was likely driving with some carelessness. Many vehicle crashes are preventable. Safety professionals dislike calling them accidents and prefer to call them crashes because they generally are not accidents. Instead, most serious vehicle crashes are the result of an intentional factor such as excessive speed, failure to stop at traffic signals and stop signs, or a general lack of caution at intersections.

Provisions to control the most common causes of FES serious accidents are outlined in NFPA 1500. For example, when drivers encounter a stop signal or sign, a stopped school bus, or an unguarded railroad crossing, they should come to a complete stop and not proceed until it is confirmed that it is safe to do so (NFPA 1500 6-2.8). Very specific SOPs governing vehicle speeds and limitations during inclement weather are needed to reduce accidents and any potential liabilities **FIGURE 9-2**.

In the 1997 Pennsylvania Supreme Court case *Schwarzbek v. City of Wauseon*, the City of Wauseon was sued by the estate of Blaine and Mildred Sanders, who were killed when their car was struck by an EMS vehicle responding to the emergency scene. The Court Justice issued the following statement, which touches on the issue of speed and emergency vehicles: "While the law—in view of the vital and urgent missions of fire department vehicles—wisely allows them certain privileges over other vehicles, it does not assign to them absolute dominion of the road…speed, which is uncontrolled, is capable of wreaking as much havoc and causing as much sorrow as fire itself" (data from Brannigan, 1997, p. 21).

In addition to standards regarding speed, NFPA 1500 contains other standards that help prevent FES crashes and injuries. For example, when operating at a vehicle crash on a highway, members should wear a garment with reflective material for visibility, use fire apparatus as a shield, and place appropriate warning

FIGURE 9-2 SOPs governing vehicle speeds reduce accidents and potential liabilities.
© Tom Carter/Photographer's Choice RF/Getty Images.

devices, including highway caution signs to warn oncoming traffic. Furthermore, the regular inspection, maintenance, and repair of vehicles are proven methods of preventing serious vehicle accidents. Requirements for creating a program to eliminate these causes are outlined in NFPA 1500. Finally, accident investigation and analysis have also been incorporated into the NFPA standard. Without feedback from analyses, safety prevention programs would not be able to adjust to new hazardous situations, information, and equipment.

Crashes Involving Ambulances

Ambulance crashes are not unusual, especially in rural areas. Between 1992–2011, there were an annual estimated mean of 4,500 motor vehicle traffic crashes involving an ambulance. Contributing factors such as maintenance, operator error and training, and safety restraints play a significant role in these crashes (Sanddal et al., 2010). Another important factor is the use of lights and sirens. Studies comparing code-3 and non–code-3 trips have shown that time saved by a code-3 response is insignificant to the outcome of the patient in nearly all cases (Sanddal et al., 2010).

NIOSH conducts research to improve safety for EMS responders. As part of the National Occupational Research Agenda, researchers at NIOSH have been looking for ways to reduce ambulance crash-related injuries and deaths. One major issue is that EMS employees often ride on the squad bench without wearing restraints, because restraint systems do not allow easy access to the patient or equipment. Without the use of restraints, personnel are at higher risk of head injuries from striking their heads on bulkheads or cabinets (Centers for Disease Control and Prevention, 2011).

In 2019, the NFPA published NFPA 1917, *Ambulance Vehicle Safety*. This standard defines stringent safety specifications for new ambulances. It addresses the issue of head injuries in the patient care compartment and the need to secure equipment that has the potential to seriously injure the provider or patient. Ambulance manufacturers should develop restraint systems designed to increase crash survivability of EMS responders and patients while accommodating accessibility to patients and equipment.

Further considerations to reduce injuries include the layout and structural integrity of ambulance compartments, hardware design, occupant restraints, and additional factors that can contribute to employee injuries and death (Centers for Disease Control and Prevention, 2011). To improve ambulance safety, federal funding is needed for ongoing research and testing of

CASE STUDY

EMS Ambulance Safety

In April 1989 Michael Montecalvo, a firefighter-paramedic and EMS educator, became the first EMS professional to be sentenced to prison for an ambulance crash. He had driven his ambulance through an intersection, killing a woman and seriously injuring her 6-year-old son. (The woman was proceeding normally through a green light at the time of the incident.) Montecalvo was sentenced to 2 years in the Ohio Mansfield State Penitentiary, but he made good use of his time by teaching the penitentiary employees first responder skills. He was released in 7 months, but as a convicted felon he was never licensed as a paramedic again. This case had a huge impact on the EMS community, because many had previously assumed that a vehicle crash in response to an emergency would not be criminally prosecuted.

ambulance vehicle design. Furthermore, EMS administrators need to develop and implement safe driving policies and procedures, monitor driving behaviors, and require proficiency training and oversight of continuous vehicle operation.

Crashes Involving Fire Apparatus

A vehicle crash can be a financial and operational disaster for the fire department, because these crashes may result in firefighter injuries and deaths, liability issues, and out-of-service time for fire and rescue apparatus. There is also the concern for civilian deaths, injuries, and property damage. Firefighters who risk their own lives to rescue a victim from a fire may unknowingly place themselves, fellow firefighters, and civilians at risk while aggressively responding to that same emergency call. When an emergency vehicle driver drives too fast for road conditions, fails to fully stop at a controlled intersection, or relies too much on their emergency lights and sirens, they are risking the lives of unaware civilian drivers and their passengers.

The NFPA has created key resources for reducing these types of crashes. NFPA 1451, *Standard for a Fire and Emergency Service Vehicle Operations Training Program*, contains training requirements for defensive driving techniques. NFPA 1002, *Standard for Fire Apparatus Driver/Operator Professional Qualifications*, and NFPA 1500, *Standard on Fire Department Occupational Safety and Health Program*, Section 4-2, Drivers/Operators of Fire Department Vehicles, contain

guidelines for the basic training new drivers should receive before being permitted to operate fire department vehicles.

It is always easier to learn a new skill correctly than to change existing habits. Nevertheless, existing drivers should also be encouraged to complete a driver-training course designed to comply with these safety standards. In-service training for existing drivers is required at least twice a year and is necessary to change the culture and tradition that encourages unsafe driving, such as speeding and not stopping at stop signs or red traffic lights. In-service training provides valuable hands-on experience and reinforces defensive driving techniques and emergency response SOPs.

Firefighter Health and Wellness

A comprehensive health and wellness program has four components that are essential for success. They are listed below:

1. Medical evaluation on a periodic basis consisting of a full evaluation and testing similar to the medical exam given for new members. In addition to identifying medical issues that would prevent the member from performing essential job functions, it provides a base line for any future changes that are a result of operating in hazardous environments.
2. Fitness evaluation and program containing the following:
   a. Medical clearance
   b. On-duty time for exercise
   c. Equipment and facilities
   d. Exercise specialists and peer trainers
   e. Fitness evaluation and enforceable pass-fail benchmarks
3. Rehabilitation program may be needed for members who do not meet the physical fitness benchmarks or as a result of an on-the-job injury resulting in a temporary physical disability. Physical therapy for rehabilitation is a priority for injuries on the job. For those that do not meet physical fitness standards, a short duration light duty assignment may be offered so the member can have adequate time for rehabilitation.
4. Behavioral health intervention may be needed for substance abuse, critical incident stress management, and other psychological issues. This will include tobacco use cessation and weight management control issues.

Joint Labor-Management Wellness and Fitness Initiative

The International Association of Firefighters (IAFF) and the International Association of Fire Chiefs (IAFC) have launched a national initiative to establish a program for medical and fitness evaluations, rehabilitation, behavioral health, and data collection. The basic design premise of this effort is to not be punitive. For example, in a punitive program, a member who is not fit for duty based on either medical or physical fitness criteria may be fined or placed on nonpaid leave. In a nonpunitive program, unfit members are placed on light duty, injured-on-the-job leave, or some other placement with no reduction in wages.

Punitive is a misnomer in this context. A good analogy would be the typical requirements for firefighters to report to duty on time. If a member is late or does not show up for the shift, it is typical for some type of disciplinary action to occur. Some would argue that an unfit firefighter is analogous to the firefighter not showing up for duty.

The next phase for an unfit firefighter is placing him or her in rehabilitation. However, several questions need to be considered in these cases. For example: How many salaried firefighters can a department afford to place in noncombat positions before it becomes a hardship? How many volunteers can be taken off combat status before it adversely affects responses? This is a complex and highly individualized problem that can have a substantial impact on the budget and operation of a department. If a reasonable implementation plan is used, however, the number of firefighters needing rehabilitation will probably be small and experienced infrequently.

When a new program is initiated, it is reasonable to allow existing firefighters who have medical and physical fitness problems an equitable time frame in which to meet standards and participate in a rehabilitation program. Also, NFPA 1500 allows phased implementation of safety requirements. For example, if a fitness test was chosen that has a pass-fail time of 7 minutes for a physical activity, the first year could start out with a time of 14 minutes for incumbents. Then, each year, the pass-fail time would be reduced by 1 minute until all members meet the 7-minute standard.

Physical Fitness Overview

The FES is a physically demanding profession that involves heavy lifting of equipment, continuous physical exertion for up to 30 to 40 minutes at a time, and exposure to toxic chemicals. Recruits are generally healthy and physically fit in departments with comprehensive physical fitness screening before hiring, but their health, especially physical fitness, can decline over time because many agencies do not require regular exercise or yearly medical examinations.

The main causes of firefighter deaths are not a secret, nor are they new. The firefighter fatality report for 2017 showed that heart disease accounts for 59.8% of firefighter fatalities. For example, in an interview about his experience as the first due battalion chief at the World Trade Center on September 11, 2001, Battalion Chief Joseph Pfiefer, New York Fire Department, described the health problems caused by the extensive stair climbing in the twin 1300 foot (396.24 m) high rise buildings: "There were messages, urgent messages of firefighters having chest pains as they went up" (data from *Firehouse*, 2002, p. 42). Another New York Fire Department officer at the scene reported similar findings, "As we got higher into the building, we heard numerous 'Urgents' and 'Maydays' from firefighters with chest pains, in need of oxygen or worse" (data from *Fire Engineering*, 2002, p. 97). Although all the other items in the safety standards are important and do save lives and prevent injuries, preventing heart disease has the potential to save a great number of firefighter lives, both on the job and in retirement.

The Need to Stay in Shape

A large U.S. study conducted by the Harvard School of Public Health suggests that firefighters face a far greater risk of dying from heart problems while battling a blaze than was thought. In the study, researchers examined a federal registry of 1,144 on-duty firefighter deaths between 1994 and 2004 (excluded were the 343 firefighters who perished in the September 11, 2001 terrorist attacks). Nearly 40% of the on-duty deaths during that period were caused by heart disease. Thirty-two percent of those deaths occurred while fighting blazes. Researchers found that the risk of suffering a cardiac-related death while putting out a fire was up to 100 times higher than the risk during down time, even though fighting fires accounts for only a small percentage of these workers' time.

A 2010 FEMA-sponsored study conducted at Saint Joseph's Hospital in Atlanta using the medical examinations of 300 firefighters aged 40 and older reported that one-third of the firefighters had heart disease that was unrelated to traditional risk factors, such as high cholesterol. According to the principal investigator, the risk of heart attack was tremendously impacted by such factors as stress and psychological pressures, diet, exercise, inherent personality, and genetic predispositions (Saint Joseph's Hospital, 2009).

Diet and exercise should be priorities of all FES administrators **FIGURE 9-3**. As noted by Dr. Linda Rosenstock, dean of the UCLA School of Public Health, "You may not be able to prevent all these deaths, but to the degree you can prevent some deaths by paying attention to underlying risk factors and better fitness programs, that's the goal" (Chang, 2007). These organization's administrators could do more to improve health by requiring annual physicals and fitness tests, and implementing wellness and fitness programs geared toward reducing heart disease risk factors such as obesity and high blood pressure.

As part of a Fire Fighter Wellness class in 2009, Memphis Fire Department training instructors calculated body mass index values for 670 firefighters ranging in age from 22 to 66 years. Based on the World Health Organization body mass index scale, 33% of these firefighters were considered obese and 5% were considered morbidly obese. Only 33% of the firefighters in this assessment were found to have healthy weight levels. "Since obesity is considered a major cardiovascular disease risk factor, many overweight and obese firefighters are predisposed to a higher incidence of adverse cardiac events than their normal-weight peers" (data from Spratlin, 2011).

In addition, excess body weight adversely affects firefighters' abilities to perform their duties. In 2011, researchers at Auburn University's Human Performance Laboratory studied firefighters in Montgomery, Alabama and found a negative correlation between weight and physical performance (Spratlin, 2011). Obese firefighters are also more likely to get sick or injured. There is concern that obesity in the firefighter demographic might only increase, as it has in the general public.

A two-pronged approach works best for reducing the number of stress- and cardiac-related firefighter deaths. First, regular medical examinations, as outlined

in NFPA 1500 and 1582, should be scheduled for all members. Any indication of heart disease should be treated without delay. In most cases, the member should be completely evaluated before being allowed to participate in emergency operations duties.

Second, an ongoing physical fitness program has many beneficial outcomes. This type of training reduces the chance of heart attack, lowers cholesterol, reduces body fat, and strengthens the heart. Individuals with stronger hearts as a result of aerobic exercise are more likely to survive a heart attack (https://www.studyfinds.org/exercise-heart-attack-survival-study/) (2017). In addition, Zhen Yan, the director of the Center for Skeletal Muscle Research at University of Virginia School of Medicine's Robert M. Berne Cardiovascular Research Center, recommends 30 minutes of cardiovascular workouts every day to prevent the respiratory disease [COVID-19]. A preventative medicine study by this research center is claiming, according to reports, the best way to prevent coronavirus complications from worsening is regular exercise (Retrieved from: https://www.foxnews.com/health/regular-exercise-help-prevent-coronavirus-complications-from-worsening. 4/23/2020).

Voluntary fitness programs have reportedly reduced injuries by 30% (Rawlyk, 2010). This benefit affects more than just the individual, who has a decreased chance of suffering a heart attack and a better chance of surviving if one does occur. Team members also benefit, becoming safer working with other team members who are physically fit. Furthermore, firefighters in good physical condition consume less air from their self-contained breathing apparatus than firefighters who are in poorer physical aerobic condition. This allows them to work for longer periods of time in a hostile atmosphere, which makes a tremendous difference, because any low air alarm activation requires the entire team to withdraw from the scene. This reduced time can be a major factor in a team's ability to extinguish a fire or search for a victim overcome by smoke.

The entire department also receives a better reputation from the public and elected officials who derive their impression of the department based on what they see of the members. Furthermore, the customer receives better service, because the results of the team are always limited by the weakest member. The public has a right to expect that all firefighters who are responding to their emergency are trained, organized, staffed, equipped, and in top physical shape. In addition, as a result of fewer injuries from members, the city and county end up paying less for worker's compensation claims, sick leave, and medical insurance.

FIGURE 9-3 Diet and exercise should be priorities in all FES organizations.
© Westend61/Getty Images.

Physical fitness should be directly related to the job, not sex or age. Predictably, the human body loses strength and endurance as it ages; however, with proper physical training a person can function at high levels of strength and endurance into his or her 60s. Discussing the Candidate Physical Ability Test (CPAT), the IAFF website comments, "It is the position of the IAFF/IAFC Joint Labor Management Wellness-Fitness Initiative Task Force that fire departments should increase the diversity of their workforce by actively recruiting candidates from throughout their communities rather than lowering candidate physical ability standards" (data from IAFF, 2011). This is in reference to a practice, to meet diversity goals, by a few departments of hiring persons who are not able to perform physical demands of the job. This practice is not well documented and may be a small percentage of departments. But, one is too many and the IAFF/IAFC position reflects this concern.

A more recent study conducted using 109 recruits from four recruit classes conducted at the University of Arizona had several interesting inferences. "Physical Fitness Intervention for New Firefighters [Tuscon, Arizona]" posted at Injury Prevention journal web site in 2015 (https://injuryprevention.bmj.com/content/injuryprev/22/3/181.full.pdf) had the following takeaways:

- "The challenge: Higher levels of fitness among firefighters are associated with a decreased risk of non-exercise related injury..."
- "This structured fitness program produced a positive return..."
- "Occupational injury frequency and workers' compensation claims costs were reduced following implementation of the Performance Fire Fighter (PFF)-Fit intervention."

This research proves that mandatory physical fitness programs result in better well-being of the firefighter, improved skills performance, and reduced worker's compensation costs.

Physical Fitness Program

Physical fitness has long been recognized as an essential trait for members of the FES. The profession requires a high level of strength and endurance, and without it many tasks cannot be done safely, promptly, and effectively. Many fire departments do not have a physical fitness program that is effective and creditable. Any program that is not mandatory will have a percentage of nonparticipants. Although resistance can be found among individuals (typically the individuals who need

it the most) and labor groups, firefighter groups are increasingly taking notice of cardiac risk.

The following recommendations can help guide the implementation, evaluation, and execution of a department's physical fitness program by the FES administrator.

- When deriving physical fitness minimum levels, the needs of the public (customers) and the safety of the firefighter must come first.
- Validating physical fitness by testing incumbents is sometimes inaccurate. There are incumbents who perform poorly because they do not participate in a physical fitness program. Also, there may have been no physical fitness standard when they were hired or accepted as a member, therefore, incumbents may be unable to perform all of the essential job functions. Most members have the potential to do better with training and meet minimum fitness standards.
- A comprehensive wellness program that includes medical evaluations must be used in combination with an exercise program, and the costs should be borne by the department. The program might allow firefighters up to 2 hours to exercise on duty each shift.
- When first implemented, a physical fitness standard for incumbents should recognize that some firefighters have not exercised for many years, so a phase-in period should be used. Most firefighters who find themselves below standards for a physical fitness evaluation can meet standards with the appropriate training and time.
- A firefighters who fails to meet the minimum levels of physical fitness (different levels for incumbents during the phase-in period) should be prevented from participating in physically demanding work. As noted by Stephen Foley in *Fire Department Occupational Health and Safety Standards Handbook*, "Members must not be allowed to begin or resume suppression duties until they pass the physical performance requirements" (data from Foley, 1998, p. 87).
- Safety and health requirements must be mandatory to work. A voluntary or nonpunitive physical fitness program guarantees that there are members in poor physical fitness. Imagine if a traffic intersection in a town had stop signs posted to prevent accidents, but the local ordinance was voluntary or nonpunitive. Some drivers would disregard the stop signs, guaranteeing some serious accidents.

Summary of Wellness Benefits to Member

1. Improved heart health is always at the top of the list. As mentioned previously, heart attacks are the number one killer of firefighters. Regular aerobic exercise helps prevent heart disease by strengthening the heart muscle, normalizing blood pressure, reducing low density cholesterol, etc. This includes stamina to perform at high levels of physical activity during emergency operations, which also provides improved pulmonary (lung capacity) function.
2. Helps achieve and maintain healthy body weight.
3. Reduces the chance of a strain or strain injuries, including the sometimes career ending back injuries.
4. Improves emotional health, including better self-esteem, reducing depression/anxiety, and sleep issues.
5. In general, exercise will enhance the FES member's ability to operate physically at emergency incidents as a result of increased stamina, strength, and flexibility.

Physical Fitness Testing

Any physical fitness test should be based on validated job performance metrics. The IAFC and the IAFF developed the CPAT in 1999 to address this issue in the FES. This test was developed to provide fire departments with a tool that would enable them to select physically capable individuals to train as firefighters. The CPAT includes the following eight sequential events:

- Stair climb
- Hose drag
- Equipment carry
- Ladder raise and extension
- Forcible entry
- Search
- Rescue
- Ceiling breach and pull

Although the CPAT is considered the best evaluation to date, the test does have limitations. For example, the test was devised with data provided by 10 metropolitan fire departments, some of which had crews of four or more firefighters. Therefore, the data may or may not accurately represent minimum physical fitness required to accomplish each task. Also, some of the tests or the background data were gathered from firefighters who were already on the job, and it can be argued that just because these individuals were already firefighters they were not necessarily physically competent to perform all critical tasks efficiently. Instead, valid studies that focus on job performance should be instituted.

Another limitation of the CPAT is that certain critical tasks were underrepresented in the test. One example is the dummy drag, officially known as Event 7. This task, which was reported as critical by 97% of those surveyed, simulates removing a victim or injured partner from a fire scene. According to the IAFC, "The candidate grasps a 165-pound mannequin by handles on a shoulder harness, drags it 35 feet to a drum, makes a 180-degree turn around the drum, and continues an additional 35 feet across the finish line" (Courtesy of IAFC). The established weight used within this test was based on the average weight of adult patients and firefighters. The established distance was based on the average distance from the living room to the front door in a single-family home. The final determination of pass-fail was made after a committee evaluated this data, using the lowest 10th percentile as a basis for passing. This means that a person passing the test can rescue only 10% of incumbent firefighters (without the weight of personal protective equipment [PPE] and self-contained breathing apparatus). This seems like a very low benchmark that can only be overcome by several firefighters helping out.

CASE STUDY

Physical Fitness Fairness

In the late 1970s, a firefighters graduated from a rookie school of a metro county fire department after an intensive 16-week course of skills and knowledge of firefighting. At that time, this department had no physical fitness test for new employees. At the first duty station, the rookie was assigned to an engine company. The officer at this station liked to think outside the box and planned to have the rookie teach a series of basic drills to the crew. He was aware that teaching the skills necessary to be a safe and effective member would help the rookie commit them to memory.

The first drill was to take the extension ladder off the engine, place it so the firehouse roof could be accessed, and climb it carrying a roof ladder. The officer observed the drill and immediately realized that the rookie was not strong enough to complete the evolution, which was a standard one-person skill. At this time the Battalion Chief was consulted for possible personnel action. Subsequently, the officer was advised there was no process to separate this employee at this time. Just make it work.

Seven years later, the same firefighter was assigned to a truck company. The new company officer had also noticed the physical issues with this firefighter. The new officer started creating a list of specific skills this firefighter could not complete either timely or safely. It was at this time that this firefighter reported a back injury as a result of lifting ground ladders and started medical treatment. Eventually, the firefighter was awarded a job-connected disability of 70% of their pay tax-free.

This account points to three possible issues regarding safety:

1. The firefighter now has a permanent painful back injury that may have been prevented if he or she had been required to successfully complete an initial pre-hiring physical fitness test.
2. As part of the company crew, the firefighter handicapped the team's ability to perform all of the necessary job functions for several years. This affects the safety of fellow firefighters and the service the public receives.
3. If the back injury was contrived, the county and its taxpayers are paying a very generous disability benefit to someone who does not deserve it.

This case illustrates the danger for departments that do not test new members. The issues listed above could have been resolved if job performance physical fitness testing had been correctly implemented. Employing firefighters who cannot pass a fitness test is not fair to the individual, the crew, the department, or the public.

Other Occupational Hazards

There are many other safety items identified in NFPA 1500 that departmental safety committees should address. In addition, safety committees should review the annual U.S. Firefighter Fatalities and Injuries report issued by the NFPA. Also, the *U.S. Fire Administration Firefighter Fatality Retrospective Study: 1990–2000* identifies the major causes of deaths and injuries from a multiyear perspective. These reports contain many clues to areas where deaths and injuries can be prevented.

Whether within a fire-based or private organization, FES personnel are exposed to a wide variety of occupational hazards, including infectious disease and blood-borne pathogens, hearing loss from exposure to sirens and loud environments, temperature extremes, a higher incidence of vehicle collisions, musculoskeletal injuries, long work hours, sleep deprivation, and inordinate physiologic and psychological stress.

Personal Protective Equipment

OSHA has created standards to protect FES personnel from some of the hazards previously mentioned. For example, OSHA's 1910.120 requires that employers provide all employees with the PPE necessary to safely do their jobs. This may include air respiratory masks, firefighting protective clothing, filter masks, radiation suits, and, when needed, quarantine suits to avoid exposure to these toxic substances **FIGURE 9-4**.

Even though respiratory protective equipment has been provided for the past 40–50 years, states and local governments have classified Heart Lung diseases as presumptively job related and eligible for worker's compensation. In addition, "A majority of states now cover firefighters for one or more cancers under workers' compensation because of presumptive legislation. In many states, the presumptive legislation contains broad or nonspecific language that can be interpreted to cover any cancer experienced by a firefighter. In other states, only specific cancers are covered. Most commonly those are leukemia, non-Hodgkin lymphoma, brain cancer, bladder cancer, and gastrointestinal cancer." (https://www.firstrespondercenter.org/cancer/toolsresources/presumptive-legislation-firefighter-cancer-state/ 2019).

In addition, the Center for Research on Emergency Medical Services' Emergency Responder Human Performance Lab has examined protective programs that are common to the fire service but seem to be lacking in EMS. The Center contends that NIOSH provides

FIGURE 9-4 Employers must provide all employees with the PPE necessary to safely do their jobs.
© Jones and Bartlett Publishers. Photographed by Glen E. Ellman.

firefighters with resources specific to heat- and work-related injuries but does not provide programs specific to EMS. As the role of the traditional EMS responder expands to include response to disasters, contagion viruses, or weapons of mass destruction, the PPE for EMS will become even more important.

EMS and Violence

There is also a psychological toll taken when, in the attempt to render aid, EMS personnel become victims of workplace violence. All too often, personnel are threatened by anxious family members who feel the response was too slow, people who feel threatened by the uniform, those with metabolic impairments or trauma that affects their behavior, or those under the influence of alcohol or drugs. Violence against EMS personnel is a real and pervasive problem. During a presentation at the 2009 Mid-Year Conference of the National Association of State EMS Officials, EMS Scholar Brian Maguire stated, "EMS personnel are seven times more likely to be killed by workplace violence than other healthcare professionals and 22 times more likely to experience injury from assault than the population at large" (data from Maguire, 2009, p. 6).

EMS personnel may become so focused on the patient and the task that they become blindsided by events occurring around them. Force protection and self-defense training are not typically covered in an emergency medical technician or paramedic school curricula. An agency or company that invests time in training personnel on how to be aware and safe can save themselves grief, the cost of a worker's compensation claim, and disability and death for their employees. One such class is a collaborative effort between EMS-1 .com, an online education resource, and Calibre Press, a provider of law enforcement training: the EMS-1 Street Survival Seminar. The class covers the following topics:

- Gang threats
- Weapons search
- Intoxicated, psychotic, and combative patients
- Specific conditions, such as excited delirium and autism
- Ruling out medical complaints

Employers have a duty to provide their employees with a safe workplace, and prepare them for instances in which hazards are not known.

Mental Health Issues (Stress)

FES personnel are constantly exposed to stress and can easily develop stress-related disorders if situations are not mitigated properly or in a timely manner. For example, chronic stress disorder can result from exposure to anxiety that occurs over time because of issues such as salary; work schedules; an unsupportive company or partner; a lack of confidence in their own skills; or general fears of exposure to communicable diseases, violence, or vehicular accidents. These issues can be exacerbated by 24-hour work schedules, lack of sleep, general burnout, or home and family issues.

Burnout is generally defined as a psychological response to prolonged work stress and is recognized as having three major components that affect work performance:

- Fatigue that can lead to unsafe actions.
- Detachment or alienation from the workplace and other members.
- Cynicism or unsympathetic attitude towards people in general. This can be especially problematic for patient care.

The major causes tend to be psychological impact from experiencing tragic injuries to patients, especially children, and lack of rest time (sleep deprivation) (https://www.usfa.fema.gov/current_events/072618 .html). Both of these are discussed in the following paragraphs starting with critical incidents.

Tragic incidents are also likely to have a psychological impact on an emergency responder. These include scenarios such as injury or death of a team member, a situation in which an error caused an injury or death of a patient, child abuse, an unsuccessful rescue, or a scene where the responder felt threatened. There are several questions that need to be asked when determining whether stress from a critical incident can be a more serious problem for a first responder:

- Can this incident be mitigated by normal coping mechanisms? Often a responder just needs time to process and adapt to the events of an incident. They may be able to reframe the issue from negative to positive or evaluate the problem from an internal perspective rather than blaming it on an external force.
- Does the person have a good social support network? Coworkers and partners are often caring and understanding of the issue. It is even better if the family is a part of the support chain. Responders without sufficient support may adopt negative coping mechanisms.
- Does the company or agency have good Critical Incident Stress Management (CISM) resources? CISM is a comprehensive crisis intervention system that includes components to manage all phases of a crisis. It starts with pre-crisis training so employees know what to expect and

understand how to access available resources. The acute crisis phase may be managed by a team of peer counselors and mental health professionals and provides interventions to individuals, crews, or larger groups. In the post-crisis phase, debriefings may occur or references may be provided for ongoing care.

If an issue is not mitigated properly, **posttraumatic stress disorder** (PTSD) can occur. This is considered by some to be a silent epidemic for FES personnel. PTSD develops from situations in which the FES member "has experienced, witnessed or was confronted with an event or events that involved actual or threatened death or serious injury, or a threat to the physical integrity of self or others and the person's response involved fear, helplessness, or horror" (data from Alexander and Klein, 2001, p. 80). The symptoms of PTSD can include re-experiencing of the event or flashbacks, avoidance, or persistent symptoms of increased arousal. Similar calls, sights, smells, or other triggers can propel the person back to the original incident and could create a dangerous situation.

Companies and agencies are wise to create their own CISM teams or partner with other agencies for resources to address stress. The agency needs to ensure that the right resources with the right training are available to members, and that a mental health professional can be accessed when necessary. For most members who suffer critical incidents, timely CISM can make a difference in whether the employee continues to be psychologically healthy, helping to ensure a long and productive career.

Sleep Deprivation

Sleep disruption is another type of hazard for FES personnel who work a 24-hour shift. "[A]t any time during the night, that person may be awakened suddenly and abruptly by a bunkroom light coming on and some form of a bell indicating there is an alarm" (Courtesy of Fire Engineering Magazine; Gagliano, 2016, p. 90). A 2009 study of firefighters noted that their heart rates "typically rise to 80 percent of maximum when the alarm is received" (Courtsey of Fire Engineering Magazine; Gagliano, 2016, p. 91). Furthermore, the inability to sleep easily and deeply may result in **sleep deprivation**, which may physically predispose an individual to becoming a stress casualty. This is especially true for EMS providers working on units that are very busy (over 2,500 calls per year).

For example, because of sleep deprivation issues, truck drivers are restricted in their time to driving the maximum of 11 hours within the 14-hour "driving window." After each shift, the driver must have 10 consecutive hours off duty (Retrieved April 7, 2019 from http://www.fmcsa.dot.gov/). These limitations are based on extensive research over many years, including investigations of thousands of truck crashes. Because similar restrictions are not in place for FES, there is a need to adopt policies from scientific comprehensive studies.

Recent FES research has suggested possible policies that would reduce the negative outcomes of a 24-hour shift. These include the following:

- Restrict members from working more than 24 hours without a 12-hour break. This is especially important when the EMS units are so busy that being able to sleep at night is usually interrupted.

- Because most departments require approval of a second job, add the provision that they cannot work the second job 12 hours before their 24-hour shift.

- Encourage or preferably require better than average physical fitness levels, which would result from a firefighter physical fitness program.

Wrap-Up

CHAPTER SUMMARY

- The safety and wellness of FES members is paramount to providing the best professional emergency service to the public.

- In addition, safety and wellness programs are of great benefit to the member.

- Contextual explanations are discussed to educate the member, including representative groups (unions) and government administration of the need for these programs because there can be opposition.

- A discussion of the issue and justification for safety and wellness programs is presented.

KEY TERMS

Body mass index A high BMI can be an indicator of high body fatness. However, percent body fat computations tend to be more accurate.

Burnout A psychological state that results from physical or mental stress that affects a member's ability to think, behave, or physically perform effectively and safely.

Candidate Physical Ability Test Test comprising eight sequential events developed by the IAFC and the IAFF in 1999 to provide fire departments with a tool to select physically capable individuals to train as firefighters.

Coping mechanism The strategies people often use in the face of stress and/or trauma to help manage painful or difficult emotions. Some good, some bad, such as overeating or using intoxicating drugs or alcohol.

Critical Incident Stress Management A comprehensive crisis intervention system that includes components to manage all phases of a crisis.

EMS Workforce Agenda for the Future A document prepared by the University of California San Francisco (UCSF), with funding from the National Highway Traffic Safety Administration (NHTSA) and the Health Resources and Services Administration (HRSA), that investigates developing and retaining a professional, well-trained EMS workforce and identifies four key research components considered critical to developing such a workforce.

Everyone Goes Home program The result of a 2004 NFFF conference that established the 16 Firefighter Life Safety Initiatives.

Heart Lung diseases This is a category of medical conditions that have been linked with the occupation of firefighting. Most firefighters have the benefit of Heart Lung presumption that assumes these medical conditions are job related and eligible for medical treatment, including justification for job-related disability and retirement.

National Fallen Firefighters Foundation Since 1990 the National Fallen Firefighters Foundation has been officially designated by the U.S. Congress as the financial support for the National Memorial to career and volunteer fallen (death) firefighters. Located in Emmitsburg, Maryland, it was conceived as a tribute to American fire service.

NHTSA A U.S. agency that develops and employs vehicle performance and safety standards, including partnerships with state and local governments. NHTSA reduces deaths, injuries, and economic losses from motor vehicle crashes.

National Institute for Occupational Safety and Health (NIOSH) The U.S. federal agency responsible for conducting research and making recommendations for the prevention of work-related injury and illness.

Occupational hazard Occupational hazards, experiences in the workplace, can encompass many types of hazards, including chemical hazards, biological hazards, psychosocial hazards, and physical hazards.

Occupational Safety and Health Administration (OSHA) A U.S. agency that is responsible for providing safe and healthful working conditions by setting and enforcing standards, and by providing training, outreach, education, and assistance.

Posttraumatic stress disorder A psychological disorder that develops from situations in which the FES member has experienced, witnessed, or was confronted by events that involved actual or threatened death or serious injury; symptoms can include flashbacks, avoidance, or persistent symptoms of increased arousal.

Punitive The imposition of an undesirable or unpleasant outcome upon a group or individual, imparted by an authority as a response and deterrent to a particular action or behavior that is deemed undesirable or unacceptable.

Safety committee A group of stakeholders with varied experience that plans and implements changes needed to comply with safety standards.

Sleep deprivation Sleep deprivation is the condition of not having enough or good quality of sleep. It can be either chronic or acute and may vary widely in severity. A chronic sleep-restricted state adversely affects the brain and cognitive function.

Stress Stress is a fact of life in which psychological pressures from the inside or outside world affect the individual, either one's emotional or physical well-being, or both. Typically, it can be helpful in short-term situations but harmful when it is chronic long term.

U.S. Firefighter Fatalities and Injuries report Annual report issued by the NFPA providing statistics on American firefighter fatalities and injuries.

Wellness The characteristic of being in good health physically and mentally.

REVIEW QUESTIONS

1. Select one major topic found in NFPA 1500 that you feel most passionate about. Identify one of your organization's SOPs related to that topic, and carefully read it. Is all of the information up to date? Research at least two sources (studies, expert opinion, etc.) that either support the SOP as it is written or provide information and data that could improve it.

2. Which of the 16 Firefighter Life Safety Initiatives do you think an individual administrator could best influence or implement in his or her own department? Defend your choice.

3. Read the case study titled "EMS Ambulance Safety." Do you think that the outcome in the Montecalvo case was just? Why or why not?

4. Briefly outline a physical fitness program you would implement in your department. Be sure to describe who would be affected, what kind of phase-in period you would include, and what resources would be needed to make the program a success.

5. Which occupational hazard do you feel poses the biggest threat to your particular work setting and employees? How does your facility address this concern, and after reading this chapter, what more do you feel could be done?

CHAPTER ACTIVITY #1: INJURIES ON THE JOB

Approximately 16% of all fireground injuries are categorized as wounds, cuts, bleeding, or bruises (Evarts and Molis, 2018). Because there are a large number of injuries in this category, a program could be initiated to prevent these injuries. Training and education would be a part of the overall plan, but the department's officers would also want to be sure that all members are wearing all their safety equipment and following SOPs.

One important item for the prevention of cuts, wounds, burns, and other injuries of the hand is the use of issued firefighter safety gloves. If a safety officer receives an injury report of a cut to the palm of the hand from a firefighter, it would be a good idea to examine the glove of the hand that was cut. It should have an identical cut and blood around the cut in the glove. Without the matching cut in the glove, it could be inferred that the firefighter was not wearing gloves when the injury occurred, and corrective action should be taken. There may be times when there is a suspicion that cannot be proved. Even in these cases, if the members know the administration is checking, they are more likely to wear all their PPE in the future.

Discussion Questions

1. Is it fair to check up on the firefighters?

2. Would it be appropriate to have the firefighter use his or her own health insurance rather than the jurisdiction's worker's compensation if the firefighter was careless or neglected to follow safety SOPs or wear safety equipment?

3. Is this an example of managing by walking around? If yes, are there other actions that could be taken using this method that would prevent injuries?

4. How could the administration be assured that all firefighters know the safety SOPs, including the use of safety equipment?

CHAPTER ACTIVITY #2: CRITICAL INCIDENT STRESS

As a paramedic at a busy suburban ambulance company, you have worked alongside one of your coworkers for over a year, when you suddenly begin to notice changes in his behavior. He always seems tired, gets angry about calls he feels are unnecessary, and seems to take advantage of long transports. You know he has experienced a number of pediatric calls with negative outcomes in the past few months and have heard him mention that he is having issues with his wife. You mention your concerns to a supervisor, but your coworker's behavior is brushed off as simply a "bad attitude." Then one day after a particularly bad call, you notice him take the station's supply of narcotics and leave. You report him to the company supervisor, and he is eventually caught by police and arrested for theft of controlled substances. The company fires him.

Discussion Questions

1. This coworker hadn't always been an angry person. What may have happened to him or her over time?

2. Was the company remiss in not listening to the initial concerns?

3. How could all of the negative outcomes have been prevented?

REFERENCES

Alexander, D.A., and S. Klein. 2001. Ambulance personnel and critical incidents: impact of accident and emergency work on mental health and emotional well-being. *British Journal of Psychiatry* 178: 76–81.

Brannigan, V. 1997. The real hazard is driving the fire truck. *Fire Chief* 41(7): 28–29.

Carter H.R. 2001. Why Are We Dying? *Firehouse* 26(8): 21.

Centers for Disease Control and Prevention. 2011. *NIOSH continues research to improve safety for ambulance service workers and EMS responders.* Retrieved from http://www.cdc.gov/niosh/docs/2011-190/

Chang, Alicia. 2007. Heart-related death risk higher while firefighters battle blazes. *Associated Press,* March 22, 2007.

Evarts, Ben, and Joseph Molis. 2018. *U.S. Firefighter Injuries – 2017.* Quincy, MA: National Fire Protection Association.

Fahy, R.F. 2010. *U.S. Fire Service fatalities in structure fires, 1977–2009.* Quincy, MA: National Fire Protection Association.

Federal Emergency Management Agency. 1999. Guide to developing effective standard operating procedures for fire and EMS: Federal Emergency Management Agency United States Fire Administration. Retrieved from https://www.usfa.fema.gov/downloads/pdf/publications/fa-197-508.pdf

Fire Engineering. 2002. Anonymous letter. *Fire Engineering* 155(9): 97.

Firehouse. 2002, July. WTC—This is their story. *Firehouse* 27(7): 42–44.

Foley, S.N. (ed.). 1998. *Fire Department Occupational Health and Safety Standard Handbook.* Quincy, MA: National Fire Protection Association

Gagliano, Anne, 2016, June 17. The 24-Hour Shift, Part 1. *Fire Engineering.* Retrieved from https://www.fireengineering.com/2016/06/17/317796/the-24-hour-shift-part-1/

International Association of Firefighters. 2011. *Candidate physical ability test program summary.* Retrieved from https://www.iaff.org/cpat/

Liao, H., R.D. Arvey, and R.J. Butler. 2001. Correlates of Work Injury Frequency and Duration among Firefighters. *Journal of Occupational Health Psychology* 6(3): 229–242.

Maguire, B. J. 2009, June 8–9. *Occupational Risks Among EMS Personnel.* Presented at the 2009 Mid-Year Conference of the National Association of State EMS Officials, Hilton Alexandria Mark Center, Alexandria, VA.

National Fallen FireFighters Foundation. 2019. *Everyone Goes Home: Firefighter Life Safety Initiatives.* Retrieved from https://www.everyonegoeshome.com/16-initiatives/

National Highway Traffic Safety Administration. 2011. *EMS Workforce Agenda for the Future.* Retrieved from www.ems.gov/pdf/2011/EMS_Workforce_Agenda_052011.pdf

Rawlyk, H. 2010. Fitter firefighters saving taxpayers money: Health program reducing injuries. *The Capital Gazette* 10.

Saint Joseph's Hospital. 2009. *Heart Disease: An Epidemic for Firefighters.* Retrieved from https://www.fireengineering.com/2009/03/17/257996/landmark-fema-study-heart-disease-is-an-epidemic-for-firefighters/#gref

Sanddal, T. L., N.D. Sanddal, N. Ward, and L. Stanley. 2010. *Ambulance Crash Characteristics in the US Defined by the Popular Press: A Retrospective Analysis. Emergency Medicine International.* Retrieved from http://www.hindawi.com/journals/emi/2010/525979/

Spratlin, K. 2011. *Firefighter Obesity: A Public Safety Risk.* Retrieved from http://www.fireengineering.com/articles/print/volume-164/issue-1/departments/fire-service_ems/firefighter-obesity-a-public-safety-risk.html

U.S. Bureau of Labor Statistics. 2019. *Occupational outlook handbook.* Retrieved from https://www.bls.gov/ooh/healthcare/emts-and-paramedics.htm#tab-3

U.S. Fire Administration. 2019. *Heart Attack Leading Cause of Death for Firefighters.* Retrieved from https://www.pennmedicine.org/updates/blogs/heart-and-vascular-blog/2019/january/firefighters-and-heart-health

LAW POLICIES REGULATIONS GOVERNANCE

TRANSPARENCY REQUIREMENTS RULES STANDARTS

COMPLIANCE

CHAPTER 10

Government Regulation, Laws, and the Courts

LEARNING OBJECTIVES

After studying this chapter, the student will be able to:

- Identify and understand the moral and ethical justifications for government oversight of the public fire and emergency services (FES) (pp 187–189).
- Understand the creation process and the importance of government regulations and laws (pp 189–192).
- Examine building codes and regulations at the state and local level, and understand their effect on FES (pp 190–192).
- Examine progressive laws and regulation trends in FES (pp 192–193).
- Recognize the influence that the Occupational Safety and Health Administration (OSHA) has on FES (pp 193–196).
- Recognize the influence that the National Fire Protection Association (NFPA) has on FES (pp 196–199).
- Understand the impact and influence that the legal system and the courts have on FES (pp 199–201).

Case Study

Fire safety experts recognized a market failure shortly after the first smoke alarms came on the market. These alarms were relatively expensive due to the low volume of purchases. Although it was recognized that this safety device was extremely beneficial, experts realized it would not necessarily be purchased if left up to individuals or building owners who did not see the benefit of spending money for these detectors. This could result in troubling consequences, for even the fire-conscious individual. For example, if a fire in a multifamily apartment building starts in an apartment without a detector, the fire will endanger the other occupants of the building, regardless of whether those occupants have a smoke alarm in their own apartment. An undetected fire in an apartment can become very sizable and hard to extinguish. In a worst-case scenario, it may trap victims by blocking the exits from other apartments.

The original effort by fire safety experts in the 1960s was to convince building and fire code organizations that these life safety devices were effective and reasonably priced. This took many years but now smoke alarms are typically found in all new residential units. Due to market failure, the smoke alarm eventually became mandated by building and fire codes and in many cases even in existing buildings. This is a classic example of a public good that had to be mandated by a government regulation.

Discussion Questions

1. Does your jurisdiction require the installation of smoke alarms in residential occupancies? If yes, which code requires them?

2. For existing residential occupancies, does your jurisdiction require the installation of smoke alarms? If no, what process could be used to adopt this type of retro requirement?

3. Does the local fire department support the maintenance of these lifesaving devices?

Introduction

In their uncomplicated form, regulations and laws are used by governments to protect the public, including employees, from actions by others that could be harmful to them. They may be described as a policy for the improvement of safety or well-being, or some altruistic goal such as fairness. To justify these regulations and laws there is the assumption that the public does not have the knowledge or ability to judge the safety or benefit of some processes or products. For example, when you buy a car you should have the assurance that when you apply the brakes, the car/truck will stop in a safe distance. Another everyday example is realized when you shop at the grocery store or go to a restaurant. For the most part, you can be guaranteed that you will not get sick from the food you consume.

Because regulations and laws are adopted by politicians or government bureaucratic employees, they are prone to have goals that are not altruistic. This is a complex subject and well beyond being discussed in this text. However, we will explain several regulations that directly affect FES.

Most FES providers are a part of either local or state governments, independent volunteer fire companies, or emergency medical services (EMS) organizations. Often, elected officials and appointed administrators of these organizations argue that there is no need for federal or state government intervention in their affairs. They protest that standards of service are best determined at the local level of government.

This is especially true when a state or federal government regulation requires the expenditure of tax dollars at the local level but is unaccompanied by state or federal funding. For example, in 2001, Robert Johnson, writing in the *Wall Street Journal*, predicted that the implementation of NFPA 1710, *Standard for the Organization and Deployment of Fire Suppression Operations, Emergency Medical Operations, and Special Operations to the Public by Career Fire Departments*, would "prompt fire departments to hire 30,000 firemen nationwide, an 11 percent increase" (Data from Johnson, 2001). Although this particular prediction was unfounded, it would have had a substantial fiscal impact on the nation's FES and taxpayers. For regulations such as these, it is not uncommon for city or county administrators and elected officials, along with FES officials, to protest the mandates. Most mandates are never fully funded or complied with because they are voluntary professional consensus standards;

however, some may at times be partially funded. For example, when the U.S. Department of Transportation issues new safety design features for roads, the federal government often provides matching funds or threatens to withhold funds to encourage states to comply.

NFPA standards can only be "mandated" if a locality or state adopts them as statute. This does not preclude holding a department liable for failing to adhere to a recognized "standard of care." In the case of NFPA 1710, the standard does not require compliance absolutely, but only that the FES department produce an "annual report" to the city/county administration and the public that designates deficiencies. That evaluation's "annual report" can be used to justify improvements funded by the jurisdiction. In the case of NFPA 1710, it is a great framework for providing a professional timely emergency service as discussed in Chapter 12, *Public Policy Analysis*.

Regardless of this issue and the fact that regulations and standards for local FES are relatively new (starting in the 1980s) in the United States, there seems to be adequate justification for the public to expect and encourage government oversight. In general, there are three factors influencing the justification for government regulation: (1) the role of government to protect its citizens when they cannot protect themselves, (2) when the public lacks the knowledge needed to evaluate a complex issue, and (3) the concern over a market failure preventing a public good or service from becoming available or widely used.

Legal Aspects of FES

The following is a general list of advice for the FES administrator:

- First and foremost, do not become paralyzed by fear of liability.
- Get the facts; do your own independent research.
- Check with more than one attorney.
- Check on the department's immunity.
- Comply with all laws, regulations, and safety standards even if they are voluntary.
- Perform all duties imposed by law or regulation.
- Be fair.
- Train, train, and train.
- Consult recognized experts (remember, not just one).
- Have access to a well-informed attorney.
- Document all actions in writing.
- Be proactive in proposing changes or new laws and regulations when necessary.

- Take prompt action on complaints.
- Follow up with appropriate discipline on violations of department standard operating procedures (SOPs).
- Know the professional expectations of the job (experience, training, and education).
- Know about relevant statutes, regulations, and court cases.
- Review legal and occupational publications.
- Treat all employees in the same, fair, evenhanded way.
- Keep customer service first when creating the department's goals.

Role of Government to Protect Citizens

It is the role of government to protect the consumer (citizens) in cases where it can be demonstrated that the consumer lacks the knowledge, background, or education to make a conscious judgment of the safety, competency, or adequacy of the service or product.

For example, governments typically regulate the contents and labeling of food products. Without a label, it is impossible for the consumer to know the health risks or contents of a can of vegetables without first sending it to a laboratory to be tested. The consumer has no readily available knowledge about how the vegetables were planted, grown, harvested, processed, and canned. Because of government health regulations, a consumer can reasonably assume that the contents are accurately described on the label and are acceptable for consumption. Without some assurance of the safety of a given food product, few consumers would continue to buy the product. Therefore, this regulation benefits the consumer and the producer.

Similarly, the state or federal government regulates most professional services, such as medicine, engineering, and the law. It is impossible for most consumers of these services to accurately assess the competency of the professionals who provide them. The customer would have to know the university each professional attended and if the academic standards were sufficient to meet a minimum standard of education, as well as information about their personal experience, credentials, and so on. Currently, these professions require a competency examination to prove mastery of the knowledge needed for the profession before they are granted a license to practice in their state. When consulting a doctor or seeking advice from a lawyer, the client must exercise a certain level of trust. Minimum license requirements reassure the public that the provider they are dealing with has

FIGURE 10-1 He has all the right safety and protective equipment, but is this firefighter competent in firefighting skills, knowledge, and abilities?
© Flashon Studio/Shutterstock.

the requisite knowledge and competency regarding his or her profession.

Can the same be said about FES? When a fire engine shows up at an emergency, can the typical customer determine the competency of the firefighters or the adequacy of the fire equipment? If a person suffers a serious injury, can he or she tell if the responding ambulance crew is proficient in providing advanced life support? Becoming a professional, competent firefighter or EMS provider requires extensive time in a specialized training program to master necessary skills and knowledge, but the typical citizen has no way to make an educated assessment of the competency of these providers. In most states, EMS providers must be certified by a state medical regulatory agency. In a few states, professional firefighter certification is required for providers; however, this is not universal throughout the country **FIGURE 10-1**. Also, generally most larger fire departments have comprehensive recruit firefighter programs that meet all of the requirements of professional benchmarks.

Market Failure Consequences

A market failure may prevent a safety product or service from becoming available to the general public. Typically, market failures are more common in products or services that can prevent injury, sickness, or death, as was the case in the smoke alarm example presented in the opening case study.

Seat belts are another example of a safety product that is now required by government regulation in all vehicles sold in the United States. At the time the regulation went into effect, seat belts were offered as an option, but very few car buyers chose them because they were uncomfortable and the likelihood of being involved in a car crash seemed low to the car purchaser. Thanks to the perseverance of safety groups, such as the American Automobile Association and insurance associations, the federal government issued a regulation requiring seat belts in all newly manufactured cars.

After seat belts were mandated in cars, it was assumed that drivers and passengers would use them. However, there was resistance; so states—encouraged by insurance companies, the federal government, and safety organizations—passed laws requiring the usage of seat belts. In most cases, regulations that make safety equipment a requirement must be paired with rules that require its use and provide some type of disciplinary action for noncompliance. Even with vigorous enforcement efforts, it is not uncommon to read about a death as a result of a vehicle crash where the occupants were not wearing seat belts.

Monopolies

When a product or service becomes a monopoly, the consumer loses the ability to choose a product based on a preferred provider or price. Some typical consumer monopoly products are utilities such as electricity, water, and sewer services. Private companies or the government can provide these, but they all work most efficiently when granted a monopoly regulated by government. For example, imagine that local cable television service has four separate providers competing for the public's business. Every telephone pole in the neighborhood would have four separate wires in addition to the electrical and telephone wires. Because of the heavy costs of the infrastructure, the government grants a monopoly to one provider and controls quality and price.

Monopolies also exist in the realm of FES. Emergency services would be very expensive if there was more than one provider in an area. In addition, because response time is such a critical part of emergency service, a given system can provide reduced response time by having all stations be part of the same company rather than having two or more providers servicing the same neighborhoods. Therefore, when an individual dials 911, they do not have a choice of fire departments or EMS; whoever is the local provider responds to the call. This inability to choose by itself justifies government oversight of FES to ensure the customer gets timely and competent emergency service.

Federal Regulations

At the federal level of government, Congress passes laws and the president approves them; however, Congress generally does not include specific guidance for implementing regulatory legislation. It usually defers to the government bureaucracy to formulate the rules of compliance with the new or revised law. Congress refrains from setting these "rules" for two practical reasons.

First, Congress normally lacks the expertise of the professional bureaucracy. For example, because of this, Congress passed a public law approved by President Nixon in 1970 that created OSHA. The law contained guidelines and a process for adopting regulations to "help employers and employees reduce injuries, illnesses and deaths on the job in America" (Data from OSHA, 2006). The adoption process required OSHA to seek out and heavily rely on subject matter experts from the industry being regulated. As exemplified by the following extract, the expertise provided by this process had a great impact on employees and employers in all occupations:

"Since then [1970], workplace fatalities have been cut by more than 60 percent and occupational injury and illness rates have declined 40 percent. At the same time, U.S. employment has more than doubled and now includes over 115 million workers at 7.2 million worksites." (Data from OSHA, 2006) However, during this same time period, the number of jobs in manufacturing has fallen, which may explain some of the reductions in fatalities.

Second, lawmakers typically do not like to place themselves in the middle of controversial regulations. Instead, there are two groups that often have more of a stake in the outcome of the regulatory process: employers and employees. In the case of OSHA safety regulations, employees are interested in safety at any cost, whereas their employers are more likely to be interested in making a profit by keeping costs as low as possible. Because of the conflicts that can arise with proposed regulations, a long period of time is often required for review and public comment before final adoption of a regulation.

The review process begins with the federal agency publishing notice of their proposed rule in the *Federal Register*, which can be thought of as the "government's legal newspaper" (U.S. Government Printing Office, 2019). This publication gives the public and other organizations the ability to comment on the proposal and participate in the federal decision-making process. Once a regulation is reviewed, the public can submit their comments either in writing or at a hearing. After this period, which usually lasts between 30 and 90 days, agencies then publish their final regulations and address any significant issues that were raised and discuss any changes made in response.

Although this review can take a long period of time, certain processes may help shorten this period. For example, issued on February 10, 1998 by the Executive Office of the President, Office of Management and Budget, Circular No. A-119 directs federal agencies to use voluntary consensus standards, such as the ones promulgated by the NFPA, instead of government-issued standards (Office of Management and Budget, 1998). The voluntary consensus standards are described as:

- Being open process
- Supporting a balance of interest
- Maintaining due process
- Providing an appeals process
- Working toward consensus that is a substantial agreement, but not necessarily unanimous support

However, many existing regulations are still based on requirements created by federal bureaucrats. Therefore, in the future, federal agencies can be expected to use voluntary consensus standards in their procurement and regulatory activities, except where inconsistent with federal law or otherwise impractical.

Offering Effective Input on Codes, Standards, and Regulations

Many national FES organizations and special interest groups routinely review the *Federal Register* and circulate items that may be of interest to the organization. Besides reviewing the information internally, members of these organizations are typically notified of proposed regulations of interest, and notices are placed on these organizations' web sites.

Controversial policy topics tend to attract the attention of strong advocacy groups who have political allegiances and commitments. These groups may focus on issues that are of benefit to them and their members. For example, consider the following:

In California (2020) legislation was proposed that would force insurance companies to cover homes in wildfire zones. This would force insurance companies to provide fire insurance for homes in the wildland urban interface where communities and property owners have "hardened" their homes against wildfire risk.

This seemingly simple issue has many interested parties, including the fire service. If the fire service knew all homes would be fire resistant and covered by fire insurance, there would not be a critical need to protect these buildings in a wildfire. There would not be a need to commit firefighters and equipment to operate in a clearly hazardous zone. Other interested parties could be: home developers and owners, insurance companies, and suppliers of construction materials and techniques to "harden" the buildings from fire. In some cases, new homes will not be built in wildfire areas, reducing real estate taxes, and so on.

The following quotes are from an article in the *Sacramento Bee* newspaper (February 19, 2020; https://www.sacbee.com/article240404781.html), "The bill, unveiled Tuesday, would require the state to develop standards for reducing fire risks. Insurance companies would then have to offer coverage to property owners who've met those standards and live in areas where the communities have stepped up as well. It's unclear how the industry would respond if the law gets passed. After the 1994 Northridge earthquake, insurers significantly scaled back their homeowners' coverage rather than be forced to include earthquake coverage in their policies. As the crisis intensified, the Legislature created the privately-funded California Earthquake Authority. All told, the costs of homeowners' insurance has tripled for thousands of Californians, including those living in lower-income rural areas."

In the case of property insurance, the industry is regulated by each state separately. If you have a mortgage, you will be required to have property insurance including damage by fire. If a property cannot be insured, it cannot be sold to a new owner and what would happen to existing owners is anybody's guess. Clearly that is a public policy issue that can only be achieved by government regulation.

When faced with political pressure from a special interest group and a lack of input from citizens, chief officers need to rely on their own ability to argue for policy and regulation changes. Although doing so might take a lot of courage, honesty, perseverance, and knowledge, the officer must remember that the primary advocate for the customer (citizen) is the FES administrator. Chief administrative officials of an FES organization should represent the public in any discussion of policies that affect service level and quality. These officials can represent the public's interest impartially. It is in these times and situations that the chief officer must use the best negotiating and consensus-building techniques possible to come to a fair and equitable solution.

If the chief officer decides to submit a proposal or comment concerning public policy, it is very important

that he or she researches and justifies a position thoroughly, honestly, and accurately. The federal government and the NFPA are looking for consensus from participants, advocacy groups, committee members, and public officials. The submitter of the proposal or comment must be able to convince the regulators that their input has merit. Proposals that are supported by hard data and backed by well-thought-out lines of reasoning are more likely to succeed.

In most cases, the submitter is not physically present when the committee or public officials review the proposal or comment. These members and officials only have the information that is submitted in writing. However, submitters who can attend the meeting are allowed to present their case personally (although permission to testify usually must be approved before the meeting). An alternative for submitters who have contacts with national organizations, committee members, or other participants in the rule-making process is to contact these members and voice opinions personally on the subject. Participants representing special interest groups in the rule-making process have more influence in the debate than the individual submitter.

State Regulations

When the federal government adopts a regulation, it may or may not be enforceable by state or local governments and their employees. Many FES departments do not need to comply with federal regulations because they are binding only on private sector employers; however, as explained in the chapter *Health and Safety*, when a state does adopt the federal OSHA regulations, the regulations then become enforceable for state and local government employers and employees. Many departments in nonparticipating states use the federal and NFPA safety standards, with the understanding that they are providing a safe working environment for their members and reducing the chance of liability. (Refer to the chapter, *Health and Safety*, for a complete list of all states that are federal partners.)

For the purposes of compliance with state OSHA regulations, volunteer firefighters are considered agents or contractors, not employees, and so they may be exempt. Again, this situation can differ from state to state. For example, in one state, volunteers are considered employees when it comes to workers' compensation protection. However, one OSHA regulation is an important exception: OSHA 29 CFR 1910.120, *Hazardous Waste Operations and Emergency Response*, was adopted in 1993 by the Environmental Protection Agency, therefore making the regulation a national mandate. Administrators should research and know what regulations, federal and state, apply to their workforce.

Building and Fire Codes

Most states and many local governments adopt building and fire codes that regulate the construction and maintenance of structures. These codes are adopted and enforced in a variety of ways. For example, in some states adopted codes cannot be changed locally, but in others local amendments to the codes are allowed.

Two national organizations currently promulgate building and fire codes: the NFPA and the International Code Council (ICC). The ICC is an organization that was created in 1994 by the three leading building code groups in the United States: (1) Building Officials and Code Administrators International, (2) International Conference of Building Officials, and (3) Southern Building Code Congress International. The ICC's goal is to produce one building code to be used throughout the United States.

The NFPA has also created a building code to complement NFPA standards and codes that previously proscribed only fire and electrical safety features for structures. Most states have opted to adopt the ICC codes, using the NFPA standards for reference. These codes can have a major impact on a fire department's ability to protect newer structures. For example, it is common for local fire protection regulations to require fire sprinkler protection for structures over a specified height or area, or that are in a certain use-group; however, one code may require this regulation and the other may not.

In addition, a few cities have their own separate building and fire codes; however, there has been a steady trend away from this practice. For example, in 2008, New York City decided to replace its own local codes with the ICC's International Building Code. In 2005, prior to the implementation, New York Buildings Commissioner Patricia Lancaster explained, "As it stands now, our Building Code is the most stringent set of construction regulations in the nation, yet its complexity is seen by many as an impediment to progress. The International Building Code will allow us to streamline the construction process while not sacrificing the effectiveness of these regulations in keeping our city a safe place to live, work, and build" (Data from Lancaster, 2005).

Building and fire codes are very complex and are best studied through a formal education process, such as by taking fire prevention courses at a community college **FIGURE 10-2**. In many states, inspectors and plans reviewers have to pass a rigorous certification process.

Zoning Regulations

Zoning regulations can be difficult to generalize based on their origin. One simple example of

FIGURE 10-2 Before construction starts, the developer must obtain a building permit after demonstrating compliance with the building and fire codes.
© DimiSotirov/Shutterstock.

a zoning regulation would be prohibiting a major manufacturing building in a residential area. Local governments generally use zoning to prevent non-compatible uses of buildings and land; however, there are many unincorporated areas and a few cities, such as Houston, Texas, that have no zoning regulations.

Some zoning regulations have their roots in fire prevention and were encouraged by insurance companies, and some are for aesthetics, such as the Washington, DC, regulation that no structure can be built higher than the Washington Monument or the U.S. Capitol building. This zoning ordinance has an unintended benefit for the local fire department—there are no high-rise buildings over 130 feet high to contend with in the city.

The best way to gain some understanding of zoning regulations is to review examples. The list below cites examples of zoning regulations that the FES administrator may come across:

- In residential neighborhoods, small convenience stores can be built only at major intersections.

- Before a major commercial development, such as a covered mall, is approved, potential locations must be detailed on a map of the municipality, and requirements must be specified for adequate and safe access to the property by the adjacent road network. The developer might have to pay for road improvements and traffic signals to gain zoning approval from the local government.

- In a congested downtown area, a masonry firewall is required on the property line of each building (as a result of a major conflagration in years past).

- If a new building is more than three stories high or three miles from a ladder truck station, that building has to be protected by fire sprinklers.
- Streets in new developments, such as gated retirement communities, must be built to a minimum width.

Local governments might also control the type of development that can occur in each geographic area by using land-use zoning maps, which are maps approved by the local zoning authority that visually identify different use areas. For example, zoning might be approved for a large commercial development, but with the condition that the developer donates a parcel of the property for a new school, library, or fire rescue station.

Union Contracts

In addition to federal and state regulations, union contracts have become vehicles to adopt standards and local regulations for FES employees. By mutual consent, and with the formal approval of the legislature and executive branch, these standards and regulations become legally binding for both parties. In essence, they are virtually the same as a law or ordinance.

Unions deserve special attention because they are an important part of the political landscape. In right-to-work states, unions may attempt to develop clout with elected officials (and those who influence them) because they cannot mandate members to join or pay dues, or mandate management to negotiate a binding contract for wages and benefits. (The chapter, *Introduction to Administration*, includes a complete listing of the right-to-work states.) In other states, unions might have mandatory bargaining rights **FIGURE 10-3**. A union contract acts just like a regulation and is binding on all parties.

The relationship between union members and the administration can be adversarial or confrontational. The main goals of a union are to maximize pay and benefits while ensuring the health and safety of their members. FES administrators, however, often strive to improve services at the lowest practical cost to the taxpayers. These goals can conflict.

For example, an administration might propose an 8-hour duty limit in response to data showing that shifts longer than 8 hours have an adverse effect on a worker's mental acuity. Theoretically, this proposal would ensure well-rested FES providers who would make fewer mistakes during emergency calls. However, firefighters who work 24-hour shifts often have a second job and, in many cases, they spend almost as many hours at the part-time job as at their primary

FIGURE 10-3 Unions are an important part of the political landscape.
© Yuri Arcurs/ShutterStock.

job. Therefore, a union might oppose this change because of the hardships it places on its members.

To facilitate change to existing services, the administrator should consider the following advice:

- Determine which decisions and policies are required by law or ordinance to be negotiated with the union.
- Delegate policies that do not require negotiation to a nonbinding task group made up of representatives from all interested parties.
- Understand the language in the existing labor contract, especially the concept of past practice.

If a decision is not a mandatory subject of bargaining, the chief does not technically have to secure the union's permission before implementation. However, the bargaining process requires good faith discussion with the union and a willingness to include the union's ideas and concerns in the final outcome. The union should be given a full opportunity to express all of its viewpoints.

Some jurisdictions might have a formal impasse process. If both parties cannot come to agreement, an arbitrator might be brought in to decide the issue. This process is called binding arbitration. Even if mandatory negotiations are not required by a jurisdiction, formal meetings between the administration and members are always a good idea. Regular contact improves member morale and is useful in implementing change, helping the organization provide better services to the public.

Tax Regulation

The ability to charge taxes and fees is a powerful tool; with it, governments can encourage or discourage behaviors. For example, home ownership is encouraged with the exemption of real-estate taxes and interest on the property owner's Federal and State income taxes. Products like cigarettes, on the other hand, may have high taxes on them to discourage consumption. Both of these are examples of the government using taxes to regulate behavior for what it deems as good outcomes.

Taxes are also used to regulate behaviors that affect the FES. For example, after a number of failed attempts to gain approval for a mandatory single-family dwelling sprinkler ordinance, one county adopted a reduction in local real estate taxes on new single-family dwellings with fire sprinklers. When a sprinkler system is installed in a new home, the cost is reflected in the total cost, which is then used to set the real estate tax. This temporary reduction offsets most of the cost of sprinkler installation.

At the federal level, the Fire Sprinkler Incentive Act was introduced in the House of Representatives and Senate in 2003. Previously, building owners were discouraged from investing in a sprinkler system due to the depreciation schedule of the 1986 Internal Revenue Tax Code: 39 years for a commercial building and 27.5 years for a residential structure. However, the recent Tax Cuts and Jobs Act includes incentives that make it more affordable for small business owners to install fire sprinklers. According to U.S. Fire Administration, "Fire sprinkler systems are the only tool that can reduce the spread of fire, heat and smoke in a building and contain the fire until the fire department arrives. This provides valuable time for employees and customers to escape a fire." (United States Fire Administration, retrieved from https://www.usfa.fema.gov/current_events/061218.html) "This incentive makes it possible for restaurants, retail outlets, night clubs and other small businesses to install, retrofit or upgrade their current fire suppression systems. Also, small businesses can deduct the interest from any loan they receive for this type of renovation." (United States Fire Administration, retrieved from https://www.usfa.fema.gov/current_events/061218.html)

Hiring and Personnel Regulations

There are numerous rules, regulations, and laws that pertain to hiring and personnel actions for employees, such as the Fair Labor Standards Act (FLSA), Equal Employment Opportunity Commission (EEOC), Americans with Disabilities Act (ADA), affirmative action, and the Family and Medical Leave Act. These regulations, laws, and acts are discussed in the chapter *Human Resources Management*.

OSHA Regulations

In 1970, Congress created OSHA to assure the safety and health of most employees in the United States by setting and enforcing safety standards and providing training, outreach, education, and assistance. State and local employees are only covered in state plan jurisdictions **TABLE 10-1**. This section discusses several OSHA regulations relevant to FES with which the administrator should be familiar.

OSHA 29 CFR 1910.146: Permit-Required Confined Spaces

This federal regulation covers employees who enter areas with limited or restricted means for entry and exit, such as trenches, pipes, tanks, and vaults. Confined spaces have the potential to trap a person and may

TABLE 10-1 OSHA State Plan Jurisdictions

Alaska	New Mexico
Arizona	New York
California	North Carolina
Connecticut	Oregon
Hawaii	Puerto Rico
Illinois	South Carolina
Indiana	Tennessee
Iowa	Utah
Kentucky	Vermont
Maryland	Virgin Islands
Michigan	Virginia
Minnesota	Washington
Nevada	Wyoming
New Jersey	

have oxygen-deficient atmospheres. When the fire department receives a 9-1-1 call that someone has been overcome and is unconscious in a confined space, the firefighters must enter the confined space to rescue the victim. Firefighters have died when entering these atmospheres without self-contained breathing apparatus. The OSHA regulation requires training, proper equipment, and written procedures for all employees operating in these hazardous areas.

OSHA 29 CFR 1910.134: Respiratory Protection

The following are specific requirements for this regulation, as noted by the International Association of Fire Chiefs and the International Association of Fire Fighters (1998):

- Personnel must use a self-contained breathing apparatus when fighting a structural fire.
- A minimum of two firefighters must work as a team inside the structure, maintaining contact with one another at all times.
- A minimum of two firefighters must be on standby outside the structure. Any tasks performed by these firefighters must not interfere with the responsibility to account for the firefighters inside the structure.

Among other requirements for training and equipment, this OSHA regulation—typically abbreviated as "two in/two out"—requires that four firefighters be on the scene before an interior structural fire can be attacked while using a self-contained breathing apparatus. Fire departments generally address this regulation by adopting an SOP. Many municipalities are increasing their staffing on engines to four firefighters to comply with this federal regulation. This requirement for four firefighters for safe operations is also found in NFPA 1500 and 1710. In 2010, fireground field experiments conducted by the National Institute of Standards and Technology (NIST) also concluded that four firefighters is the best number of firefighters for safe and effective fireground operations.

If a department staffs its engine companies with fewer than four persons, the first crew on the scene has to wait until another company (or volunteer/call firefighters) arrives at the structural fire to make up the minimum of four firefighters. In these situations, the only opportunity to start extinguishment is using an outside attack. **FIGURE 10-4** This practice can be successful if the room where most of the fire is located has a window or door to the outside—a layout common in residential occupancies. For the safety of any occupants who remain inside, the attack should be done

FIGURE 10-4 Outside fire attacks can be very effective when an interior attack would be dangerous or there are fewer than four firefighters on the scene.
© Volodymyr Baleha/Shutterstock.

with a solid stream, which tends not to push the fire, heat, and smoke into other areas of the structure where victims may still be present. This is especially true when using a wide pattern fog nozzle according to NIST Research #1938 (2016). In fact, what is called hydraulic ventilation is the technique of using a wide pattern fog nozzle from inside the building to push heat and smoke out a window or door and is very effective.

A Summer 2013 article by Underwriters Laboratories had the following observations on the use of outside fire attacks: "Firefighters are being challenged by different fireground hazards due to today's more open floor plans and the use of synthetic materials in furniture and building products. These changes have made structure fires more deadly than ever before and call into question traditional firefighting tactics. Innovative fire attack tactics matter because, although they may go against traditional practices, they represent a more effective way to make the fireground safer for both building occupants and firefighters." (Innovating fire attack tactics, summer 2013, UL.COM/NEW-SCIENCE, retrieved from https://ulfirefightersafety .org/research-projects/impact-of-fire-attack-on -firefighter-safety-and-occupant-survival.html).

OSHA 29 CFR 1910.1030: Occupational Exposure to Bloodborne Pathogens

This regulation requires the employer to reduce or eliminate employee exposure to diseases that can be transmitted by bloodborne pathogens or other potentially infectious materials. Any agency involved in patient care is required to have a comprehensive education and control program for personnel who may become exposed to bloodborne pathogens or other potentially infectious materials. This includes providing all personal protection equipment necessary to protect personnel from bloodborne pathogens and offering hepatitis B vaccinations at no cost to the employee. These requirements are typically covered in an FES agency SOP.

Standard precautions generally include gloves; approved eye protection; and, in cases where it is warranted, a disposable gown. It is also prudent to have waterless antibacterial hand cleaners available until crews can reach soap and water. Repeated handwashing is the best defense against disease transmission in general.

OSHA 29 CFR 1910.120: Hazardous Waste Operations and Emergency Response

This regulation is based on a 1986 draft of NFPA 472, *Standard for Competence of Responders to Hazardous Materials/Weapons of Mass Destruction Incidents*. It is generally a good idea to base training and certification on the latest edition of NFPA standards, because OSHA has not updated its hazardous materials regulation since 1989. OSHA recognizes and accepts these more recent standards as equivalent to the original rule.

This regulation was also adopted by the Environmental Protection Agency; therefore, it covers all persons who would have contact with hazardous materials in all states, including those not covered by OSHA state plans. All firefighters and other emergency responders who respond to hazardous materials incidents, such as police and EMS personnel, must be trained to a minimum level of competency.

For firefighters, this minimum competency is the operations level and is usually incorporated in recruit firefighter training programs. First responders, such as police officers and EMS support staff who need to protect themselves but do not take emergency action during a hazardous materials incident are required to meet the awareness level, which is the lowest level of training and competency.

To gain a better understanding of this regulation and comply with its mandates, the NFPA *Hazardous Materials Response Handbook* provides an in-depth explanation of this standard and several useful appendix articles dealing with issues of organizing a hazardous materials response team.

OSHA 29 CFR 1910.156: Fire Brigades

This OSHA regulation, first adopted in 1980, has been largely ignored by the public fire service. Many of its requirements are nonspecific and therefore a challenge to interpret and implement. Also, the title *Fire Brigades* seems to confuse readers, making them believe it is not applicable to public fire departments; however, "fire brigade" is just another name for "fire department," although used more commonly in other countries and industries.

Items required by this safety regulation include:

- An organizational statement
- Training and education similar to fire training offered in such states as Maryland, Georgia, and Washington
- In-service training
- Protective clothing and respiratory protection at no cost to employees

In the appendix, the regulation also goes on to suggest:

- Preplanning
- A physical fitness program
- Fire officer training and education

Most of these same items can be found in NFPA 1500, *Standard on Fire Department Occupational Safety, Health, and Wellness Program*. NFPA 1500 contains more specific language regarding requirements for safety policies, procedures, and equipment, and therefore many find it easier to comply with and enforce its requirements. OSHA 29 CFR 1910.156: Fire Brigades is also sometimes used in conjunction with the General Duty Clause that is explained in the next section.

OSHA's General Duty Clause

This clause requires each employer to furnish a place of employment that is safe from hazards that are likely to cause injury, sickness, or death. OSHA and court precedent have shown that in the absence of specific or up-to-date OSHA regulations, a national consensus standard, such as NFPA 1500, can be used to assess compliance. In state plan jurisdictions, numerous violations for failure to comply with nationally recognized safety standards have been issued using the General Duty Clause.

For example, in the early 1990s firefighters in a large suburban county filed a complaint with their state OSHA agency to stop the practice of firefighters riding on the back step of fire engines. Even though there is no specific requirement in the federal OSHA Fire Brigades regulation, the state agency cited the General Duty Clause along with NFPA 1500 when it filed a violation against the county. The practice was halted immediately and the proposed fine was eliminated.

Chief officers should always keep in mind the General Duty Clause. At a minimum, departments should consider complete compliance with NFPA 1500 for the FES agency. Otherwise, the chief and the organization could be open to civil liability or state OSHA citations.

NFPA Codes and Standards

The NFPA was formed in 1896 by a group of insurance companies who met to address inconsistencies in the design and installation of fire sprinkler systems. According to the NFPA, "At that time there were nine different standards for piping size and sprinkler spacing, and these business people realized that unless these discrepancies were resolved, the reliability of these sprinkler systems would be compromised" (Courtesy of National Fire Protection Association; NFPA, 2019). The group created a standard for the uniform installation of sprinkler systems, and the NFPA standard-making process began. The organization cites their mission as follows: "to reduce the burden of fire on the quality of life by advocating scientifically based consensus codes and standards, research, and education for fire and related safety issues" (Courtesy of National Fire Protection Association; NFPA, 2019).

The NFPA home page contains several links to information under the heading "Codes and Standards." There is also a link entitled "Documents Accepting Public Input" that allows the review of a proposed standard or changes to an existing standard, so comments regarding these proposals can be prepared and submitted **FIGURE 10-5**. The electronic version of *NFPA News* also contains a list of codes and standards that are in the process of being revised.

NFPA 1500, *Standard on Fire Department Occupational Safety, Health, and Wellness Program*

NFPA 1500 was first issued as a standard in 1987. At that time, many in the FES industry and city or county administrators proclaimed that it would be too expensive and impractical to implement. Some even believed that the standard would be the end of many smaller departments, which would collapse under the substantial economic burden that came from safe operations and equipment. Although most fire departments have not yet completely complied with NFPA 1500, many have made substantial changes to provide a safer work environment for firefighters. As with any change in the FES field, safety improvements under NFPA 1500 have been evolutionary rather than revolutionary.

FIGURE 10-5 An FES administrator should keep up to date with publications such as *NFPA News*.
Courtesy of National Fire Protection Association (NFPA) 2020.

NFPA 1500 can be viewed as an umbrella document that adopts by reference many other NFPA standards, such as:

- NFPA 1001, *Standard for Fire Fighter Professional Qualifications*
- NFPA 1021, *Standard for Fire Officer Professional Qualifications*
- NFPA 1403, *Standard on Live Fire Training Evolutions*
- NFPA 1582, *Standard on Comprehensive Occupational Medical Program for Fire Departments*
- NFPA 1901, *Standards for Automotive Fire Apparatus*
- NFPA 1971, *Standard on Protective Ensembles for Structural Fire Fighting and Proximity Fire Fighting*
- NFPA 1981, *Standard on Open-Circuit Self-Contained Breathing Apparatus (SCBA) for Emergency Services*

These standards, in addition to providing for firefighters' safety, also provide firefighters with the training, tools, and protective clothing needed to operate effectively and professionally in mitigating dangerous uncontrolled fires and other emergencies that threaten to destroy property and injure or kill occupants.

One aspect of this standard that still has not been defined clearly is physical fitness. Although firefighting requires high levels of strength and endurance, the NFPA committee that was assigned the responsibility for defining this standard—the Fire Service Occupational Safety and Health Committee—has not come to a consensus on specific levels and definitions of minimum levels of physical fitness.

However, the 2018 edition of NFPA 1582 outlines a comprehensive fitness evaluation in Chapter 8, Annual Occupational Fitness Evaluation for Members. The evaluation describes itself as nonpunitive. To understand why this would be an ineffective policy, as described in the chapter on *Health and Safety*, consider an evaluation for alcohol for a driver after a major vehicular crash or a traffic stop by law enforcement for dangerous driving behavior. If the laws about drunk driving were similarly nonpunitive, there would be no pass/fail percentage such as a blood alcohol concentration of 0.08% and no consequences for failing the test. Removing the penalty would likely result in an increase in deaths, injuries, and crashes by drunk drivers.

If this fitness evaluation is implemented by your department, be sure to adopt measurable pass/fail levels and enforce compliance. Anything less will result in failure of the policy.

NFPA 1710, *Standard for the Organization and Deployment of Fire Suppression Operations, Emergency Medical Operations, and Special Operations to the Public by Career Fire Departments*

This standard, which was adopted in 2001 and has been revised several times, requires minimum levels for response times, staffing, and training. This standard probably has more impact on FES than any other standard, with the exception of NFPA 1500. The word "career," as used in NFPA 1710, does not necessarily mean that a department is fully paid, although in most cases members are salaried. Career fire departments include departments that rely on firefighters who are on-duty at the fire station. Obviously, completely paid departments belong in this definition, but many combination departments can also use this standard, especially if they are in the process of increasing their on-duty personnel. However, this standard can also cover partially

and fully volunteer organizations that schedule their firefighters to cover shifts, although this is not common.

NFPA 1710's definition of "member" allows volunteers who are on-duty at the station to count toward the minimum number of firefighters on fire companies and the total first alarm assignment. In theory, it is possible for an all-volunteer company that assigns its members to specific duty times to comply with NFPA 1710. The key is that the firefighters and officer of the company have to be on-duty in the station and trained to minimum levels of competency in accordance with the NFPA Professional Qualifications Standards.

There are many legal implications of not complying with NFPA 1710. However, even if a department's liability exposure is limited, it is still a good idea to use the NFPA standard as a benchmark for planning for public fire protection in a community. The citizens of a community deserve to know what level of public fire protection they are receiving.

Four-Person Companies

The preferred method of complying with NFPA 1710 is to provide on-duty, in-station staffing of three firefighters and one officer per company. The NFPA 1710 document contains many references that support this requirement, and chiefs and elected or appointed officials are encouraged to review them if they do not understand. One of the main reasons why a four-member company works best is that each firefighter is prepared to perform an assigned role. Firefighters who arrive individually have to fill a particular role that might be unfamiliar to them. Even with well-trained firefighters, this can result in an uncoordinated effort.

Response Time Requirements

The NFPA 1710 response time requirements for a structural fire state that the first due engine must arrive within 4 minutes (an additional 80 seconds are allowed for turnout) or the full first alarm assignment must be made within 8 minutes (an additional 80 seconds are again allowed for turnout). For record-keeping purposes, the response can be considered successful when either time target is met. These response times were set with the understanding that engine companies might be out of position when notified to respond, such as during preincident planning, fire prevention education, physical fitness, or training exercises. It is not uncommon for engine companies to be out of the station but in their first due areas. In addition, high call volumes due to EMS calls may find the company unavailable, servicing a medical call. In some jurisdictions with areas that are undeveloped or that have low population densities (Rural), the 4-minute response time is not justified.

Response time performance should be measured at the 90% level (i.e., 90% of responses are made in less than or at the maximum time limit). This flexibility allows for exceptions because of the nature of a situation. For example, when a company from a distant station answers a call in the first due area of a company that is already responding to or working at an emergency incident, it is nearly impossible, or at least unreasonably expensive, to provide anything close to 100% compliance.

The number of times that a company is busy when a second call comes in is in direct proportion to call volumes (i.e., the greater the number of calls per year, the greater the amount of busy time and percentage of calls not meeting the maximum response time standard). Circumstances vary but, for the average company, volume in the range of 2,000 to 3,000 calls annually can frequently lead to situations in which the maximum response time is not met. The Insurance Service Office Municipal Fire Grading Schedule once contained the requirement that a second company would be needed if more than 2,500 calls were responded to annually.

As exemplified by the following list, departmental policies and SOPs can help solve—or exacerbate—the increasingly common problem of excessive responses:

- Dispatch policy can sometimes lead to responses to alarms that may not be necessary. For example, some departments send a full first-alarm assignment to an automatic alarm, whereas others send only one engine.

- A vehicle crash may generate the automatic response of a paramedic unit, an engine, and a heavy rescue squad in one jurisdiction; in another jurisdiction, only the paramedic unit responds unless additional information indicates a serious accident with occupants trapped.

- Other than the required companies that must be dispatched to a structural fire (totaling 14 firefighters), chief officers should review all dispatches that send more than one unit and calculate the percentage of times each unit was actually used by call type.

- Adding a second unit in the same station cuts the call volume substantially. For maximum efficiency, only one unit should be dispatched on any fire call, leaving the other unit (engine) to answer a second call in the first due area within the expected 4 minutes.

Modifications to the Standard

The standard originally adopted in 2001 has been updated several times: 2004, 2010, 2016, and most recently 2020. In summary, these changes go beyond the original response to a single-family dwelling unit that is still the basic deployment requirements. Two new factors must be considered in determining minimum fire department response to fires: population density and occupancy.

There are now four categories of population density: Rural, Suburban, Urban, and Dense Urban. Each has its own specific requirement for staffing and total number of firefighters. In addition, the type of occupancy can be the determining factor using either: Rural Single-Family, Open-Air Strip Shopping Center, Apartment (garden style usually less than four stories in height), and High-Rise. The relationships between these diverse categories are complex and lend themselves to an intense study by using a community risk assessment.

One example of the complexities and judgments that may be needed might be a high-rise apartment building (fully protected by automatic fire sprinklers) in a First-Due Response Zone that contains mainly single-family dwellings and some small retail stores. Would this one building dictate the staffing and response to this entire area? There are no absolutes in using this standard other than the basic 4-minute response time with four firefighters. Chapter 12, *Public Policy Analysis* has an overview of the process of making these sometimes intuitive subjective decisions on staffing.

NFPA 1710 will continue to have a distinct impact on fire and EMS services in the future. It is a planning document that can help identify and quantify weaknesses in fire protection and EMS. Quantifiable comparisons help citizens and both elected and appointed officials learn which parts of their communities have adequate fire and EMS coverage and which parts are beyond response time standards.

NFPA 1720, *Standard for the Organization and Deployment of Fire Suppression Operations, Emergency Medical Operations, and Special Operations to the Public by Volunteer Fire Departments*

This standard, which was adopted in 2001, has been revised several times and requires minimum levels for response times, staffing, and training. Because it adopts by reference many of the same NFPA standards as NFPA 1710, requirements for safety, training, and equipment are identical.

The major difference is in the response time requirements and the practice of using fire companies with a minimum of four firefighters per company. Where NFPA 1710 uses a 4-minute response time for the first due engine company and 8 minutes for the entire first alarm companies, the NFPA 1720 use requirement is divided by population density. For example, 1720 uses a minimum of 9 minutes for Urban (density of greater than 1,000 people per square mile) areas for 15 members to arrive on the scene. There is no mention of travel time for the first due engine, only the entire first alarm response.

In volunteer fire departments, it is the typical practice to not require any members to be in the fire station, but to respond to either the station or the emergency location when alerted. There are many different response scenarios such as one volunteer firefighter drives the fire engine to the scene and the others travel by personal vehicle to the scene to staff the fire company. Whereas the NFPA 1710 fire company has each of the four members of the company in an assigned position and function when they are dispatched, the volunteer company assigns the firefighters a position and function when they arrive on the scene. Some argue that the complete four person fire company from the time of dispatch provides a more coordinated and safer operation on the scene.

National Highway Traffic Safety Administration

In the 1960s, it seemed clear that the quality of prehospital care was an important determinant of survival from sudden injury and preventable deaths from highway trauma. It was for this reasoning that federal funding for the standardized training of ambulance attendants and purchasing ambulances was provided through the Highway Safety Act of 1966. In 1969, the Highway Safety Bureau came into existence, funding the development of the National Standard Curricula, which set the precedent for the way EMS education would be standardized for the next 50 years.

In 1970, the Highway Safety Bureau transformed into the National Highway Traffic Safety Administration (NHTSA), with the goal of reducing the number of vehicle crashes, injuries, and deaths. The NHTSA collected data, investigated safety defects, reduced the threat of drunk drivers, promoted the use of safety belts and airbags, and provided consumer information on motor vehicle safety topics. It has also been instrumental in implementing a nationwide 9-1-1 emergency reporting system; setting certification standards for emergency medical responders, emergency medical technicians, and paramedics; and setting standards for medevac helicopters and shock trauma centers **FIGURE 10-6**.

FIGURE 10-6 NHTSA provides safety regulations for medevac aircraft.
© Karl R. Martin/ShutterStock.

Legal and Court Issues

Legal issues and opinions are in a continual state of change as a result of new legislation and legal decisions from numerous judges throughout the country. When reviewing decisions from a court, keep in mind the hierarchy of the judicial system in this country. Unless the Supreme Court of the United States has made a determination about a legal issue, the issue can always be appealed for another opinion. For example, a lawyer may explain that a circuit court awarded a large settlement to an employee as a result of some action by an FES organization. If the administration is using this court ruling to determine if a similar policy at another department is illegal or defensible in court, the circuit court's decision should have less of an influence on the administration's actions than a decision from a higher appeals court. Out-of-court settlements have even less influence.

In addition, each lawyer, attorney, or counselor may have a different opinion about a particular law or regulation. A critical emergency service to the public can be compromised if a department is persuaded by legal advice to reduce professional standards for its members. Legal actions can have major and undesirable consequences for the department. The chief officer should be wary of any attorney who wants to settle quickly and not take the issue to court (Edwards, 2010). Public service is a very serious responsibility that should never be taken lightly by the administration. In some situations, it is wise to seek a second opinion and to do independent research. Contact peers and check with professional organizations. Look for articles that may have been written about the subject. Do not rely on one source. Many individuals have their own agendas.

Chief officers can take several measures to protect the department and its members from legal problems.

When applicable, follow the requirements of an appropriate professional standard or regulation (e.g., NFPA or OSHA). Ensure that all the department's policies treat all people equally. Demand that any legal advice be supported by case histories from courts at the appeals level or higher. Ask the lawyer to present all sides of the argument, not just the legal precedent that supports the lawyer's opinion or the department's policy.

The Court System

The better the FES administration understands the court system and the legal process, the better it can protect the interests of its organization and the public it serves.

Federal and state courts have three separate levels: (1) the district court, (2) the appellate court, and (3) the Supreme Court **FIGURE 10-7**. Judges in the court systems interpret existing laws and regulations to come to their decisions. If a decision made in a district court is appealed, the case moves on to the appellate court. The decision made in the appellate court then becomes known as a precedent, or common law, and may be used by other judges in similar cases. For the most part, lower appellate court decisions can set precedents only in the district or state in which they are issued; it is possible for appeals courts to make decisions that are in conflict with appeals courts in other districts.

The U.S. Supreme Court, whose decisions are binding nationwide, has final authority (Figure 10-7). Its decisions become precedent for the entire nation. If the Supreme Court refuses to hear a case, its refusal indicates that it supports the lower court's decision. If the

CHIEF OFFICER TIP

Legal Counsel

Chief officers should not wait for a legal crisis to meet the lawyers representing their department. By developing and maintaining a relationship with their lawyers, chief officers can get ahead of any legal crises that may occur. Regular conversations allow the chief officer and lawyer to familiarize each other with information on legal issues and trends, court cases, and other organizational developments.

Data from Edwards, S. T. 2010. *Fire service personnel management*. 3rd ed. Upper Saddle River, NJ: Brady/Prentice Hall Health.

Supreme Court makes a decision based on the U.S. Constitution, this decision can be changed only by changing the Constitution, which is difficult, if not impossible, to do. Although there is a process for adding to and changing the Constitution, it is used only in rare cases because it requires a three-quarter super majority vote by Congress and three-quarters of the States approval.

Attorneys

As is true for many areas of management, when it comes to legal issues, it is a good idea to seek help as soon as possible. The chief officer also needs to make sure to get advice from an attorney who has the appropriate background and best interests of the FES organization.

Attorneys might not understand the internal workings of an emergency organization. Chief officers should educate them, starting with the basics. For example, the FES organization may have a strict policy that any new applicant may not have any record indicating that he or she cannot be trusted, such as a criminal record, excessive driving citations, or poor credit. Chief officers would need to explain to the attorney that there are many occasions when members are in private homes during an emergency call, and the organization cannot tolerate theft of private property or any other criminal activity.

Assure yourself that the attorney and elected officials in a case involving the department that may be lost at the district court level will support an appeal. Judges may make decisions based on their personal beliefs about the law or the specific circumstances. This is known as "legislating from the bench." Do not compromise when the issue is very important to the organization and its ability to provide quality service to the public. Always be prepared to appeal a misguided or inaccurate ruling. Make sure the other side is aware of the department's resolve. Showing this resolve can sometimes keep the department and the chief officer out of court.

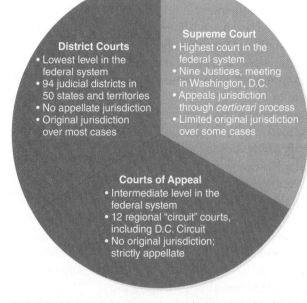

FIGURE 10-7 U.S. Supreme Court decisions are binding nationwide.

If the organization or the municipality is too small to have a staff attorney, or if the attorney does not have enough time to provide quality service, ask the municipal administrator or mayor to select an outside attorney. This is especially helpful if the loss of the case will have a serious adverse impact on the organization. If the present attorney does not seem to be meeting the department's needs, the administrator should insist on changing attorneys.

Administrative Rule Making

Administrative rule making allows the public administrator to function as a lawmaker. As public administration expert Gerald Garvey states, "The opportunity for administrators to function, in effect, as lawmakers often arises because the legislators themselves may lack the interest, the information, or the political courage to settle every last provision of a new public program" (Garvey, 1996, p. 138). For example, when the Florida legislature adopted a state-wide minimum building and fire prevention code in 2001, it did not specify the exact model building code to adopt but instead delegated that authority to the Florida Building Code Commission, which was made up of government representatives, subject matter experts, and stakeholders. The Florida State Fire Marshal was then given the authority to adopt, modify, revise, interpret, and maintain the Fire Prevention Code, which subsequently was based on NFPA 1, *Fire Code*, and NFPA 101, *Life Safety Code.*

Garvey goes on to explain, "The rule maker's role—one of the most important in public administration—requires administrators to function as lawmakers, or at least in a quasi-legislative mode. That mode casts appointive officials in a role which strict democratic theory once reserved for elected representatives sitting in formal legislative bodies" (Garvey, 1996, p. 139). These administrative rules are sometimes issued under the signature of the head administrative official, such as the secretary of labor or the state fire marshal. However, a special team of experts within an agency or a task force of stakeholders generally is assigned to see rule making through from start to finish. This team then provides their recommendations to the administrative official. The FES administrator must be aware of these officials including the committees and their agendas. Ignorance is no excuse for not attempting to influence these officials. Generally, public input is requested either in writing or through verbal testimony at scheduled meetings throughout the jurisdiction. The FES administrator's participation and influence can many times make a big difference in the outcome.

Many meetings and numerous hours are spent in consultation and negotiation with individuals and representatives of groups whose members will be affected by the rule. Public hearings are held, in theory, to solicit the input of all concerned parties. However, special interest groups, such as big business or labor groups, may concentrate their influence during the rule-making process, and their efforts sometimes allow them to have more of a voice than the public. Although appointed officials should be above local politics and should use system-wide planning to approach the problems of interdependent communities, it is not uncommon for vote-conscious politicians to be influenced by prominent business leaders, voters, and special interest groups in their election districts.

For this reason, many controversial regulations are delegated to the administrative rule-making process. This process goes a long way toward moderating special interest influence and leveling the playing field because, in most cases, it is open to public input and scrutiny. Still, the chief officer must remain cognizant of the possibility that special interest groups want to advance their own agendas and should, therefore, advocate for all those affected by the rule. In the FES business, there is a slow and steady movement away from the excessive political influence from special interest groups and elected officials. Instead, progress is driven by national standards, mutual and automatic aid agreements, consolidations, and college-level education for FES administrators.

Wrap-Up

CHAPTER SUMMARY

- How do government regulations, laws, and courts affect FES?
- Understanding federal and state regulations
- How do union contracts affect FES administrations?
- Comprehending hiring and personnel regulations and laws.
- How do safety standards such as OSHA and NFPA affect FES?
- Finally, an overview of legal and court impacts on FES

KEY TERMS

Building and fire codes Government policies that regulate the construction and maintenance of structures.

Community risk assessment Community risk assessment is the identification of fire hazards and planning an appropriate suppression response force to mitigate medical and other emergencies. This all-hazards approach provides the opportunity to have a much bigger impact on life safety and property loss.

Environmental Protection Agency The mission of the U.S. agency, EPA, is to protect human health and the environment.

Federal Register The U.S. government's legal newspaper.

International Code Council (ICC) An organization created in 1994 by the three leading building code groups in the United States to produce one building code to be used throughout the country.

Liability A liability is something a person or company owes, usually a sum of money.

Mandate To officially require (something). To make (something) mandatory.

Market failure This occurs when there is an inefficient allocation of resources in a free market. Market failure can occur due to a variety of reasons, such as monopoly (higher prices and less output), negative externalities (overconsumed and costs to third party), and public goods (usually not provided in a free market).

Monopoly Exclusive ownership through legal privilege, command of supply, or concerted action of a product or service.

Regulation A rule or order issued by an executive authority or regulatory agency of a government and having the force of law.

Special interest group A group with a shared interest in advancing a specific area of knowledge, learning, political view, financial interest, or technology where members cooperate to affect or to produce justifications and solutions to support their point of view.

Supreme Court The highest court in the federal judiciary of the United States of America. It has ultimate (and largely discretionary) appellate jurisdiction over all federal and state court cases that involve a point of federal law.

Zoning Regulations created to prevent noncompatible uses of buildings and land.

REVIEW QUESTIONS

1. Do you believe that state and federal governments should be more or less involved in local FES organizations and operations than they are today? Why? Provide at least three specific examples to support your opinion.

2. Compare and contrast the administrator's role with respect to special interest groups and unions. What considerations are the same? Different? Do you think these concerns change depending on the size or type of FES organization?

3. For national catastrophic incidents is it a good idea to expect mutual aid resources provided by local governments to meet skills, education, and national standards for competency and safety?

4. How does your local FES organization measure up to OSHA and NFPA training and safety standards and regulations? Provide two specific examples.

5. This chapter provides a general list of advice for the FES administrator at the beginning of the chapter. If you had to add a bullet to this list based on your own experience, what advice would you give?

CHAPTER ACTIVITY #1: NFPA COMPLIANCE

Using the chapters *Training and Education*, and *Health and Safety*, complete the NFPA 1500 Worksheet found in Appendix B of the standard. For this activity, it is not necessary to complete all columns; just complete the columns marked "Compliance," "Partial Compliance," and "Compliance with Administrative Action." Contact an official representative of a department who will help provide copies of written official documents supporting compliance, including SOPs and policy directives.

Remember, for an accurate understanding of a section or paragraph, you may need to do some research, starting with a review of the NFPA handbook. If you are not convinced that a requirement has been explained accurately, contact another knowledgeable person for an opinion. An e-mail to an NFPA staff person is always a good option (be sure you mention your title and organization).

CHAPTER ACTIVITY #2: LEGAL ADVICE

Find an article from an EMS periodical that discusses a recent appeals or Supreme Court decision. Then find two more articles that discuss the same issue either before or after the court's decision. Summarize the articles, comparing and contrasting the sides being taken. After you think you have a good understanding of the issues and the court decision, find an attorney who will provide additional legal advice regarding the issue. (Explain that it is a school project. Many attorneys will give a student a limited amount of time without professional charges.) After writing down a synopsis of the attorney's advice, express your own personal thoughts about the case. Limit the answer to no more than two pages.

REFERENCES

Edwards, S. T. 2010. *Fire service personnel management* (3rd ed.). Upper Saddle River, NJ: Brady/Prentice Hall Health.

Garvey, G. 1996. *Public administration: The profession and the practice*. New York, NY: St. Martin's Press.

International Association of Fire Fighters, International Association of Fire Chiefs. 1998. *IAFF/IAFC two in/two out questions and answers*. Retrieved from http://cms5.revize.com/revize/satellitebeach/City%20Boards/Citizen%20Budget%20Committee/2018%20Resources/US%20Dept%20of%20Labor%20OSHA%20Firefighers%20Two-in%20Two-out%20Regulation.pdf

Johnson, R. 2001. Lines of duty: as blazes get fewer, fire fighters take on new emergency roles—Cities complain about costs, but safety group's plans may help fuel the trend. *Wall Street Journal*. Retrieved from https://www.wsj.com/articles/SB981497980155540530

Lancaster, P. J. 2005. Press release. New York City Department of Buildings.

National Fire Protection Association. 2016. *1710, Standard for the organization and deployment of fire suppression operations,* *emergency medical operations, and special operations to the public by career fire departments*. Quincy, MA: National Fire Protection Association.

National Fire Protection Association. 2019. *History of the standards development process*. Retrieved from https://www.nfpa.org/Codes-and-Standards/Standards-development-process/How-the-process-works/History-of-standards-development

Occupational Safety and Health Administration. 2006. *All About OSHA*. Retrieved from https://www.osha.gov/Publications/about-osha/3302-06N-2006-English.html#Introduction

Office of Management and Budget. 1998. *Circular No. A-119, Federal participation in the development and use of voluntary consensus standards and in conformity assessment activities*. Retrieved from https://www.whitehouse.gov/sites/whitehouse.gov/files/omb/inforeg/inforeg/revised_circular_a-119_as_of_1_22.pdf

U.S. Government Printing Office. 2019. *Your one stop site to comment on federal regulations*. Retrieved from https://www.regulations.gov/

CHAPTER **11**

Ethics

LEARNING OBJECTIVES

After studying this chapter, the student will be able to:

- Understand the ethical justification for government oversight of the public fire and emergency services (FES) (p 207).
- Understand the creation, effect, and importance of a professional organization's code of ethics (pp 207–208).
- Identify commonly cited justifications for lying (pp 209–211).
- Understand the impact that lying and other ethical issues have on fire and emergency services (p 212).
- Examine the role of the administrator as a caretaker of public trust (pp 212–214).
- Examine how ethics affect cost-benefit analysis in the FES (pp 214–215).

Case Study

An Example of a Tough Ethical Choice

During his first week as fire chief for the Prince George's County Maryland Fire Department, now retired Chief Steven Edwards faced a serious ethical dilemma that could have cost him his job. A citizen group in the county was holding a fundraising event for the county executive. The chief learned that the event, which was already advertised and had sold hundreds of tickets, was being held in a vacant store that had significant fire code violations. He contacted the event planners, but they were very upset and did not want to cancel the event. Still, the chief believed he had no choice and contacted the county executive to explain that the event could not be held in the building. To his surprise, the county executive thanked the chief for presenting him with this information and wound up moving the event to a tent close to the original site. In this situation, the chief *and* the county executive displayed courage and personal integrity, making a tough decision in the name of ethics (Edwards, 2010).

Data from Edwards, S. T. 2010. *Fire Service Personnel Management*. 3rd ed. Upper Saddle River, NJ: Brady/Prentice Hall Health.

Discussion Questions

1. There is nothing more frightening to a chief administrator than to run afoul of a political boss. Would you have another suggestion on how to handle this situation?

2. Imagine you were this fire chief and you did nothing about these fire code violations. Would you just pretend it did not happen? Would your lack of being ethical by not enforcing the fire code be spread around the department's personnel?

3. Your personal ethics can and do influence your behavior in your public official position. What would you do about the following ethical issues:

   a. You are in line at the grocery store and you notice the person in front of you drop a large quantity of money. Would you keep it or give it back to the person who dropped it?

   b. As your groceries are being tabulated, you notice that an item was missed and will now be free for you. Because this is a large corporation, you say nothing. Is this ethical?

Ethical Behavior

The *American Heritage Dictionary* defines ethics as a "principle of right or good conduct" or "a system of moral principles or values" (*American Heritage Dictionary*, 2005). Ethical behavior is guided by that system of moral principles. Although using one's ethics to make appropriate moral decisions might seem straightforward, increasingly complex situations and ever-changing policies can complicate decision-making. In addition, most people have a set of values derived from family, friends, experiences, and religious beliefs. Some believe that only those actions that are illegal, such as robbing a bank, are wrong. However, illegal actions are only a subset of the larger category of potentially unethical actions.

Some experts have even argued that most of our likes, dislikes, and habits are established in or before kindergarten. Robert Fulghum in his book, *All I Really Need to Know I Learned in Kindergarten* (2003) lists the following examples of the things learned early in life: share everything, play fair, don't hit people, put things back where you found them, don't take things that aren't yours, say you're SORRY when you HURT somebody, warm cookies and cold milk are good for you, and take a nap every afternoon. Not everyone has the same experiences or cultural background as Fulghum describes. So this chapter's discussion gives the reader a way to judge ethical choices without the normal prejudices most of us have internalized from past experiences. Unethical behavior often involves some form of lying, such as deliberately presenting a false statement or action as true or hiding factual information with the intent to deceive. This can be a difficult issue to discuss and study because many people hate being called a liar and either rationalize their behavior or deny their actions are lies. No one is perfect and everyone has made mistakes. However, it is hoped that this chapter offers a new way to govern future decisions and behavior by administrators. The trick to success is to learn from one's mistakes; to learn, one must identify and admit the mistake and commit to correcting the error in the future.

Administrators of FES agencies and organizations make many decisions each day, and every decision has potential ethical consequences. Because many of these decisions are ultimately made by the chief officer, the officer needs to consider whether he or she is making morally correct decisions. This is not always easy to determine, especially for the decision maker.

Professional Ethics

Ethical behavior includes personal and professional conduct. For example, personal ethics may include such behaviors as complying with laws, complying with societal norms, telling the truth, and having concern for others' well-being and respect for their freedom and independence. Professional ethics, however, may include such behaviors as complying with company policies and procedures, telling the truth to the public and elected officials, protecting the confidentiality of coworkers and those one serves, not mismanaging financial resources either unintentionally or illegally, and performing due diligence for the safety of the workers and the public.

Many professions are under the oversight of organizations that regulate the ethical behavior of their practitioners. For example, states may create boards made up of practicing professionals, such as doctors, lawyers, or engineers, to regulate a specific profession. As noted by philosopher and ethicist Sissela Bok in the book *Lying: Moral Choice in Public and Private Life*, these boards should expose any members of the profession who act illegally or unethically; however, this is rarely the case (Bok, 1999). A colloquial expression to describe this relationship is "the fox watching the hen house."

Although FES does not have any organizations that specifically watch for unethical actions, a few fire organizations do have the potential to make judgments on professional conduct. These organizations include the National Fire Protection Association, the International Association of Fire Chiefs, the National Volunteer Fire Council, and the International Association of Fire Fighters. Some of these organizations, such as the Florida Fire Chiefs' Association (FFCA), use a suggested code of ethics to guide members' conduct. The FFCA's Code of Ethics cites the following principles (FFCA, 2012):

- Maintain the highest standards of personal integrity; be honest and straightforward in dealings with others; and avoid conflicts of interest.
- Place the public's safety and welfare as well as the safety of members above all other concerns. Be supportive of training and education, which promote safer living and occupational conduct and habits.
- Ensure that the lifesaving services offered under the members' direction are provided fairly and equitably to all without regard to other considerations.
- Be mindful of the needs of peers and subordinates and assist them freely in developing their skills, abilities, and talents to the fullest extent; offer encouragement to those trying to better themselves and the fire service.
- Foster creativity and be open to consistent innovations that may better enable the performance of our duties and responsibilities.

Another detailed Code of Ethics can be found at the National Fire Academy web site: https://www.usfa.fema.gov/downloads/pdf/code_of_ethics.pdf and is worth taking into consideration.

Similar to this example, the National Association of Emergency Medical Technicians adopted a Code of Ethics in 1978. This Code acknowledges the emergency medical services (EMS) practitioners' obligation to make ethical patient-care decisions and to avoid opening their organization to liability. As stated within the Code of Ethics, "Professional status as an emergency medical technician and emergency medical technician-paramedic is maintained and enriched by the willingness of the individual practitioner to accept and fulfill obligations to society, other medical professionals, and the profession of emergency medical technician" (Data from Gillespie, 1978). The Code also goes on to list a number of specific ethical behaviors to follow, including conserving life and alleviating pain, respecting the dignity of all patients, maintaining confidentiality, and upholding medical standards and laws.

EMS organizations, many of which are private companies, are expected to train to a professional standard and to investigate incidents where unethical acts may have occurred. Most organizations are held accountable to local government oversight, which is also responsible for investigating any incident involving unethical behavior or a departure from city- or county-approved protocols or policies. If the oversight entity finds wrongdoing, the incident might then be referred to a state-level EMS authority that would perform its own investigation and decide whether or not to take disciplinary action, such as revoking an EMS provider's license.

In situations where the failure of an administrator, supervisor, or organization to comply with a safety regulation results in a death or serious injury, state or federal Occupational Safety and Health Administration (OSHA) agencies step in to assess unethical conduct. Although there have been some past examples of chiefs being dismissed after a firefighter death, generally in cases of noncompliance, the organization—not the individual—is found at fault. This is one situation where the administrator cannot be too careful. Seminars that cover these safety regulations for all officers is a must to assure that everyone is fully aware of the requirements. Ignorance is no excuse.

CASE STUDY

The following case study was taken from actual events but the names and all of the circumstances are not factual.

In the fire service, there is nothing more prestigious than the fire chief becoming the President of the International Association of Fire Chiefs (IAFC). For that reason a fire chief of a large fire department approached his boss and explained what he was planning. He mentioned that it would be a prestigious benefit for the municipality having the best fire chief in the country. He needed approval and commitment of funds to send a team of fire officials to several national conventions. These funds were unbudgeted and unapproved costs. The mayor subsequently made arrangements for the funds to be made available for the fire chief's mission.

This team was needed to do the personal contacts for convincing many individuals, groups, and associations to provide backing in the form of endorsements and some donations. The position was the 2nd Vice President and was the entry point for the progression to President two years later. The process of using taxpayer's funds to pay for this project is somewhat questionable, but not uncommon.

At each conference the fire chief's financial staff would reserve and pay for hotel rooms for the team. It also was common practice for the candidate to arrange an open bar reception to influence voters. This was not an inexpensive endeavor. The funding for these open bar receptions was paid for by a kickback from the hotels. When making the hotel reservations the chief's representative would agree to a fee per night of x dollars. This was at the high end of the normal hotel room costs, but because these reservations were for a large number of people, the representative would demand a kickback, which was used to buy liquor and even a buffet. Clearly this is unethical. The justification: for the greater good and prestige of the fire chief and municipality.

Discussion Questions

1. Was there a significant difference in ethical behavior between the mayor's funding involvement and the fire chief's staff arranging the kickbacks? If yes, is one more egregious than the other?
2. Do you know of any other spending of public funds that would be open to kickbacks? List a minimum of two.
3. Could these same circumstances occur when purchasing fire equipment or apparatus in your department?

Why People Lie

Researchers report that extroverts and coworkers are some of the most likely culprits of lying (Feldman, 2005). The tendency of coworkers to lie to each other is an important phenomenon to understand in the FES field. For example, consider the fire chief who is surrounded with staff who are "yes individuals." This practice prevents the discussion of controversial policy issues and hinders the delivery of honest feedback. Such lack of communication is unhealthy to the organization.

However, the most insidious types of lies are not those that are classified as boasting or are told in the name of discretion or politeness. The most harmful lies involve leaving out the truth or stating something misleading or absolutely false. These types of lies have the potential to destroy trust and intimacy in personal and professional relationships.

For example, a universal complaint concerning many politicians is that when they are running for election, they make many promises that are not kept once elected or they make deals with large donors for favors (quid pro quo) when elected. This has resulted in very low approval ratings for many politicians. For April to June 2020, a Real Clear Politics survey found only 25.8% of those surveyed approved of the U.S. congress job. This may be a reflection of the promises not kept or kept for large donors.

Therefore, it is important to understand why someone would lie in the first place. According to University of Massachusetts psychologist Robert Feldman, the act of lying is attached to a person's self-esteem: "We find that as soon as people feel that their self-esteem is threatened, they immediately begin to lie at higher levels" (Data from Feldman, 2005, p. 373). It is with this notion that lying can be viewed as a person's attempt to see himself or herself as they would like to be seen by others. "People are so engaged in managing

Habits of Dishonesty

The study entitled "2012 Report Card on the Ethics of American Youth" discusses the results of a 2012 study conducted by the Josephson Institute in which the ethical habits of nearly 23,000 high school students were examined. According to the study it revealed some disturbing habits of dishonesty for the workforce of the future, including the following alarming statistics (Josephson Institute, 2014):

- 36% answered yes to " A person has to lie or cheat sometimes in order to succeed."
- 55% admitted to lying to a teacher.
- 52% of students admitted to cheating on a test, with 38% saying they had done so more than once.

Although these numbers may be concerning, what is even more surprising is the students' perspective on ethics. "Some 93 percent of students indicated satisfaction with their own character and ethics, with 81 percent saying that 'when it comes to doing what is right, I am better than most people I know'" (Josephson Institute, 2014). Keeping in mind that these same high school students are now elected or appointed officials, members of the FES organization, and even officers, administrators must use a zero-tolerance policy for lying for themselves and the members of their organization. This kind of behavior can be changed with perseverance and leading by example.

Finally, one issue that has become a hot topic is systemic racism that is asserted to be prevalent in the American culture. "I am prejudiced against certain groups" was asked in the survey, and 77% of the high school students reported that they were not. What is not known is the race, religion, and ethnicity of the students. Therefore, when you considered all the possibilities of prejudice this seems like a good result. The main prejudice reported as systemic racism is the claim of the police prejudice against blacks even though real world statistics do not back up this perception.

In a July 2020 *USA Today* analysis of the statistics, they concluded that: "…the evidence does not support the charge that biased police are systematically killing Black Americans in fatal shootings. Much of modern policing is driven by crime data and community demands for help. The African American community tends to be policed more heavily, because that is where people are disproportionately hurt by violent street crime. In New York City in 2018, 73% of shooting victims were Black, though Black residents comprise only 24% of the city's population" (Ref: https://www.usatoday.com/story/opinion/2020/07/03 /police-black-killings-homicide-rates-race-injustice-column/3235072001/). However, there are numerous other possible prejudices such as religion, age, weight, race, gender, sexuality, ethnicity, etc. and these were not addressed specifically in the survey for the 23% who reported prejudice against a group. For administrators, they must guard against these prejudices and require sensitivity training for members. Even though there are differences between members, they must all work together without personnel conflicts.

how others perceive them that they are often unable to separate truth from fiction in their own minds" (Data from Feldman, 2005, p. 375). For people to keep fact and fiction separate, they must consciously commit to telling the truth in every situation.

Justification for Lying

Imagine a woman is in her kitchen at home when two very young children come running through the kitchen door. They tell her that a crazy person is chasing them with a knife. The woman quickly directs the children to hide in a closet. Almost immediately after the children close the closet door, a man bursts through the kitchen door with a large machete in his hand. He asks the woman in the kitchen if she has seen two children. The woman lies to him and tells him she has not seen any children. The intruder goes away.

Is this person justified in lying? The rational answer is yes. However, most situations are not so "black and white" **TABLE 11-1**. Trying to determine if a lie is justifiable is often very difficult, and opinions can vary drastically. The following sections address a few commonly cited justifications that may not pass the test as ethical.

"It's for the Public Good"

Administrators who see themselves superior by means of birth, wealth, training, or education often feel they have a right to make decisions for the public. Bok

TABLE 11-1 Examples of Ethical Dilemmas
■ Should a paramedic lie to a patient who asks, "You won't let me die, will you?"
■ Should professors and supervisors exaggerate the excellence of job applicants when giving references to give the applicants a better chance at being hired? Is this fair to the potential new employer or other applicants?
■ Should journalists write false or misleading statements, or fail to tell the whole story, if doing so supports their personal views on a subject that they strongly believe is right?
■ Should a fire chief exaggerate the consequences (e.g., more deaths and property loss) of not funding the FES budget properly?
■ Should a police officer use fabricated evidence to help convict a known criminal?

FIGURE 11-1 Members and officers need to use truthful statements when justifying their beliefs to the public and elected officials.
© Patricia Marks/ShutterStock.

explains, "Convinced that they know the truth—whether in religion or politics—enthusiasts may regard lies for the sake of this truth as justifiable. They may perpetrate so-called pious frauds to convert the unbelieving or strengthen the conviction of the faithful. They see nothing wrong in telling untruths for what they regard as a much 'higher' truth" (Reproduced from Bok, 1999, p. 7). In practicing this belief, however, administrators are left with only their own consciences to judge whether their choices are morally right.

For example, suppose an incumbent mayor is running for reelection and is convinced that his continued leadership is in the best interest of the city. Most likely, all of his staff and political party also support this notion. Under these circumstances, the mayor might hide an indiscretion, fail to be completely frank about his position on a controversial issue, or support a popular proposed law or allocation of funding even if he knows doing so would not be in the best interest of the city and most of the citizens.

One effective technique administrators could use is to look at issues through the eyes of different viewpoints, such as the young and the old, minority members, men and women, members of the emergency service's agency, customers, and elected officials. They should then consider how the people in each of these roles would feel about choices being made for them. There is an idiom, "you need to walk a mile in someone's shoes," that illustrates the need to understand another person's perceptions, experiences, or motivations before making a judgment for or about them. Administrators must also keep the "Golden Rule" in mind, which is the principle of treating others as one would wish to be treated. This can be difficult, because many members of the FES industry have strong beliefs regarding what is needed for emergency operations and safety.

Consider the chief at an incident where a fire death has occurred. The chief is asked by a newspaper reporter, "What could have been done to prevent the death?" The chief could ignore several prevention factors (e.g., smoke alarms and sprinklers) and answer that having four firefighters on each engine would have averted the death. Although this answer may open up a conversation for additional staff funding, which would ultimately lead to better service by the department, it would not be entirely truthful. In the emergency services field, there is often more than one way to prevent a tragic incident and, to tell the whole truth, the chief officer should mention all solutions. Although beliefs regarding service needs may be well founded, members and officers still need to use truthful statements when justifying these beliefs to the public and elected officials **FIGURE 11-1**.

"It's in Self-Defense"

There are situations in which lying seems justifiable when the alternative would be to use physical force. As considered by Bok, "If to use force in self-defense or in defending those at risk of murder is right, why then should a lie in such cases be ruled out?" (Reproduced from Bok, 1999, p. 41). Therefore, when physical force seems like the only means of self-defense, a person could also consider whether lying would accomplish the same objective. One example was discussed previously under "Justification for Lying." The woman in that example could have used physical force to stop the assailant with the machete, which would have accomplished the same result as the lie she used.

"It's a Miscommunication"

Chief officers should be very careful that all communications are received accurately. A person may be accused of lying when instead something actually went wrong within the communication process. This is especially true with verbal communications. A person may have meant to say one thing, but inadvertently said something with a completely different meaning. If an officer conveys the circumstances of a fire and the service's response to it, but fails to do so clearly, the message may be misunderstood and people may believe that the chief has lied about the event or the department's response to it.

Another type of miscommunication might occur when what the listener has already assumed to be true differs from what is actually said. In this case, the listener might hear what he or she wants to hear. In the case of emergency incidents, prompt communications after a major event helps prevent the public from being informed by gossip and rumor, which might then color their perception of the official story.

For example, after a major fire in an apartment complex, occupants complain about the slow fire department response. Several occupants had been forced to jump from upper floors and have sustained serious injuries after waiting a long period of time for the fire department to arrive. The fire department's on-scene public information officer immediately called the dispatch office to inquire about the response times noted in the computer-aided dispatch system. After checking the system, it is determined that the response actually took place within 4 minutes, clearly an acceptable time range. Once the public information officer repeated the actual response time in a press conference, the complaints stopped.

So what happened? This was an afternoon fire and a major incident, with heavy flames and smoke showing from numerous windows. After checking with several witnesses and asking if they called 9-1-1, it is determined that nobody actually called, because everyone thought that someone else would have already called. To correct the misinformation, more than the specific facts were needed. These facts had to be paired to a plausible explanation. This was doubtless a case where quick and accurate communications saved the fire department's reputation.

CASE STUDY

Propaganda is a form of lying that is typically used by governments and special interest groups to convince the uninformed audience to support their ideas or policies. For example, a reported horse race in Moscow during the Cold War was arranged between the best U.S. and Union of Soviet Socialist Republic (USSR) horses. The results of the race found the U.S. horse coming in first, beating the USSR horse. However, for the USSR citizens, the Pravda newspaper reported: The USSR horse came in second and the USA horse came in next to last. In reality a true statement but clearly misleading. This is actually a good example of propaganda.

This points out the need to attain more than one source for any information to assure yourself that you are indeed getting the whole truth. In any political situation this is imperative given today's highly political environment. Many examples of this unethical technique are commonly used every day around the world, including the United States. In totalitarian countries or those countries that do not have at least two independent media sources, the citizens normally hear only one side of a discussion or claim. This is a complex subject well beyond being able to be fully explained in this text. Therefore, if you have an interest in not being deceived by this type of lie, a good read is *Nineteen Eighty Four (1984)* by George Orwell.

Discussion Questions

1. Do FES administrators and their organizations have to be careful not to unintentionally use propaganda?

2. What areas of an FES organization could be prone to misuse of this technique?

3. Would a personnel goal of not believing anything you hear the first time (sometimes two or more) help in being able to identify propaganda?

4. If you take the approach of challenging everything you hear does that help to identify propaganda?

Consequences of Lies

Denial of the Public's Freedom to Choose

When we talk about taking away someone's freedom, most people think of a dictatorship or a prison where freedom is removed by physical force. But deceit and lying can also result in the loss of freedom. Furthermore, deceit does not have to be a direct communication of something that is untrue; it can also be the withholding of information that others have a right to know. For example, before the information became public knowledge, a well-known tire manufacturer kept many reports of vehicle rollover crashes involving its tires a secret from the public. Without knowing this information, the public and government safety agencies were unable to choose the safest tires for cars.

Now consider a common situation in the fire service. When the general public sees the fire engine responding with its red lights and siren operating, they cannot see how many firefighters are on board; furthermore, they often have no knowledge of how many firefighters are even needed to do a safe and effective job. Even though the minimum number of firefighters on the truck has been established by a national standard (e.g., National Fire Protection Association 1710) and bolstered by OSHA's safety regulation, the FES administrator can easily mislead the public and elected officials into thinking that whatever level of service is being provided at the local level is appropriate. This may result in fire engines responding with only one, two, or three firefighters, which necessitates more than one fire company for a fire to be safely attacked within a structure. In many cases, this level of service (i.e., number of stations and staffing) is justified by such factors as cost avoidance, unique local situations, or tradition, none of which meets the test of truthfulness.

Similarly, members of the public often assume and expect that the response of an ambulance means that advanced life support care is provided. Although there is a distinct difference in the scope of practice between a basic life support emergency medical technician and an advanced life support paramedic, the uniform often looks very similar. Should the service be obligated to clarify if it offers only basic life support service? Just like the engine company that does not publicize it has only two crew members, the ambulance company may not attempt to change what the public perceives about their available level of service.

In cases such as these, it may be easy to get caught up in addressing the needs of members or elected and appointed officials and fail to remember the primary goal of serving the public. A chief officer may not attempt a change because the change would be resisted by specific groups or individuals, such as the union, volunteer members, the mayor, or the city council; however, the officer should continue to try, using appropriate preparation, leadership, and courage to gain acceptance.

Damage to Professional Reputation

With the media's global reach, everyone has instant access to news about fraud, corruption, and cheating. It seems that the more immoral the act and the more famous the perpetrator, the bigger the story. It is now very difficult for public officials to cover up inappropriate behavior. If administrators have enemies or members in their organizations who simply do not approve of their leadership, impropriety can quickly be turned into ammunition.

"Paradoxically, once [the person lying] is no longer trusted, he will be left with greatly decreased power—even though a lie often does bring at least a short-term gain in power over those deceived" (Reproduced from Bok, 1999, p. 26). Some studies indicate that it may take up to 2 years for others to acknowledge any positive changes in behavior made by a person or an organization. As noted by social psychologist Roy Baumeister, "Bad impressions and bad stereotypes are quicker to form and more resistant to disconfirmation than good ones" (Data from Baumeister et al., 2001, p. 232). After an individual or an organization loses the confidence of the public and its own members, it becomes very difficult to ever regain that trust.

It should be noted on a personnel basis, the official may suffer harsh consequences. If found out about using lying, deception, and/or withholding important information, they may be fired or suffer other punitive actions. With this also comes issues with loss of income and loss of respect from family members, colleagues, and friends. It can be a nightmare for the individual and is not worth the chance of being exposed.

Moral Obligations of Public Roles

There is no reason to think that individuals in public roles are exempt from traditional ethical and moral requirements, or that the end always justifies the means. The person who accepts the position of chief officer of an FES department or EMS organization has a special obligation to the public, and to the organization's members and elected officials. The FES or EMS chief officer is directly responsible for the service the public receives and for the safety of the members. Also, in many cases, the chief officer is among the few who have the technical expertise to judge the adequacy of the service provided. Right or wrong, the officer has the power to fool the public and politicians into believing the organization is providing great service when in fact the organization and its members may not be operating at 100% effort and efficiency.

For example, a common judgment used to assess their FES organization adequacy by the public is the condition, age, and cleanliness of the apparatus. In reality, it would be better to have an old (although functional) apparatus with a well-trained and physically capable team of emergency responders. The critical element to efficient safe outcomes of emergencies is predominately the members.

Duty to Obey

There are many times during a chief officer's career in government when the officer feels pulled in several directions by superiors, such as a mayor or city administrator. People entering authoritarian systems might no longer view themselves as acting for their own purposes, but acting rather as agents for executing the orders of another person. "[A]dministrators typically work in environments where the presumption of the obligation to obey is powerful—so powerful that, in an extreme case, the administrative environment may promote a willingness on the part of officials to give up all sense of personal responsibility" (Data from Garvey, 1996, p. 328). Chief officers must be careful in these situations, because they can end up involved in a civil or criminal action.

To offset this tendency, number 4 of the 16 Firefighter Life Safety Initiative states "All firefighters must be empowered to stop unsafe practices." The normal behavior of firefighters is to obey all orders without question. A major goal of recruit training is to teach a firefighter to be capable of timely, effectively, and efficiently functioning at emergencies. They should also be educated in the practice of recognizing unsafe situations and orders that must be called out. This is sometimes characterized as the "culture" that must be identified so the individual can have the courage to resist the duty to obey when necessary.

Although obedience may be a deeply ingrained behavior for many public officials, it is important to remember that blind obedience is a powerful force that can override training in ethics, common sense, and moral conduct. For example, the chief officer receives a direct order from the mayor to eliminate all overtime. The officer knows that overtime is needed

to maintain minimum staffing levels, but he obeys the order without a fight, possibly fearing loss of his job. A leader of an FES department or EMS organization must find a way to stay independent by nurturing a relationship with superiors that allows for open, honest communication. Although most of this frank communication must take place in private, the chief officer should be able to approach his or her superior in any setting without the fear of reprisals.

Unions, volunteer organizations, and business owners have all used their influence over the years to change or attempt to change fire- and EMS-related public policy. These groups were specifically created to look out for the best interest of their members. Administrators need to remember their obligation to the public because many of the groups attempting to influence the behavior or policies of the organization have their own personal self-serving agendas. Officers must always try to do what is ethically right, using some of the techniques and suggestions in this chapter as a guideline.

Fairness Among Members

Fairness is equated by many as being the "right" or ethical thing to do. This is not always achievable, but when possible, chief officers should treat all members equally and fairly **FIGURE 11-2**. Every person should have an equal opportunity to succeed in accordance with his or her abilities and hard work.

For example, in any FES organization, some members perform at a higher job performance level than their peers. A chief officer could easily start treating these few outstanding workers better than their peers. However, the officer should remember that it takes a

FIGURE 11-2 Absolute balanced justice is not always achievable in organizations because it is applied by people who are not perfect, but it should be the goal strived for by administrators.

© Aerial Mike/Shutterstock.

team of players to win the game. All players contribute to the best of their ability if they are properly motivated. The job of a leader is to get all members to operate near their maximum performance level, whatever that level may be.

Only in those cases where an member's behavior is unacceptable should a different treatment be administered. In these cases, the other members are watching, and if they see an employee getting away with substandard performance, some will lower their own performance levels. Everything a chief officer says or does is a signal to all members of what to expect in relation to job performance. If the officer allows a few self-motivated overachievers to carry the bulk of the work or does not discipline those not meeting minimum performance standards, that officer is not treating members impartially or fairly.

Ethical Tests

The following is a list of methods to consider when evaluating an ethical dilemma:

- Assume the public is in the room and has adequate knowledge to review and discuss the issue (especially if the decision must be made secretly, a discussion using this assumption provides needed input).

- Follow the Golden Rule: treat others as you would want them to treat you.
- Consider the viewpoint of everyone who will be affected: age, gender, sexuality, race, religion, income, disability, ethnicity, etc.
- Examine the validity of the "avoiding harm" and "greater truth" rational. For example, "Recently, medical experts argued that we need to lock down the populous to reduce transmission of a virus." Other scientific, psychologist, and economic experts disagreed because of the greater harm to the economy and peoples' psychology. This debate should have been in the public and any adverse consequences (especially unintended) fully evaluated. Be cautious in these debates of occasionally invalid justifications.
- Investigate from top to bottom the "produces greater benefit" justification.
- Be cautious of the "means are justified by the end." This is almost always false. The means must be ethical.
- Consider the fairness of the decision.

In addition to the considerations listed above, keep in mind the following quote from Mark Twain: "When in doubt, tell the truth. It will confound your enemies and astound your friends" (Data from Twain, 1906).

Wrap-Up

CHAPTER SUMMARY

- What does ethical behavior involve?
- Why do people lie?
- Are there justifications for lying?
- Are there consequences for lying?
- Is there a special duty to be honest for public officials?

KEY TERMS

Code of ethics Ethical codes are adopted by organizations to assist members in understanding the difference between right and wrong and in applying that understanding to their administration actions.

Ethics A principle of right or good conduct or a system of moral principles/values.

Fairness Treating people or ideas equally or in a way that is right or reasonable. However, this is a very

subjective concept and can sometimes be clarified by an ethical evaluation of the person or idea by considering how one ought to act, morally. One father explained to his children, who wanted to be fairly judged (the "that's not fair" declaration), that if life were fair to him, he would have won the lotto several times. Life is not always fair but administrators should do all they can to be as fair as possible in their supervisory responsibilities.

Lying Deliberately presenting a false statement or action as true or hiding factual information with the intent to deceive.

Obedience The act or an instance of obeying; dutiful or submissive behavior such as following orders of a supervisor at the scene of an emergency.

Oversight Review by entities such as the government or professional organizations to regulate ethical behavior.

Personal ethics May include complying with laws and societal norms, telling the truth, showing concern for others' well-being, and respecting others' freedom and independence.

Professional ethics May include complying with company policies and procedures, telling the truth to the public and elected officials, protecting confidentiality, and ensuring the safety of FES workers and the public.

REVIEW QUESTIONS

1. Answer each question posed in Table 11-1 according to your own moral and ethical code, and the information gleaned from this chapter.

2. Compare and contrast personal and professional ethics, and provide at least two examples of each. Do you think one is more important than the other?

3. Can you think of any other justifications for lying not covered in this chapter? Describe a scenario in which this justification could be used by a hypothetical unethical member of the FES, and describe the potential effects for the department or organization.

4. Do you expect the superior officials or politicians to be ethical in all their actions? List two examples of individuals that you think may have had ulterior motives other than being truthful in their actions.

5. List three motives people may have to act other than ethically in administrative responsibilities.

CHAPTER ACTIVITY #1: AUTOMATIC SPRINKLERS

A fire chief is scheduled to testify at a city supervisor's workshop regarding a proposed fire code amendment requiring installation of automatic sprinklers in all newly constructed parking garages.

An automobile fire in a parking garage several months ago prompted this proposed fire code amendment. Fire companies had a difficult time advancing hose lines to the area of the parking garage where the fire was and, subsequently, the fire spread to cars on each side of the original fire. After the fire was out, a reporter from a local newspaper interviewed the chief and asked why the fire companies seemed to have so much trouble putting out the fire. Without much thought, the chief pronounced that if a sprinkler system had been installed in the parking garage, the fire would have been contained with smaller property loss.

Sometime later, the chief received a report from a battalion chief indicating that the fire company had been very slow in advancing the hose line at the fire. The battalion chief then went on to recommend that the department should institute a "back to basics" program. The battalion chief also noted that the number of structural fires in the city was down, causing the proficiency of the companies to diminish.

In addition, while researching his testimony for the city supervisor's workshop, the chief discovered that the model building codes did not require sprinklers in garages if there were appropriate exterior openings on each level. The chief also discovered articles in fire service publications based on real-life experience and controlled burns where valid cost-benefit justification for sprinklers could not be established.

The fire chief had been in his present job for several years and had a reputation as a very knowledgeable leader. In other presentations and appearances before public hearings, the chief had gained a lot of support for the department's opinions, goals, and ideas.

Discussion Questions

1. Make a list of options the chief might choose based on his situation and explain the ethical considerations that each option entails.

2. Would a full public apology to the city council along with requesting the code change be rejected be appropriate for the chief?

3. Name three specific skills that would be needed to extinguish the car fire in a garage.

CHAPTER ACTIVITY #2: CALL RESPONSE

Your FES agency has just experienced a 45% increase in emergency responses after implementing a new policy that requires an engine company to respond to all calls for medical assistance, not just life-threatening medical calls.

Discussion Questions

1. Based on a cost-benefit analysis, are the citizens getting more for their tax dollars under the new policy?

2. How would you defend the following situation: A first due engine company was slow to respond to a structural fire where a 5-year-old girl died. At that time, the first due engine was busy at a medical call scene where a 45-year-old construction worker had broken his arm. After providing first aid and splinting the arm, they were waiting for the arrival of a transport ambulance. Take into account all ethical considerations when answering a question from the grieving parents; What took you so long to get to our house?

3. With this new activity level, the officers of your engine companies are now reporting that they are not able to complete their in-service fire prevention inspections and physical fitness programs. How might this situation affect your organization's ability to meet its goal for quality service to the public?

4. Would you consider a change in response policy that required an engine company response only in those truly life-threatening emergencies (e.g., heart attacks, serious trauma, trouble breathing) to reduce the call volume?

CHAPTER ACTIVITY #3: RIGHT OR WRONG

Charlie is a 54-year-old homeless man with coronary artery disease. He calls 9-1-1 almost every day, especially in the winter when he is looking for a warm meal and a dry bed. Every time the ambulance crew serving the primary service area in which Charlie lives transports him to the busy county hospital, they know that they will wait at least 20–30 minutes for a bed. The crew also knows that they are going to get grief from the emergency department staff when they arrive with Charlie.

On a particularly busy, rainy day, the crew finds Charlie at a telephone booth with the usual complaint of chest pain. Jack, the senior medic, tells Charlie this is a bad day to be fooling around and that the emergency department is full. Charlie says he is hungry and agrees to sign an Against Medical Advice form if Jack gives him $20 for food. George, the new intern at the hospital emergency department, questions whether they should have transported as the patient requested or at least done an assessment. Jack replies, "It's no different than any other day. The guy is just a system abuser."

Two hours later another crew responds to find Charlie in cardiac arrest. He cannot be resuscitated. When the family finds out a crew had seen him earlier and did not treat or transport, they file a lawsuit.

Discussion Questions

1. Was Jack's decision to talk Charlie out of transport ethical?

2. Should the intern have reported his discomfort to a supervisor?

3. Can Jack and his crew be liable for patient abandonment even with a patient signature?

REFERENCES

American Heritage Dictionary of the English Language, 4th ed. 2005. Retrieved from http://education.yahoo.com /reference/dictionary/?s=ethics

Baumeister, R. F., Bratslavsky, C. Finkenauer, and K. Vohs. 2001. Bad is stronger than good. *Review of General Psychology,* 5(4), 323–370.

Bok, S. 1999. *Lying: Moral Choice in Public and Private Life.* New York, NY: Vintage Books.

Edwards, S. T. 2010. *Fire Service Personnel Management,* 3rd ed. Upper Saddle River, NJ: Brady/Prentice Hall Health.

Feldman, R. 2005. Deflecting threat to one's image: Dissembling personal information as a self-presentation strategy. *Basic and Applied Social Psychology, 27*(4).

Florida Fire Chiefs Association. 2012. *Code of ethics.* Retrieved from http://www.ffca.org/i4a/pages/index .cfm?pageid=3287

Garvey, G. 1996. *Public administration: The profession and the practice.* New York, NY: St. Martin's Press.

Gillespie, C. 1978. EMT oath and code of ethics. Retrieved from www.naemt.org/about_us/emtoath.aspx

Josephson Institute. 2014. *The ethics of American youth: 2012.* Retrieved from https://charactercounts.org/wp-content /uploads/2014/02/ReportCard-2012-DataTables.pdf

Twain, M. 1906. Retrieved from http://www.twain quotes.com /freefind.html

CHAPTER 12

Public Policy Analysis

LEARNING OBJECTIVES

After studying this chapter, the student will be able to:

- Comprehend the different types of policy analysis (pp 221–222).
- Examine the role of the administrator as a leader and facilitator in policy analysis (pp 220–221, 222–223, 226–229).
- Understand the impact and influence of policy analysis on the budget and on resource allocations (pp 222–226).
- Examine the necessity of consensus building and the political process used to implement policy changes (pp 226–229).
- Understand the importance of the public policy presentation and the different elements involved (pp 229–231).
- Understand the use of statistics in policy analysis (pp 231–234).
- Analyze and research case studies and other historical experiences (p 234).

Case Study

Firefighter Heart/Lung Disease Study

Studies of firefighters and the incidence of heart/lung illnesses conducted by the Medical Research Council (London, England) concluded that firefighters had a much higher incidence of these illnesses than the general public (Douglas et al., 1985). These studies were used to justify most of the heart/lung presumption retirement and health benefits that are common throughout the country. Later, that same data was reanalyzed and normalized for cigarette smokers. The new studies found that firefighters who do not smoke had a slightly lower statistical chance of heart/lung disease than the general public, most likely as a result of hiring fit employees and the fact that fire and emergency services (FES) professionals tend to retire at younger ages. However, the firefighters who smoked had an incidence rate much higher than the smoking public.

The results pointed to a synergistic outcome from the combination of cigarette smoking and firefighting, rather than the initial conclusion that connected the illness to all firefighters. As this example points out, if more than one variable can cause the outcome, all potential variables should be studied. Because of these findings, many municipalities require their firefighters to not use any tobacco products. This makes it more likely that any heart/lung illness is the result of firefighting, not cigarette smoking.

Discussion Questions

1. Does your department have a no tobacco use at all times policy? If yes, how do they enforce it?

2. Because the majority of tobacco use is done off the job, is the no tobacco policy fair?

3. What administrative process would you use to adopt the no tobacco policy? Please explain in detail.

Future Planning and Decision-Making

The final product of policy analysis can be a comprehensive "Master Plan" or simply a budget enhancement request to the city/county council on one specific program. And in rare cases, the Master Plan can include entire government agencies or just public safety. More common, many political jurisdictions use these plans for land use that includes commercial growth control. The more comprehensive Master Plans may call for the input from many diverse individuals and organizations with an interest in the outcome. Typically, the more complex, controversial, and costly plans will have implementation periods such as 5, 10, or even 20 years. This is done for the validity and reasonableness of the recommendations including the phase that requires support for the funds needed for implementation.

On the operational side of public service, emergency services personnel often take immediate action and complete emergency operations in minutes or hours. In the public policy arena, however, decisions may not be so immediate and a long-term plan may take years to implement.

The concept of future planning in an FES organization has been encouraged through such documents as the U.S. Fire Administration's (USFA) *Urban Guide for Fire Prevention & Control Master Planning* (1977) and several courses at the National Fire Academy (executive planning and strategic analysis of community risk reduction). Many of the short- and long-term planning functions in an FES organization are relatively simple; advanced planning may allow one to anticipate needs and forestall problems for the future. The following are items for the chief officer to consider when planning for the future:

- If a project or policy in question is relatively simple, it may be possible to anticipate needs and forestall problems through advanced planning. For example, the basic measure of emergency service is response time. The determination of the appropriate response time can be complex, but once the maximum acceptable time is agreed upon, analysis to determine coverage areas for existing and proposed stations is fairly straightforward.

- If the policy area being studied makes it unwise to delegate broad discretionary authority to the frontline workers, the major policy decisions must be made at the department level so that each unit provides consistent, reliable emergency service to the public.

- If the implementation effort fits unambiguously within the responsibilities of an existing agency, the appropriate organization to be used for the job should not become the subject of discussion.

- In the case of FES operations, it is not a lack of knowledgeable experts at the company level that is a problem; it is a lack of adequate time and big-picture policy analysis that is needed to weigh all the pros and cons to make a valid decision. That is why standard operating procedures (SOPs) are extremely important in creating a safe working environment and attaining consistent quality service. The majority of emergency actions taken by first arriving companies need to be scripted to guarantee positive results each time. These decisions are best accomplished by a thoughtful unhurried analysis conducted through the task force process.

- If there is a high level of confidence in the causal theory underlying the policy, planners can predict the effects of executing changes even in the absence of additional study and research.

In many cases, a plan or policy can be implemented less painfully by incremental adjustment. This method works by making a small change and assessing the results. Incremental adjustment tends to be a safe way for policy makers to proceed with their decisions without the risk of making big changes that may have unforeseen consequences. This method spreads out a large financial impact over several years, allowing for small increases in tax revenues each year. Also, this can allow for adjustments to the plan if the outcomes are different than anticipated.

Exercise caution when implementing public policy; leading change is not a simple process. The notion that policy makers exercise, or ought to exercise, some kind of direct and determinant control over policy implementation might be considered an erroneous belief of conventional public administration and policy analysis. However, once a new policy is identified, techniques discussed in the chapter *Leading Change* will help implement change and overcome resistance. One successful method for determining future plans is to use public policy analysis—also known as strategic planning.

Strategic Planning

Strategic planning is normally conducted at the department level. This type of planning identifies the department's vision, mission, and strategies. The USFA document *America Burning* (1973) contains helpful guidelines for the vision and mission of a modern FES organization. Generally, this type of planning is conducted by an appointed task force of FES officials and senior members. In some cases, an official master plan is produced with input from the public, other agencies, and appointed or elected officials of the municipality. Strategic planning can help identify where the organization is at the present time and where a new vision can take the organization in the future.

To implement the strategic plan, tactical plans should be prepared and monitored on a monthly basis. For example, suppose a department has just received additional funding for a new station and part of the justification for the funding was to reduce response time to a specific area of the jurisdiction. The strategic plan lists a specific response time as one of its goals; the tactical plan lists the details. In this case, a new fire station is needed. After the tactical plan is implemented, a monthly check of response times to emergency incidents should be maintained and a formal report made to brief the executive and elected officials on the results. At the appropriate time, a public summary of this new program would be prepared to inform citizens, politicians, and the press of the impact of the new spending of tax dollars.

There are also operational everyday items that need to be accomplished that can indicate the need for

Strategic Plan	It is the policy of the department to provide excellent fire suppression services to all city residents using NFPA 1710 as a guide to acceptable response time.
Tactical Plan	An area of the city has been identified as having excessive response times when compared with the Strategic Plan standard and in comparison to other areas of the city. A new fire station is subsequently funded and constructed. Note: see Chapter 4, *Leading Change*, for a process for implementing any cultural changes that may be required by the Tactical Plan.
Feedback Analysis	Response times are recorded after each fire call. At a prescribed time interval, a formal report of the new improved response times is assembled. The chief administrator prepares a press release and makes arrangements to submit the report to the mayor and city council.

Rational-Comprehensive Decision-Making

In the book *The Policy-Making Process* (Lindblom and Woodhouse, 1993), the concept of rational, comprehensive decision-making is explained. This method is based on logical reasoning, accepting significant change, and favoring the solution that completely resolves the issue rather than temporarily fixes it. The FES administrator can use this decision-making process to:

1. Identify the problem.
2. Rank goals.
3. Identify all possible alternatives that solve the problem.
4. Perform a cost-benefit analysis (if possible).
5. Perform a comprehensive analysis of various ways of solving the problem.
6. Select alternatives that accomplish goals.

Data from Lindblom, C. E., and E. J. Woodhouse. 1993. *The Policy-Making Process*. 3rd ed. Upper Saddle River, NJ: Prentice Hall.

policy changes. These situations may not need a formal decision-making process, only a slight adjustment. For example, it is common to have minimum staffing levels that must be maintained daily. There may also be guidelines for the number of people who may be on annual leave for any particular day. If staffing levels fall below the minimum level, personnel must be called back on overtime or transferred from other stations. It is very important to document the exact reasons for unexpected expenses, such as overtime. At the tactical planning level, this information can be used to suggest changes in departmental policies; e.g., if the timesheets show an increase in overtime as a result of members using sick leave, an evaluation of leave policy and decreasing the number of personnel authorized to be off on annual leave may be appropriate.

The Role of Budgeting Within Public Policy Analysis

The budget and planning processes are a continuous and inseparable cycle. Plans should dictate the items placed in the budget; the budget is the process of funding the plans. When the planning process is completed, the administration should determine the financial resources needed to achieve the goals. Creative strategies for funding may be available and

should be pursued (refer to the chapter on *Financial Management* for examples of some of these strategies). Visionary economical actions are always a good way to gain respect from elected officials.

If the numbers in the budget become the focus of the policy analysis, administrators may find they are simply adding an incremental percentage to each line item in the proposed budget each year. If a strategic plan is followed, however, the chief may decide to move funds from one part of the budget to another to fund a new goal, priority, or objective, or they may find a justifiable need for a significant increase in the budget.

For medium to large FES departments and emergency medical services (EMS) organizations, budget preparations should start at the lowest feasible level of the organization. For example, the chief may direct the fire prevention division to prepare its own budget based on an incremental increase over last year's budget. In addition, division chiefs may be asked to submit justifiable enhancements as separate budget items or a budget based on a percentage decrease, including identification of programs that could be eliminated. There might be as much competition for the budget dollars internally as there is externally. Internal conflicts can be avoided or at least reduced by including division chiefs in the preparation, justification, and prioritization of the budget; this helps them to see the big picture (strategic plan) and better support the unified budget (see Figure 3-2 for an FES organizational chart).

In addition, including the division chiefs in the process gives the administration many options to fine-tune the budget. Using groups of people to study and make recommendations about the organization's goals and objectives can help to ensure the administration is not blindsided by the lack of appropriate information and expert advice when analyzing a management or functional problem. The chief may be able to eliminate funding for some items while at the same time adding funds for new programs. Also, after the decision has been agreed to by consensus, the FES administrator can expect to receive support from all the managers.

After the division chiefs have prepared their budgets, the chief administrator may then ask them to prioritize between adding back the proposed percentage cuts and funding the enhancements that have been requested. After this step is complete, the chief administrator must combine the proposed budgets from all divisions and have the entire organization prioritize any decreases and enhancements. After this prioritization, it is up to the FES administrator to balance funding by choosing decreases, adding funding for those enhancements that warrant support, and requesting additional funds for other enhancements that the administration believes are justified.

Measuring Outcomes

To carry out a public policy analysis, outcomes must be measured. An outcome is a measured benefit to stakeholders, created as a direct result of dollars spent by the organization. However, many of the measurements that are kept by the traditional FES organization are measures only of workload. For example, if you have an active public fire safety education program and a fire sprinkler ordinance, you may be able to reduce the risk of fire or control it before it reaches flashover. So these measures will indicate a better outcome of less fires, injuries, and deaths. In this example, the measure (e.g., the number of people receiving fire safety education or the number of new fire sprinkler installations) is based on the common sense realization that the results (outcomes) will be prevention of deaths.

Although the FES organization may keep accurate statistics on the number of emergency incidents to which they respond, that does not show the actual outcome. This concept is further explained by research analyst Jennifer Flynn (2009, p. 3):

> Performance measurement relies on the evaluation of achieved outcomes, compared to desired outcomes. For the fire service, the desired fire protection outcomes—which are not easily measured—include fires prevented or suppressed, and ultimately the human life and property preserved. … Performance measures are the quantitative or numerical representation of activities and resources that help evaluate whether the goal is met.

Consider the department that responds to a small trashcan fire in a high-rise building and extinguishes the fire quickly. Could the department then claim to have saved the entire dollar value of the high-rise building? What about if a complete floor was on fire and the department was successful at extinguishing the fire before it extended to any other floors? As illustrated, assessing the dollar value of properties saved is very difficult, if not impossible.

It is also rare to be able to document saving a life. If firefighters were to search, find, and remove an unconscious person from a building on fire, they would be able to claim an unquestioned save. In the fire prevention arena, however, lives are more often saved from built-in fire protection equipment, such as smoke detectors that allow occupants to escape without injury. Many of these types of saves are never reported to the fire department. Therefore, outcomes for an FES organization must be measured using other more easily verifiable items.

Because some outcomes are based on relative or subjective items, the accuracy of the outcome can often be evaluated by looking at efficiency, responsiveness, or equity. For example, a newly purchased computer-aided dispatch system might reduce dispatch time by 50%. This measured outcome is an indication of greater efficiency, greater responsiveness, and equity because it equally serves everyone who calls for emergency assistance. In general, many outcomes of response time improvements cannot be measured directly (e.g., three lives saved last week); however, relative improvement can be measured. A better outcome is the result of arriving sooner, starting emergency operations sooner, and using professionally trained and well-equipped responders.

If it can be demonstrated that response times can be reduced by changing existing procedures (e.g., streamlining the call-taking process) or adding new resources (e.g., increasing the budget for new staff or new facilities), the argument can be made that improved service to the public has been accomplished. Even though the outcome cannot be measured directly, this is a valid approach to reporting achievements (outcomes).

Staffing Levels

Personnel make up over 80% of the career FES budget; this section provides a detailed discussion of the critical need for this line item. In addition, the minimum number of personnel needed for safe emergency operations is addressed in the Occupational Safety and Health Administration's Respiratory Protection Standard 29 CFR 1910.134 (two in/two out) and National Fire Protection Association (NFPA) 1500, *Standard on Fire Department Occupational Safety, Health, and Wellness Program*. The Occupational Safety and Health Administration (OSHA) and NFPA agree that a minimum of four firefighters must be at the scene of a structural fire before an interior attack can be initiated. Therefore, one key measure of an adequate level of emergency service has to be the time needed for four trained firefighters to arrive on the scene. It is believed that anything less than this number could cause delays in attacking the fire or unsafe actions by firefighters.

Several studies have been conducted relating efficiency—measured by the time required to complete fireground operations—to staffing levels. The most recent of these studies, conducted by the National Institute of Science and Technology (NIST), investigated "the effect of varying crew size, first apparatus arrival time, and response time on firefighter safety, overall task completion, and interior residential

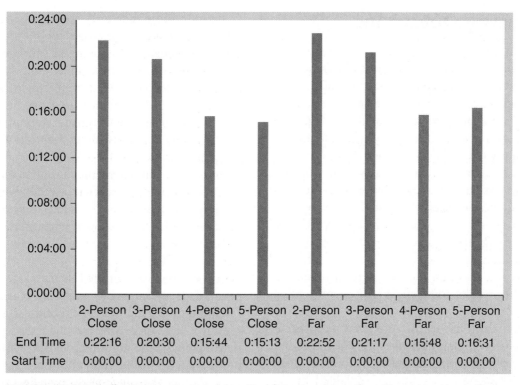

	2-Person Close	3-Person Close	4-Person Close	5-Person Close	2-Person Far	3-Person Far	4-Person Far	5-Person Far
End Time	0:22:16	0:20:30	0:15:44	0:15:13	0:22:52	0:21:17	0:15:48	0:16:31
Start Time	0:00:00	0:00:00	0:00:00	0:00:00	0:00:00	0:00:00	0:00:00	0:00:00

FIGURE 12-1 Overall scene time to complete critical fire ground tasks for extinguishment, ventilation, and search for victims.

Official contribution of the National Institute of Standards and Technology; not subject to copyright in the United States.

tenability using realistic residential fires" (Data from National Institute of Science and Technology, 2010, p. 9). Of the 22 critical fire ground tasks performed in this study, the four-person crew was the clear winner by 25%–30%, completing tasks 5.1 minutes faster than the three-person crew **FIGURE 12-1**.

Providing on-duty crews is the key to a fast, coordinated response to an emergency incident. In many cases this means career personnel, but volunteer departments can establish on-duty crews by assigning a duty time to each of their members. This has worked successfully in a small number of volunteer departments, and the trend seems to be spreading. In areas where public funding is not available for career firefighters and volunteers cannot provide complete coverage of the station with on-duty members, staffing must rely on alerting volunteer members by outside sirens or pagers to respond to the emergency incident. Generally, these volunteer departments are in areas that serve 2,500 or fewer residents.

As a general guideline for planning purposes, if the fire company or EMS unit responds to more than 2,500–3,000 calls per year, planners should look very closely at the need for a second unit, either fire or EMS. Actual response data should be used, including the time out of service spent either responding to or working at emergency incidents. If time out of service totals more than 10% of the available time during

the busiest 8-hour shift, it might be time to plan for a second unit. In some cases, these high call numbers could be the result of a dispatch policy that sends too many unnecessary units to each call. For example, a department may send a full commercial structural response to an automatic alarm and then find that during weekday business hours it is rare to need more than one engine. One other technique that is more commonly seen for EMS transport units is to staff additional units only during the busiest hours. The New York City Fire Department also used this technique by staffing part time ladder companies during busy times.

Response Times

The number of fire stations serving an area has the greatest impact on response time for an FES department. In addition, the more stations, the greater number of firefighters needed. Some jurisdictions use their existing fire station spacing as the local benchmark. Paid firefighters are used to staff units that respond immediately, keeping response times to a minimum. Some volunteer-staffed companies also may respond immediately if an appropriate number of members are in the station. A few volunteer organizations require members to stand by in the station to provide an immediate response.

NFPA 1710, *Standard for the Organization and Deployment of Fire Suppression Operations, Emergency Medical Operations, and Special Operations to the Public by Career Fire Departments*, was first adopted in 2001 and has since set a benchmark for response times. According to the standard, the first due company has 80 seconds to start its response after dispatch, and 4 minutes for travel time to 90% of emergency incidents.

The NIST report mentioned in the section on staffing levels does not validate the 4-minute response time benchmark as clearly. In this study, NIST examined only two response times: 3 minutes and 5 minutes using computer modeling technology. A substantial difference in the effectiveness of the fire companies was not found (NIST, 2010, p. 48). Because achieving 4-minute vs. 5-minute response time takes substantially more fire stations to cover the same area (36% more), a cost-benefit analysis would find that the municipality would have to fund 36% more stations and staffing. Therefore, cost-benefit analysis appears to support a 5-minute response time for the first due engine from the considerable extra cost of more fire stations and staff. For many jurisdictions, the question is always can they afford the higher number of fire stations and staffing for a questionably marginally improvement in

extinguishment and victim rescue time. The NIST results show no advantage when considering the intensity of the fire and prediction of victim rescue between the 4- and 5- (late) minute response times **FIGURE 12-2**.

Some consultants and planners use travel times for locating stations but do not consider the staffing of the companies. As previously mentioned, the number of firefighters that should be on the scene before an interior attack on a hostile fire is four.

The NFPA standard also dictates that a minimum of two people trained at the emergency medical technician and paramedic level and two people trained at the emergency medical technician level should arrive within the established time of 8 minutes on all advanced life support emergencies. NFPA 1720, *Standard for the Organization and Deployment of Fire Suppression Operations, Emergency Medical Operations, and Special Operations to the Public by Volunteer Fire Departments*, also addresses response times for volunteer fire departments, allowing 9 minutes for "urban" areas (greater than 1,000 people per square mile), 10 minutes for "suburban" areas (500–1,000 people per square mile), and 14 minutes for "rural" areas (< 500 people per square mile) 80% of the time. There is no limit on turnout time, which is a weakness of the standard.

Crew Configuration	Rescue Time	Fire Growth Rates		
		Slow	Medium	Fast
2 Early	12:47	.12	.72	1.49
2 Late	14:47	.35	1.37	2.56
3 Early	9:03	.01	.11	.40
3 Late	11:03	.04	.36	.84
4 Early	9:10	.01	.11	.42
4 Late	11:10	.05	.38	.91
5 Early	8:57	.01	.10	.38
KEY	White	89% or more of population may be capable of effecting their own escape if they are able.		
	Yellow	Potential for certain sensitive populations (such as children and the elderly) to become incapacitated.		
	Red	More than 50% of the population would be incapable of effecting their own escape.		

FIGURE 12-2 Fractional Effective Dose (FED) as a function of deployment configuration and fire growth rate.

National Institute of Science and Technology. 2010. *Report on Residential Fireground Field Experiments*. Retrieved from http://www.nist.gov/customcf/get_pdf.cfm?pub_id=904607

The headline of a 2014 news report from San Diego, California (*The San Diego Union-Tribune*), read "City tries to speed emergency response" (Garrick, 2014). The study made the argument for several two-person crews to reduce response times in a few areas of the city. However, as noted in the previous section on staffing levels, most credible sources state that four firefighters should ultimately respond to the scene before an interior attack can begin. Therefore, response time should be measured only after these four firefighters arrive on the scene. It may seem understandable for a department with understaffed fire companies to report a response time when the first fire engine arrives if it is also planning and budgeting to increase the staffing to four. However, if there is no plan to increase staffing, response time should be reported when four firefighters arrive, which would be almost twice as long as the response time of a four-person crew, because at least two fire companies must be on the scene.

If a four-person response time is used for reporting, it can also be used to justify staff increases. For example, if the department has minimum staffing of two or three firefighters, the response time is the arrival at the scene of the second engine, which in many cases is twice as long as the response time of a fully staffed company. The maximum response time with understaffed companies is actually at the physical location of the two stations. For properly staffed companies it is exactly the opposite. Remember, the standard is a maximum response time. This is a difficult concept to comprehend, but critical to justifying adequate safe staffing.

Professional Competency

Professional competency is an additional area that relates to better service. EMS responders have nationally recognized certification for training and education to ensure equal competency across the United States; however, individual departments and states require different levels of training for firefighters. The NFPA report *Fire Service Performance Measures* states that the competency of firefighter training "is a proxy for quality of service provided. It is assumed that a high percentage of responders with completed training and certification are providing high quality service when responding to calls" (Data from Flynn, 2009, p. 23).

Performance Measures for Fire and EMS

The following is a summary list of performance measures taken from the 2018 NFPA Urban Fire Forum (ref: https://www.nfpa.org/-/media/Files/Membership/member-sections/Metro-Chiefs/Urban-Fire-Forum/2018/2018UFFPerformanceMeasures.ashx?la=en):

- 90% of emergency 911 calls processed in not more than 64 seconds, or 95% of calls processed in not more than 106 seconds.
- 90% of all fire and special operations responses turned out in less than 80 seconds.
- 90% of all EMS responses turned out in 60 seconds.
- First responding company arrival at the scene with minimum of 4 persons = 90% of the time in 4 minutes.
- Full alarm effective response force assembled on scene of low- and medium-hazard structure fires = 90% of the time in 8 minutes.
- 480 seconds or less response time for the arrival of an advanced life support (ALS) unit at an emergency medical incident.

The NFPA Urban Fire Forum is made up of metropolitan-sized department fire chiefs from primarily the United States, but it also includes a few from around the world. Each department must have at least 350 paid firefighters. They have no formal regulatory authority but act as an advisory expert group for large metro fire departments. The above stated performance measurements are identical to those contained in NFPA 1710 *Standard for the Organization and Deployment of Fire Suppression Operations, Emergency Medical Operations, and Special Operations to the Public by Career Fire Departments.*

For comparison purposes, the London Fire Brigade protects over eight million residents and many more tourists. In 2013 they completed a comprehensive evaluation of response times and adopted a 6-minute response time for the first due engine company. As previously stated, cost-benefit analysis suggests a 5-minute response time is acceptable. However, in heavily populated metro cities, the shorter response time of 4 minutes may be appropriate due to many structures with large numbers of occupants.

Consensus Building

To be successful at maintaining the department's budget and adding enhancements when needed, administrators must understand the approval process. In general, if the administration can make arguments that are backed up by common sense and solid research, there is a good chance of receiving approval for the budget request. If a national consensus standard can be used—such as those of the NFPA—to support the budget request, it is easier for elected officials to justify their vote.

For example, a chief may propose a new program to certify all officers to NFPA 1021, *Standard for Fire Officer Professional Qualifications*. Certification might require a training program, incentive pay, and overtime pay to cover classroom attendance. The selling point for this proposal is that, when finished, the city or county has a more professional fire department. This is the kind of proposal that the municipal administration and elected officials can clearly understand.

Still, there has been some controversy over the use of national consensus standards in the fire service. In the EMS field, responders are now certified to national standards for emergency medical technician and paramedic; in the fire service arena, however, national standards for training and safety have been available for a relatively short period of time. Although these national standards can provide an opportunity to justify the existing budget or gain approval for a budget increase, local FES officials often want the freedom to provide the level of public protection that is determined necessary by the local community. With the use of national standards, there is the understanding that a group of experts reached consensus to determine the level of protection needed.

At the local level, consensus building must start with the municipal administrator or elected officials. A chief should seek their full advice and consent. Keep in mind that it has been reported that city/county administrators would like the FES Chief to:

1. Be a team player.
2. Become politically astute.
3. Create a personal relationship.
4. Be honest.
5. Not get out in front of your city/county manager.
6. Not surprise the city/county administrators or elected officials.
7. Communicate, communicate, communicate.

Also include influential private citizens and any administrators who give trusted advice to the chief administrative officers, such as a budget director or public safety director, in the consensus-building analysis. The task is to build support among these influential people. The chief officer should begin by preparing a list of all those people and groups who could influence the final decision, such as firefighters' unions or volunteer fire associations in **TABLE 12-1**

The chief should rate each individual or group's potential impact on the final decision. For example, in

Person or Group	Potential Influence (high, medium, or low)	For, Against, or Neutral
TABLE 12-1 Consensus Building Example		
Chief administrative officer		
Mayor		
Power elite member (*your choice*)		
Firefighters' union		
Local newspaper		
Police chief		
Budget director		
The public		
City attorney		

a mayor-council form of government, the mayor and the most influential private citizen members should be rated as having a very strong potential influence. Next, each person or group's support of the proposal should be assessed. Again, if the proposal were to hire additional firefighters, the union president would be rated as strongly supportive.

After this list is completed, the chief should seek out those people who are very influential and will more than likely support the request. In private one-on-one discussions, these likely supporters should be briefed on the proposed request and asked for their input before their support is requested. These discussions should also take place with those individuals who are deemed neutral on the issue. The object is to bring these people to a point where they mildly support or do not oppose the proposals.

Finally, the people who will most likely oppose the proposal need to be approached. The object is to present the best case for support of the proposal. The hope is that these people will become neutral on the issue. Discussing the issue with these individuals also keeps them from feeling like they have been defeated

AN EXAMPLE: COMMON SENSE STATISTICS

FYI, this is a true story; however, all specific names have been removed. Several graduate students at a major research university decided to study the relationship between response time and property loss as a result of accidental single family structural fires. They were formulating this research project to satisfy the requirements for a dissertation in a Ph.D. program.

They contacted a county fire department that served the university to ask for help. They were looking for digitized computer data on structural fire calls. This department had been keeping fire data to be used for the National Fire Incident Reporting System (NFIRS) since the late 1970s. The fire officials were very happy to allow the students to analyze the data because they had been struggling themselves with determining an adequate response time. This planning information is vital to fire station locations.

When a few months had passed, the students contacted the fire officials and asked for a meeting. They stated that the results did not make sense to them and they wanted some insight. After processing the data through a sophisticated computer statistical program, the results showed a slight negative correlation between average response and property loss. Essentially, the longer it took the first due company to arrive, the lower the fire loss!

They discussed several possible explanations. First, this county fire department was the result of a consolidation of many volunteer fire companies, with a total of 46 fire stations. Where each station was built was not the outcome of a scientific planning process, but happenstance. As a result, there were stations that covered large areas and populations and some with small areas and populations. For example, the top four populations per station were: 40,000, 34,000, 33,000, and 31,000. These are very large numbers and these areas generated many fire calls. On the other end of the spectrum, the bottom four populations per station were: 2,000, 5,000, 6,000, and 6,000 (all urban areas).

At this time, this department did not have any standardized in-service training program. It could be assumed that firefighters assigned to the busy stations were handling many fires and gaining from this experience, therefore these stations had longer average response times because they covered larger first due areas. Thus, what the students had uncovered was the greater efficiency of the firefighters with more real world experience.

Ultimately, fire loss can be affected by several independent variables such as response time, staffing levels, and the training and experience of firefighters, to name a few. Presently, there are no known studies of field data to analyze response time and property loss correlations. To say it another way, at the present time, there is no actual proof that lower response times result in lower property loss or fewer fire deaths, although this would seem intuitively obvious.

CHIEF OFFICER TIP

Political Action Plan

A document from the International Association of Fire Chiefs (IAFC), entitled *How to Develop a Political Action Plan* (2012), is available for members of the IAFC at the organization's website (www.iafc.org). Many of the suggestions in this document can be used in the everyday political process to facilitate approval of a new policy or spending initiative. The document also has some good ideas for approval processes that involve a vote by the local town or city. A document from the International Association of Fire Fighters (IAFF) advises that political goals should be as follows:

- Explaining the value of the department or organization and of the services provided
- Discussing the value of the changes sought
- Creating public pressure on political leaders
- Producing any necessary votes by the city/council or commission

by an organized conspiracy of their opponents, without a fair chance to influence the decision. Each person or organization should be personally contacted. If they do not feel that they have been fairly treated, these individuals or groups may become a permanent opponent.

This detailed process should be necessary only for potentially controversial and expensive enhancement items in the proposed budget. For every new tax dollar the department receives in its budget, an equal amount must be taken from another agency's budget or generated from increases in taxes.

Incentive Pay for Paramedics

An EMS administrator proposes incentive pay for employees who have been certified as paramedics. The union supports the change, but the city officials oppose it. The public has no opinion because the change was not made known to the public by the local press. The municipal administration and the labor organization had come to an agreement that funds for this incentive pay could be taken from a recently approved, successful Risk Watch program that the chief officer championed last year and that had previously been successfully funded. The NFPA Risk Watch is an educational program for young children intended to reduce and eliminate loss of life and property damage from all types of accidents including fires. Although administrators have a responsibility to their organization to enforce its rules, support its mission, and serve its members, in this situation it is the responsibility of the chief officer to speak up and advocate for the public interest. In this example, support of the existing Risk Watch program would be a good idea. Then lobbying for additional funding separately for the new incentive pay as an enhancement would seem to be justified.

Discussion Questions

1. Does your department have an incentive pay program for paramedics?
2. Are there other certifications that should be part of an incentive program?
3. Complete Table 12-1 to help guide consensus building for the approval of the new policy program.
4. What process would you use to adopt this policy? Please explain in detail.

Baltimore City
FIRE STATIONS & RESPONSE AREAS

FIGURE 12-3 Four-minute response contours.
Courtesy of City of Baltimore, Office of the Mayor.

to reduce response time in one section of the city, the administration should determine the response time in the other surrounding areas **FIGURE 12-3**. It can then be argued that all residents in the city deserve equal fire and rescue protection (similar response times) and that the proposed funding increase would work to increase the percentage of city residents who fall into the adequate response time area.

The following sections discuss suggestions for handling public policy presentations as professionally and successfully as possible.

Public Policy Presentation

The next step is for the chief administrator to prepare a presentation to be delivered in public for the elected and appointed municipal officials. These officials act as arbitrators of the competing agencies' requests for public funds.

The administrator's presentation should be easy to understand and include common sense arguments. For example, if a department is requesting funding for a new fire station and firefighters to staff the station

Consider the Audience

When preparing a presentation, administrators should consider: What is the audience's real knowledge of FES operations? Many appointed and elected officials have no background in FES. Do these officials understand the jargon or acronyms that are typically used? If not, ensure you use common language instead of specialized terms that could alienate your audience.

Before scheduling a policy or budget hearing appearance, consider arranging a tour and orientation for the officials at one of the department's stations. Adapt an educational program used for children, having the officials try on the protective clothing and the self-contained breathing apparatus. Explain the environment that the fire and EMS personnel operate in when performing emergency operations. Use worst-case—but plausible—examples to illustrate those incidents when the department is pressed to its maximum effort.

Chief officers who are unsure of elected and appointed officials' understanding of the budget request and its impact on service efficiency should explain the details in their presentations along with emphasizing the improvements. A time-tested sales technique is to sell the product (improve emergency service) before informing the customer (legislators) of the cost. Also, marketing experts recommend communicating the message at least three times and preferably up to twenty. Never assume anything. Elected and appointed officials need to be educated in the nuances of FES operations. If in doubt, ask questions of the legislators' understanding but remember to not embarrass the legislators in public meetings. Consider giving private in-person meetings with each legislator. Again, inform the chief city/county administrator of the necessity to talk privately with legislators. The administrator must be onboard with the budget request, and that is where you must start your sales pitch.

Practice

As with any presentation, the chief should be very knowledgeable about the subject. At the first level of preparation for the presentation, the chief should become familiar with the entire budget and justifications. One technique that has merit is to ask the staff to review the budget and write down questions that the appointed and elected officials may ask. For questions that require in-depth knowledge to answer, an appropriate staff member should prepare a briefing document for review. This person should also be present at the budget hearing to fill in details or correct any misunderstandings.

If a chief lacks presentation experience, it is a good idea to practice the presentation in front of a friend or staff member or videotape the simulated presentation and review the recordings **FIGURE 12-4**. There are two different types of situations to practice: the prepared presentation and the response to the questions at the public hearing. Practice helps the chief be confident, relaxed, and persuasive.

Be Positive and Cheerful

Tell the audience, in simple terms, the positive aspects of the proposals. From practice, the chief should be able to maintain a positive, confident appearance that reflects professional ability and credibility. State the

FIGURE 12-4 Chief officers must be accomplished public speakers.
© Joe Amon/Denver Post/Getty Images.

justification in positive terms. For example, point out that the department can provide better service with the approval of the enhanced budget requests.

Maintain Eye Contact

Reading is distracting and prevents the speaker from maintaining eye contact. The presentation has more impact and is more believable if the speaker gives the presentation extemporaneously from bulleted notes, rather than reading through pages and pages of material. Maintain as much eye contact with the elected officials as possible; many people relate eye contact to honesty and sincerity. Scan the officials without staring, which usually involves eye contact of about 6–8 seconds. A glance at their eyes and facial expressions may give an indication of their reaction to the presentation, and the presentation can then be adjusted accordingly.

Use Down-to-Earth Ideas

Whenever possible, use an analogy or a real-life story, perhaps including a humorous anecdote, to help explain a theoretical or abstract proposal. "A picture is worth a thousand words" also applies to the use of analogies.

Use Visual Aids

Include visual aids such as handouts, charts, slides, and videos **FIGURE 12-5**. However, make sure they do not distract from the main presentation. If using films or videos, be careful not to use too much of the time allocated for the presentation. If necessary, review the film or video and limit it to no more than 3–10 minutes. Selective use of video aids can increase the understanding and attention of the elected officials.

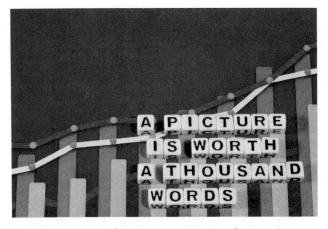

FIGURE 12-5 Carefully choose a slide that fits into the presentation and sends the accurate message. In addition, most Power Point presentations have several slides with bulleted items that are an outline of the presentation.
© Michael R Ross/Shutterstock.

Involve the Audience

Appropriate techniques for involving the audience may vary, so it is helpful to have a good understanding of procedural rules and the respect that is expected by the elected officials. At a public hearing, elected officials strive to positively influence the public and media who may be present. A direct confrontation or disrespect can cause a loss of support.

Research the background of each elected official. A thorough understanding of the audience and their past experiences increases the chance of involving them positively in the discussion. In addition, the chief should have personal one-on-one meetings to get to know all the officials before the presentation.

At the end of the presentation, ask for questions. It is at this point that the chief officer really has to think on his or her feet. Remember, the elected officials may have prepared questions for the hearing. Answer truthfully and as simply as possible. Keep to the facts and professional opinion. If the answer does not come to mind, say so and offer to locate the answer after the hearing.

Using Statistics

Statistics can also be useful for creating and justifying policy analysis and recommendations. National FES statistics from 2008 to 2017 reflect the following trend: fire incidents per million population decreased 12.3%, civilian deaths per million population increased 2.4%, and civilian injuries decreased 31.2% (USFA, 2019). The chief officer should be aware, however, of trends that might affect these statistics in the future. For example, consider the effect the following trends might have on these national fire statistics:

- Smoke alarms have been installed in 96% of U.S. homes; however, approximately 60% of fire deaths occur in homes where the smoke alarm was not installed or was installed improperly, was dysfunctional, or was missing the battery (Ahrens, 2019). To further add to the complexity of statistics regarding these devices, homes protected with working smoke alarms might have a fire that is detected early and extinguished by occupants, therefore never resulting in a call to the fire department.

- The greatest percentage (21%) of fatal fires is started by misuse of smoking materials. In 1965, 42.4% of the adult population smoked; in 2017, that figure had dropped to 14.0%. In addition, as of 2011, all states now require cigarettes to meet a fire safety standard that is designed for the cigarette to self-extinguish after a few minutes left unattended. This decrease in smokers and more stringent standards has most likely had a large impact on fire deaths over the past years and will continue to do so in the future.

- Mandatory installation of residential sprinklers continues to raise strong opposition from homebuilders and elected officials. As a result, fire sprinklers protect few single-family homes; however, NFPA 101 *Life Safety Code* and the *ICC Residential Code* now require residential sprinklers for new one- and two-family dwellings, which will increase the adoption of sprinkler requirements over time.

- The addition of first-responder medical service by fire departments is very common. In most cases, citizens are receiving emergency medical treatment a lot faster because the response times of fire companies are shorter than for the typical EMS unit. However, this trend may also cause an increase in response times for fire calls if the closest fire company is working at an EMS incident when a fire is reported in its first due area.

As with the concept of outcomes, statistics in the FES can be misused and misunderstood. For example, consider the structural fire data shown in **FIGURE 12-6**. The data show that an increase in the average dollar loss per incident equates with an increase in the number of firefighters who respond to the incident. This makes sense; a large fire resulting in extensive property damage requires a large number of firefighters on scene, and a fire that results in minimal property loss requires only a few firefighters to extinguish the fire.

A city manager using this data and graph to determine how to reduce fire loss might notice that reduced

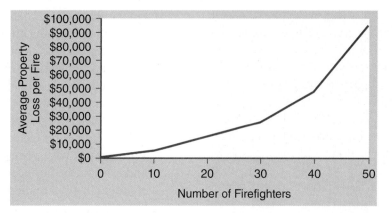

FIGURE 12-6 Average property loss per fire versus number of firefighters responding.

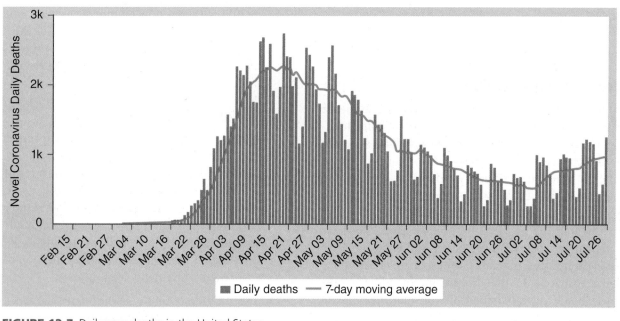

FIGURE 12-7 Daily new deaths in the United States.

Source: https://www.worldometers.info/coronavirus/country/us/ (October 27, 2020).

Courtesy of Worldometers.info; Retrieved from https://www.worldometers.info/coronavirus/country/us/ (July 29, 2020).

property loss is associated with fewer numbers of responding firefighters. The city manager may then conclude that sending fewer firefighters to fires results in smaller losses. Although the data are correct, the analysis lacks a sensible approach to interpret the results. This is generally called "common sense."

As explained above, statistics from surveys and studies can be interpreted and analyzed in many different ways. If administrators or their staffs do not have a good understanding of statistics, they can unintentionally be misled by some of these reports. Helpful in understanding a statistical report are visual presentations; bar graphs, pie charts, histograms, stem & leaf plots, dot plots, scatterplots, and time-series graphs. In many cases to understand the visual presentation, it would be a good idea to find an expert to explain these different statistical methods. Also very important is to

recognize "outliers" and remove them from the data set. Many are the result of errors in data collection or recording. A visual graph or plot will help identify these "outliers."

The bar and line graph in **FIGURE 12-7** demonstrates some great advantages to visual statistics. First notice the bar graph varies wildly by day. Researchers normalized the data by using a 7-day average and placed the subsequent line graph over the actual daily data points. This clearly shows the deaths starting in early March 2020 and peaking in mid-April 2020. Because the original goal of the lockdown was to assure that the hospitals would not run out of beds, the line graph clearly shows a substantial reduction in deaths; therefore, less seriously ill virus patients. For medical experts, policy officials, politicians, and the public, this indicated that this public policy objective was being accomplished.

A good example of a possible "outlier" is noted on October 21, 2020 of the bar graph (Figure 12-7) where the deaths seem to "sky rocket." Checking the data from this one day, we found one instance of a possible mistake. In October it was being reported the deaths attributed to the seasonal flu are substantially down; maybe they are mistakenly being counted as Covid 19 deaths? This looks like an "outlier," either accidental or purposeful.

The following observations can be made from this data set that may be used depending on which side of the political fence the reporter is on:

1. The seven day average is trending up and this is problematic.
2. October 21st numbers are a spike in the deaths and very concerning.

3. October 26th numbers are substantially down compared to October 21st and there should be optimism.
4. Compared to the highest average death rate in April, the death rate is down about 63%.

This indicates success in confronting the virus.

This points out the difficulty in using statistics. They can be used in various arguments to support a person's viewpoint. A person's ethics are very important in reporting statistics. In any case, the administrator or trusted staff should access the raw data and question all observations for truthfulness.

When it comes to surveys, there are several items the chief officer can look for in determining veracity and reliability. The first is that the survey was truly

---

Facts and Figures

Total Quality Management

What does it mean to the public if a business or agency gives less than 100% effort at all times? In the book *Root Cause Analysis: A Tool for Total Quality Management*, this question is examined. The following calculations are based on the assumption that businesses lowered their goals by only 0.1%, to 99.9% effort (Wilson, Dell, and Anderson, 1993):

- 16,000 pieces of mail would be lost per hour.
- 20,000 incorrect drug prescriptions would be prescribed each year.
- 500 incorrect surgical operations would be performed each week.
- 50 newborn babies would be dropped at birth by doctors each day.
- 22,000 checks would be deducted from the wrong account each hour.

It is important that the source providing the statistics is credible and reliable. This does not guarantee that a mistake will not be made in the study, but it can give a strong indicator of whether the statistics are correctly reported. Some organizations or individuals collecting statistics may have a particular interest in a study being conducted, also called a bias. Although many reported studies are accurate, the chief should be aware that some may be structured with a narrow focus toward a preferred outcome.

Also, consider the organization(s) sponsoring or funding for the study. The largest number of these studies are created for the use of lobbyists, marketing groups or companies that provide a service or product. Almost all issues have some type of advocacy group trying to

guide the outcome to the advantage of their clients. Even government organizations and research schools can be influenced to conclude the appropriate "truth."

Consider the following example:

"How about this: The National Oceanic and Atmospheric Administration has made repeated 'adjustments' to its data, for the presumed scientific reason of making the data sets more accurate.

Nothing wrong with that. Except, all their changes point to one thing — lowering previously measured temperatures to show cooler weather in the past, and raising more recent temperatures to show warming in the recent present." (Reproduced from The Stunning Statistical Fraud Behind The Global Warming Scare, 2018, © 2000-2020 Investor's Business Daily, Inc. Retrieved from: https://www.investors.com/politics/editorials/the-stunning-statistical-fraud-behind-the-global-warming-scare/ There is no smoking gun in this example, but this looks suspicious.

Each valid statistical conclusion from a random sampling study should have a confidence level reported. For example: "There is a 95% certainty that the average age of firefighters in this country is between 36 and 39 years old." Or, a statistic may report that there is a plus or minus variance of some percentage. In any case, consider each study and its conclusion using common sense before incorporating it into your public policy analysis. Statistics and other mathematical descriptions of problems in the public sector can be very helpful for understanding a problem, analyzing possible solutions, and finalizing the plan to solve the problem.

Reproduced from Wilson P. F., L. D. Dell, and G. F. Anderson. 1993. *Root Cause Analysis: A Tool for Total Quality Management*. Milwaukee, WI: ASQC Quality Press.

randomized; most surveys sample a small number of people and then generalize the results for a large population. One effective method of selecting participants for a telephone survey is to use a random number generator (on a computer) to select telephone numbers from all participants in the area being surveyed. Furthermore, a sufficient number of respondents are needed to ensure that the standard of error is small, which is expressed as a plus or minus percentage error after the result. Although not everyone in the population can be surveyed, the more people included, the better chance of receiving an accurate prediction of the outcome for the larger group. At minimum at least 250 observations are required for a low standard of error. Many surveys use greater numbers primarily to guarantee a random selection and gain the confidence of the general public.

Case Studies

In addition to statistics, researching case studies can be a good approach to studying, reviewing, and analyzing administrative and policy problems. Administrators can use case studies to learn from others' experiences. When dealing with problems relating to humans, administrators should consider subtle factors, such as how people feel about an issue or how they will behave, which is relatively easy to learn about using case studies. In many cases, without being able to prove these factors scientifically, as in a controlled statistical study, administrators must treat these human behavior traits as facts. For example, in the case of smoke alarms, a human behavior problem is that many smoke alarms that have been installed are nonfunctional because of missing or dead batteries; this is very troublesome to fire protection professionals. Case studies are anecdotal evidence and may not meet the same rigorous standard as randomized studies; however, they can also be very powerful at persuading others that the policy analysis is valid or for understanding the perspectives of others.

Program Budgeting

Senior public administrators and politicians demand analytical, quantitative justifications to support budget requests and policy decisions. Formal policy analysis can provide the justification needed when the leader of a fire department or EMS organization is faced with scarce resources.

Program budgeting—sometimes referred to as the Planning, Programming, and Budgeting System

(PPBS)—is not a formal part of the budget process but rather a separate policy analysis. This budgeting system describes measurable outcomes and gives detailed costs of every activity or program that is to be carried out in a budget. If justifications for the budget requests are supported by irrefutable quantitative outcomes or expected results, the FES administration has an advantage to bringing increased funding to the agency; however, with politics there are no guarantees. In addition, because of its time-consuming effort, it often results in opposition from managers and therefore is not widely used. However, the administration must make its best effort to get the resources needed to accomplish the organization's goals, and program budgeting is one of the tools that can be used to accomplish this task.

Cost-Benefit Analysis

Because it is often difficult to measure benefits or direct outcomes from FES programs, city or county administrators seldom rely on cost-benefit analysis to provide guidance in allocating public funds and developing public policy.

In private industry, the balance of cost to benefit is at the heart of any plan of action, and the financial bottom line is one of business owners' and stockholders' most important considerations. However, for environmental, safety, and health services and regulations, there may be many instances where a certain decision might be right even though the regulation's benefits do not outweigh its costs. There are good ethical reasons not to assign dollars to the value of life.

For example, if the chief proposed a new fire safety regulation that required the installation of sprinkler systems in all new one- and two-family dwellings, the opposition would probably produce a cost-benefit analysis showing that the costs (installation of the system) outweigh the benefits (savings in fire insurance, property damage, and life loss). Although the difference in these values has become almost insignificant as the cost of residential sprinkler systems continues to fall, the opposition would probably use the highest costs it could find to help support its argument that sprinklers would cost substantially more for the home buyer. In this case, the chief should point out the benefits of the sprinkler regulation that cannot be quantified, such as saved lives, prevented injuries, and preservation of the home and the irreplaceable personal items in it. If an administrator were to survey the public about the value of FES, there would probably be a great variance in the valuation from person to person, situation to situation, and group to group. What makes cost-benefit analysis even more difficult for FES organizations is the realization that the outcome—saving lives and property—is in most cases the result of a complex set of circumstances. The following is a hypothetical set of circumstances that could lead to a death in a house fire:

1. Smoke alarms were not required by the local fire prevention regulations to be installed in existing dwellings.
2. Home fire inspections along with fire safety education were not provided by the local FES organization.
3. The occupants of the dwelling did not see any value in installing smoke alarms.
4. The occupants of the dwelling did not recognize the hazard of careless smoking in the home.
5. The first arriving FES unit was understaffed or personnel were not adequately trained.
6. When the victims were eventually rescued and brought to the front yard, the FES organization was not equipped or trained to provide EMS.

It could be argued that if any of these six elements were different, lives could have been saved. This makes it almost impossible to construct a valid cost-benefit analysis. Although some of these items are clearly more expensive than others to correct, it would be good to implement programs to correct all the deficiencies indicated by each item. Anything less than 100% effort in all areas leaves open opportunities for preventable life and property loss in this example.

Program Analysis

One method that can be used in place of cost-benefit analysis is program analysis. This systematic process examines alternative means, methods, or policies to accomplish the goals and objectives of the fire department or EMS organization. This type of analysis accepts the use and reporting of intuition and judgment, and statistical studies and sensitivity analysis. This method should be used only for those problems that have a big impact on the services the department delivers such as new fire stations or major reductions in funding of the department.

Program analysis works best when a chief appoints a task force representing all members of the department. This technique is used to make sure that all possible solutions are discussed and considered. There may be cases when citizens, special interest groups, and elected officials are also requested to participate.

This participation can be used as part of a strategy to gain support for a new program.

The FES or EMS administrator leads the task force and defines the problem to study. For example, a report on response times in a city notes that one geographic area has substantially longer response times to emergency calls than the averages in other parts of the city. A task group is called together, the problem is outlined, and the group is asked to complete an analysis and recommend several solutions. This group is given the following guidelines:

- Brainstorm. Allow all ideas to be listed and considered. Identifying the problem can be the most critical point in this process. Often, a symptom is confused with the real cause. All solutions should be analyzed. Have the group consider all traditional and inventive ways to solve the problem.

- Evaluate. Remember to test each solution for political sensitivity, human behavior problems, and common sense. Review any public preference surveys that have been conducted (see the chapter on *Customer Service*). Surveys can be a strong justification to support the plan.

- Implement. If the selected solution requires additional funding, consider all options, such as increasing taxes, consolidating or relocating existing stations, charging fees for service, or establishing mutual aid with an adjoining jurisdiction. In addition, a plan that contains phases is very useful to the decision-maker. The administrator can then focus on obtaining resources for part of the plan each year, rather than dealing with the entire financial impact in the first year. For example, the plan may call for the construction and staffing of three new fire rescue stations. The administration may be able to use its skills in negotiation, compromise, and political sensitivity to gain approval to construct one new fire station every 2 years, with full implementation of the plan in 6 years.

If the FES or EMS administration is not successful in gaining approval or financing for the plan or change initiative, the plan should be stored for use in the future if an opportunity surfaces.

Policy Analysis Reference Sources

One of the most comprehensive outlines of public FES is contained in NFPA 1201, *Standard for Providing Emergency Services to the Public*. This document, which was created in 2010, references the following sources:

- NFPA 1500, *Standard on Fire Department Occupational Safety, Health, and Wellness Program*
- NFPA 1710, *Standard for the Organization and Deployment of Fire Suppression Operations, Emergency Medical Operations, and Special Operations to the Public by Career Fire Departments*
- NFPA 1720, *Standard for the Organization and Deployment of Fire Suppression Operations, Emergency Medical Operations, and Special Operations to the Public by Volunteer Fire Departments*
- Insurance Service Office (ISO), Public Protection Classification Service; Fire Suppression Rating Schedule (contact the ISO at ISO Customer Service Division, 545 Washington Boulevard, Jersey City, NJ 07310–1686)
- Center for Public Safety Excellence; Commission on Fire Accreditation International (CFAI); http://www.publicsafetyexcellence.org/agency-accreditation/about-accreditation-cfai.aspx

These sources should all be reviewed and incorporated into policy analysis. In addition, for the emergency management function, NFPA 1600, *Standard on Continuity, Emergency, and Crisis Management*,

should be consulted. Note that each NFPA standard may contain references to other standards and reference sources that need to be referred to while developing a policy analysis. The following sections further discuss NFPA 1710 and the ISO documents.

NFPA 1710

NFPA 1710, *Standard for the Organization and Deployment of Fire Suppression Operations, Emergency Medical Operations, and Special Operations to the Public by Career Fire Departments*, was first issued by the NFPA in 2001. There was opposition to its adoption from many special interest groups. For example, the International City/County Management Association stated in an article that several managers believed the standard was "being specifically cited in their communities as a means to leverage more resources for only one aspect of the fire service: staffing and deployment" (Data from International City/County Management Association, 2002).

The IAFC and the IAFF originally published separate documents in 2001 explaining their interpretation of this standard. There were some differences. However, in a 2002 joint analysis of the standard, these two groups came to an agreement and jointly published a new document titled *NFPA 1710: Implementation Guide* (2002). This guide states the following (Reproduced from International Association of Fire Chiefs and International Association of Fire Fighters, 2002, pp. 1–9):

> The NFPA 1710 standard could be found highly relevant to the question of whether a jurisdiction has negligently failed to provide adequate fire or emergency medical protection to an individual harmed in a fire or medical emergency. To prevail in such a claim, the individual would have to show that the jurisdiction failed to provide the level of service required by the standard, and that this failure was a cause of an injury, death, or property damage.

The guide is an excellent source for any analysis on staffing and deployment of fire and rescue services.

ISO Fire Suppression Grading Schedule

The ISO gathered 5 years' worth of data, reviewing the cost of fire claims around the country for homeowners and commercial property insurance. The data showed that "the communities with better classifications [Classes 1–4] experienced noticeably lower fire losses than communities with poorer classifications" (Data from ISO, 2001, p. 3). Because better ratings are very sensitive to reduced response distances, adequate staffing, training, and water supply, it is not surprising that departments that have these resources can more effectively extinguish fires.

The NFPA *Fire Protection Handbook* points out one very important factor to keep in mind when reviewing a recommendation from the ISO for public policy analysis: "The purpose [of ISO grading] is to aid in the calculation of fire insurance rates and is not for property loss prevention or life safety purposes" (Courtesy of NFPA; NFPA, 2008, pp. 7–40). Therefore, if the department is looking for recommendations to improve emergency services to the public, use the ISO recommendations as part of the analysis, but not exclusively.

The ISO measures fire company coverage with a response distance calculation. If the area served by the station is within 1.5 miles of the station, the department gets full credit for this item. This is a numerical evaluation and partial credit is allowed if the station is over 1.5 miles from the protected area. However, using only distance can be misleading and result in big differences in response time from area to area. For example, a fire station that is located in an urban area that has heavy traffic or a volunteer station that relies on members traveling to the station before responding will have substantial increases in response times.

There are two advantages to using ISO ratings. The first is that the evaluation is free to the municipality, although there are costs if the municipality attempts to improve its rating. The other advantage is that a chief can characterize any development that would improve the public fire protection rating as likely to reduce insurance costs to the owners of property in the city. However, only owners of certain structures receive substantially reduced insurance premiums. Properties protected by automatic sprinklers—such as large industrial complexes—already have lower rates. Single-family dwellings are also rated in standard policies that do not see lower premiums unless the rating is dropped several classes. For example, in Florida, the insurance for single-family dwellings and commercial structures protected by automatic sprinklers is essentially the same for Classes 1–7. In addition, several national jurisdictions with populations of over 200,000 have their insurance rates based on loss experience, and the ISO rating has no effect.

The problem with using ISO for policy analysis, especially in cost-benefit calculations, becomes evident when the financial pluses and minuses are considered. Any improvements for the department are funded by tax dollars. Any benefit to the citizens and property owners comes from lower insurance premiums. The two may not seem connected or noticeable; however, a

TABLE 12-2 NFPA 1710 vs. ISO Class 1 Requirements

	NFPA 1710	ISO Class 1
Engine company travel distance	2–2.5 miles*	1.5 miles
Staffing	4 firefighters	6 firefighters

*Note: Because NFPA 1710 uses response times, the distance will vary depending on items such as road layout and traffic conditions.

business owner may pay less insurance but have to pay higher property taxes to fund improvements in public fire protection.

It is incumbent on the department to educate the business community and elected officials about the advantages of a more effective FES department. Many jurisdictions have used lower insurance premiums to attract new commercial development; however, as can be seen from the ISO fire loss data, complying with ISO recommendations does make fire departments more effective.

TABLE 12-2 further compares the requirements of NFPA 1710 to the requirements in ISO (to gain the greatest credit in the rating calculations). In general, a department using NFPA 1710 for deployment and staffing has a lower budget than a Class 1 ISO department because fewer stations are needed to meet the response times (distances) and fewer firefighters are needed per company. The calculations for costs using ISO are very complex, so it is hard to generalize for other protection classifications, although some fire experts have commented that they think the break-even point in a cost-benefit analysis for an ISO rating is somewhere around Class 4.

Changing Social Perspective

Administrators who use a group to make recommendations on policy should be aware of the attitudes and values of members of the organization when conducting policy analysis. Because of such management techniques as total quality management and Six Sigma, many new members are under the impression that they will be a part of every decision-making and planning process.

Members and their support groups today may feel entitled to the benefits of the job and may not be open to discussions about sacrifices that are necessary to implement change. In the past, most FES members lived in the community they served. This is no longer true;

many members—volunteer and paid—live miles from the districts they serve. These commuters may not be as committed to the municipality they serve because they do not live there. Several conflicts of values can be seen in these groups; for example, group commitments versus individual needs, respect for authority and obedience versus participation and democratic rule, a melting pot of cultures and races versus diversity and appreciation for differences, and materialism versus value gained through new approaches and experimentation.

Therefore, the administrator must be prepared to guide any policy task group toward the strategic goal, which is ultimately the best emergency service for the tax dollars spent. Keep in mind, it's still the FES administrator's responsibility to make every member, regardless of their generation, feel involved. You also need to incorporate members into the FES culture and make them feel valued. There may be conflicts between the member's wants and the citizen's needs; the administrator must make it clear that the public comes first.

Empowering Employees

Empowerment has been praised by many management gurus as the way to put policy analysis at the worker's level; however, the chief officer should not adopt and implement these or any business management methods without first conducting a detailed study. Empowerment could cause havoc if used by front-line emergency service workers at the company level. Should paramedics choose the medical protocol on an ad hoc basis for each call? Should the members of each engine company be allowed to change the location and setup of equipment on their engine? With no consistency or adherence to generally accepted practices and procedures, each shift, crew, station, or battalion would do its own thing. This encourages "freelancing," which invariably creates a safety issue.

If there is a need for policy analysis or change in operations for emergency service, empowerment is not a good idea. However, if, for example, a chief assigns an officer the responsibility to supervise, plan, and create a new collapsed building and confined space rescue training area, this officer may be empowered with the authority to do whatever is necessary to complete the project. The officer may reach out to the private sector for donations of equipment, materials, and even help with the construction. The officer might then be empowered to mount a plaque in a prominent place at the training area, thanking the donor for the help.

Wrap-Up

CHAPTER SUMMARY

- A discussion of public policy and how it fits into FES planning.
- How does the availability of public revenue affect the analysis of public policy?
- What are "outcomes" and how are they measured?
- Public policy presentation to elected/appointed officials, the public, and members of the FES organization is discussed including the significant elements.
- Professional standards for measuring and defining FES service are discussed and should be the core of any FES public policy analysis.

CHAPTER ACTIVITY #1: THE QUINT CONCEPT

One metropolitan fire department has adopted a new strategy for equipping and staffing its fire companies. The quint concept is named after a fire apparatus that is constructed to replace an engine and ladder company with one piece of equipment containing hose, water, a pump, and an aerial ladder. Some experienced fire officers have observed that this type of apparatus should have a crew of six or seven to fully use the quint's potential; however, this department uses only four firefighters for each company. The senior officials of the department are quick to point out several tactical advantages that mainly focus on their ability to direct many elevated streams quickly on major defensive strategy fires. In actual experience in another major east coast fire department, the quint companies were staffed with six firefighters. It was typical for these companies to default to engine company duties and leave all truck company duties, such as ground ladders, ventilation, search and rescue, etc., for other dedicated ladder companies to accomplish.

However, the original pressure for the new policy may have come from other sources. This city was faced with a substantial shortfall of tax revenues, and the elected officials were pressuring the fire department for reductions in personnel. The fire department had also not been able to purchase any new fire apparatus for many years.

Generally, many older central cities in this country are protected by fire departments that have a long tradition and history. These older departments typically have some stations that were constructed very close to each other at a time when horses pulled the steam pumpers, but this spacing is no longer necessary or justified with motorized apparatus.

In addition, most of these cities have a greater number of ladder companies than typically found in newer cities or suburban areas. In the past, there was justification for these ladder companies to rescue trapped occupants from windows in taller buildings, but the enforcement of modern building and fire codes, including automatic sprinklers and fire-resistive exit stairs, has eliminated many of these firetraps.

Before changing to quints, the department operated 30 engines and 10 ladders out of 30 stations, totaling 40 companies. After the conversion, the department ended up with 30 quints and 4 ladders totaling 34 companies. Slightly over 15% of its fire companies were disbanded. Even with this major cutback in personnel and companies, the department managed to keep all fire stations open by placing one quint engine in each of the 30 stations. It has been argued that the original pressure for this change was not a better way to extinguish building fires, but to comply with a major cutback in the fire department's budget.

Discussion Questions

1. How would you research the facts in this case to assess their accuracy?

2. If the original justification was budget cutting, is this new system actually giving better or equivalent fire protection?

3. What harm would there be from not informing the public of all the circumstances behind the new policy decision?

4. Do customers believe FES members are their protectors and trustworthy, and therefore blindly trust policy decisions?

5. How does reduced revenue for the city affect the customers' demand for FES services?

CHAPTER ACTIVITY #2: SMOKING BAN

Many fire departments provide automatic retirement benefits for firefighters who develop any heart/lung illness. One major cause of these illnesses in the general public is smoking cigarettes. An article describing the results of studies by the Medical Research Council (London, England) indicated that cigarette smoking has a greater harmful effect than fire fumes on measures of pulmonary function. "The prevalence of respiratory symptoms was higher in smokers than non-smokers and increased with the number of cigarettes smoked" (Douglas et al. 1985). In addition, a 1973 study by H. D. Peabody discovered that cigarette smoking and exposure to carbon monoxide interacted in a synergistic fashion to substantially lower pulmonary function, reducing the ability of the firefighter to perform heavy work for sustained periods of time (Peabody, 1973).

In March 1978, the city of Alexandria, Virginia, stopped hiring smokers as firefighters because smokers accounted for a disproportionate number of early retirees from the fire department. Although the argument to support banning cigarette smoking by all firefighters seems strong, there are some practical and legal questions that need to be resolved when implementing this type of policy. For example, many unions or individuals would argue that this is an infringement on personal rights especially when they are off duty.

Discussion Questions

1. Is it reasonable to adopt the policy of not hiring smokers and requiring all new firefighters to sign a condition of employment that they will not smoke while employed as a firefighter? Annual medical examinations could test for nicotine as part of the enforcement process.

2. Should the city fund a recognized program for smokers to quit and encourage firefighters to attend?

3. Should a policy that prohibits smoking in the fire stations be adopted?

4. Would this new policy save the city or county money by reducing costs for medical insurance, reducing sick leave, and in general providing a more physically fit firefighter? Provide references.

CHAPTER ACTIVITY #3: GENERAL DISCUSSION QUESTIONS

Choose a major program in an FES services agency and assume the executive official or the elected legislature of your city has cut the department's proposed budget for this program. As the FES administrator of the department, answer each of the following questions with consideration to justifying the budget.

Discussion Questions

1. Should you mount a campaign to support your proposed budget using family, friends, and the firefighters' union?

2. Should you work quietly within the system to determine the political feasibility of fighting publicly?

3. Should you fight just enough to show your employees that you care, but not enough to do political damage to yourself? (How would you determine what is "just enough"?)

4. Should you leak information to the media about the disastrous consequences of the budget cut?

5. Should you accept the cut and prepare to be creative about how you implement it?

6. Should you be prepared to point out problems with the budget cut when the next catastrophic fire or emergency incident occurs?

7. How would the old adage, "You don't want to win the battle and lose the war," affect your actions?

8. What influence would knowing that your boss has the authority to fire you have on your actions?

CHAPTER ACTIVITY #4: PROBLEMATIC PATIENTS

An 84-year-old woman named Hazel Henry is known by the agency as a "frequent flyer." She has limited income and lives alone in an old house that is badly in need of repair. She lost her husband a year ago and has no other family close by. Her daughter has tried to convince her to move into a senior home, but she refuses.

Hazel has mobility issues and a number of medical problems, including diabetes, high blood pressure, and a heart condition. She does not drive and her doctor is a considerable distance away. She falls frequently, is often hypoglycemic, and has the occasional hypertensive crisis and chest pain. She calls your fire-based EMS resources at least once a week. She loves your crews and sometimes calls just because she wants the company.

Discussion Questions

1. Because your local and regional policies require response and transport for any patient that requests it, patients like Hazel tax your minimal resources. Is there anything you can do to prevent this?

2. What kind of programs could help create efficiency for your service as well as benefit the patient?

3. How will you know if this type of program would be a viable option in your community?

KEY TERMS

Case studies A research process that tells a story about an administrative issue. After it is completed and verified, it is used many times as a lesson in public policy curriculums. Simply, these case studies describe either the failure or success of an administrative program.

Consensus building Usually used in a formal committee as the process to determine the conclusion of a discussion on an issue. The voting can be anywhere from a simple majority to complete agreement.

Cost-benefit analysis Cost–benefit analysis is a systematic approach to estimating the strengths and weaknesses of alternatives by identifying the real complete costs versus the benefits in terms of economic outcomes.

Efficiency A determination that a process has an advantage to others resulting in a better outcome for the amount of input. For example, my car is more efficient than yours because it gets better gas mileage. Many times subjective determinations.

Empowerment Allowing of the authority, right, or ability to perform various acts or duties without a supervisor's approval. More commonly found in the retail or service industries where immediate satisfaction of the customer is important.

Incremental adjustment A method of implementation that works by making a small change and assessing the results.

Outcome A measured benefit to stakeholders, created as a direct result of dollars spent by the organization.

Performance measures Generally, are measures that are quantifiable and are essential to manage anything. Used as a planning input or to assess a person's contribution to the organization or their meeting some performance measure of job accomplishment.

Policy analysis A method used in government and nonprofit organizations to examine and evaluate the available options to implement the goals of laws, regulations, and agency policies. Generally, there is no profit involved, just service to the public, such as FES organizations.

Program analysis A systematic process that examines alternative means, methods, or policies to accomplish the goals and objectives of the FES organization.

Program budgeting A budgeting system that describes measurable outcomes and gives detailed costs of every activity to be carried out in a budget.

Strategic planning A type of planning that identifies the department's vision, mission, and strategies.

Statistics The field of scientific study that concerns the collection, organization, analysis, and presentation of data as opposed to opinions. One simple example would be the percentage of people who have blue eyes. If you could count everyone, this would be a hard number. However, many times statistical inferences are used, such as polling 1,000 people. The study may report 10% have blue eyes with a plus or minus of 2%. This is a mathematical technique.

Tactical plans Tactical plans are the very actions that are necessary to carry out the strategy. For example, the strategy may be to travel to a distant city. Some of the tactics could be fill the car with gas, check all fluids, get directions including rest and food stops, etc.

REVIEW QUESTIONS

1. List at least four resources that can help the FES administrator during policy analysis, and provide an example (hypothetical or from your experience) of how each one can be used.

2. Review the suggestions for handling public policy presentations. Which do you think will be easiest for you? Hardest?

3. Research and identify a scientific study related to your area of FES. Read about the study and then critique its usefulness for public policy use. What major areas does the study aim to illuminate? What other studies on the subject would enhance its conclusions? Do you detect any potential bias based on its authorship?

4. Research and identify a case study related to your area of FES. Read the case study and then critique its usefulness for public policy use. What major areas does the case study aim to illuminate? Would a scientific study on this topic strengthen the case or not? Do you detect any potential bias based on its authorship?

5. What are the pros and cons of employee empowerment? Give an example of one situation in which empowerment can support public policy, and one in which it would hinder it (different from those discussed in the chapter).

REFERENCES

Ahrens, M. 2019. NFPA's *Smoke Alarms in U.S. Home Fires*. Retrieved from https://www.nfpa.org//-/media/Files/News-and-Research/Fire-statistics-and-reports/Detection-and-signaling/ossmokealarmstables.pdf

Douglas, D. B., R. D. Douglas, D. Oakes, and G. Scott. 1985. Pulmonary function of London firemen. *British Journal of Industrial Medicine, 42*(1), 55–58.

Flynn, JD. 2009. *Fire Service Performance Measures*. Quincy, MA: National Fire Protection Association.

Garrick, Gary. 2014. City [San Diego] tries to speed emergency response. Retrieved from https://www.sandiegouniontribune.com/sdut-san-diego-fire-fast-response-crew-pilot-2014may11-story.html/.

Insurance Service Office. 2001. *ISO's PPC Program: Helping to Build Effective Fire-Protection Services*. Retrieved from https://www.verisk.com/insurance/about/faq/the-public-protection-classification-ppc-program/

International Association of Fire Chiefs. 2012. *How to Develop a Political Action Plan*. Retrieved from www.iafc.org.

International Association of Fire Chiefs and International Association of Fire Fighters. 2002. *NFPA 1710: Implementation Guide*. Fairfax, VA: International Association of Fire Chiefs.

International City/County Management Association. 2002. *NFPA 1710: Your Community May Be Next*. Retrieved from http://www.icma.org

Lindblom, C. E., and E. J. Woodhouse. 1993. *The Policy-Making Process*. 3rd ed. Upper Saddle River, NJ: Prentice Hall.

National Fire Protection Association. 2008. *Fire Protection Handbook*. 20th ed. Quincy, MA: National Fire Protection Association.

National Institute of Science and Technology. 2010. *Report on Residential Fireground Field Experiments*. Retrieved from http://www.nist.gov/customcf/get_pdf.cfm?pub_id=904607

Peabody, H. D. 1973. San Diego Fire Department Health Survey. In: *Survival in the Fire Fighting Profession: Second Symposium on Occupational Health and Hazards of the Fire Service*. University of Notre Dame, Indiana: John P. Redmond Memorial Fund, International Association of Fire Fighters.

U.S. Fire Administration. 1973. *America Burning*. Retrieved from www.usfa.fema.gov/downloads/pdf/publications/fa-264.pdf

U.S. Fire Administration. 1977. *Urban Guide for Fire Prevention & Control Master Planning*. Washington, DC: U.S. Department of Commerce, National Fire Prevention & Control Administration, National Fire Safety & Research Office.

U.S. Fire Administration. 2019. *Fire in the United States 2008–2017* (20th ed.). Retrieved from http://www.usfa.dhs.gov/downloads/pdf/statistics/fa_325.pdf

Wilson P. F., L. D. Dell, and G. F. Anderson. 1993. *Root Cause Analysis: A Tool for Total Quality Management*. Milwaukee, WI: ASQC Quality Press.

CHAPTER 13

The Future

LEARNING OBJECTIVES

After studying this chapter, the student will be able to:

- Comprehend the creation, effect, and importance of progressive change (pp 244–245).
- Understand the impact and influence that the future will have on fire and emergency services (FES) administration (p 246).
- Examine and comprehend the effect that outside influences have on the future (pp 246–248).
- Identify the responsibility and accountability of administration to constantly strive for improved service to the public (pp 248–249).
- Understand the importance of predicting future trends in the public FES (p 249).

Case Study

Even though many fire departments (and their governments) are experiencing economic difficulties, consolidation or mergers continue to be viewed with apprehension. Actually, this would be the most challenging future change for a government or FES agency on many levels.

Consolidation or mergers usually result in a savings for the city and/or county and its residents. In addition, these larger fire departments tend to be more efficient and effective because of their greater resources and funding levels. Consolidation or mergers number in the thousands and many are found in the Midwest and upper South, which have the highest concentration of large consolidated city-county governments in the United States. In fact, the largest fire department in the country is a consolidation (since 1898) of five counties, each of which is also known as a borough: the Kings, Richmond, Queens, Bronx, and New York counties form the New York City fire department (FDNY). This example seems to suggest that anything is possible. Even though there is the potential for many pitfalls and very detailed planning must precede these efforts, consolidation is an alternative that should not be overlooked for departments that are required to cut their budgets or that have shortages of volunteer members.

Discussion Questions

1. Looking into the future, what departments in your area could be consolidated? Or, are there governments in your area that have been consolidated?

2. Make a list of their present funding, staffing, and equipment.

3. Make a list of five issues that could begin a consolidation discussion and provide a short validation of each.

Persistent Change in the Past and Future

Taking into consideration that the U.S. fire service is more than 280 years old, many changes in FES are relatively new. For example, the federal fire focus of the U.S. Fire Administration and the National Fire Academy (NFA) is only slightly more than 30 years old. The first National Fire Protection Association (NFPA) fire service professional qualification standard, NFPA 1001, *Standard for Fire Fighter Professional Qualifications*, was adopted in 1974, and many more recently adopted standards are still finding their way into the training and operations of the fire service.

In the free enterprise business community, where turning a profit is the number one goal, the urgency for change is intense. There is always the possibility that the company might go out of business if it does not keep up with the latest and greatest consumer products and services. In contrast, although there are some outside pressures for change in the FES organization (e.g., regulations and standards for operations and safety), most of the urgency for change must come from within, leaving the future in the hands of the chief officer. In many cases, this requires competent leadership and personnel who are comfortable with change.

In the FES industry, the typical employee and organization itself receive little feedback on their performance. This lack of feedback is not the result of poor management but rather the result of a lack of performance measures. To measure future performance, valid outcomes directly resulting from input are needed. Leadership expert John Kotter further explains this notion: "The combination of valid data from a number of external sources, broad communications of that information inside an organization, and a willingness to deal honestly with the feedback will go a long way toward squashing complacency. An increased sense of urgency, in turn, will help organizations change more easily and better deal with a rapidly changing environment [future]" (Data from Kotter 1996, p. 163).

One recent example of a document, addressing this issue of measurable outcomes is NFPA 1710, *Standard for the Organization and Deployment of Fire Suppression Operations, Emergency Medical Operations, and Special Operations to the Public by Career Fire Departments*. Adopted in 2001, this standard addresses the issues of response times, deployment capabilities, and staffing in FES, which are all easily measured. It also contains requirements for training, operations, and safety through the reference of other NFPA standards. This standard now provides the FES community with legitimate measurements of service delivery that have been validated by consensus of subject matter experts using scientific research and extensive experience, allowing FES administrators to set goals for the future.

New positions and promotions should go to those who believe in these future goals and are comfortable with change. Individuals should be compatible with the intimate family atmosphere of the typical FES company. This characteristic can be difficult to assess, but it is not impossible. Psychological testing is becoming more and more common in the public safety professions. Professional consultants can help devise tests to screen applicants for personality traits that are desirable in a new member. Incumbents who aspire for promotions can also be evaluated periodically based on a set of known qualities.

In the past, departments have typically attracted and accepted risk-takers who were physically strong and preferred to interact with others who had similar personalities. Although these traits are still valued, compassion, open-mindedness, and being a team player are also now essential. Moving forward, departments should also look for diverse individuals who enjoy and feel comfortable interacting with the public. These individuals should be tolerant of others and want to help the public. They should also be able to deal with the rapidly changing world and the effects these changes have on the FES.

Technology and Research

Competency at the emergency scene is the result of three separate functions: (1) personnel; (2) equipment; and (3) command (standard operating procedures). To have FES organizations equipped with the newest and best technology is only one part of providing quality emergency services to the public; additional training and improved physical fitness also dramatically improve performance at an emergency incident.

Still, technology seems to be a never-ending source of change, some good, some bad, and some unnecessary. Examples of new technology in FES are 1.75-(44-mm) and 2-inch (51-mm) hoses, large diameter supply hoses, positive pressure ventilation, automatic defibrillators, the National Incident Management System (NIMS), compressed air foam, infrared cameras, and GPS navigation **FIGURE 13-1**. Major research participants, such as the National Fire Academy and the U.S. National Institute of Science and Technology, are studying respiratory protection, dermal protection, thermal imaging cameras, and other personal protective equipment (PPE) technologies. Other studies focus on firefighter safety and health as it relates to

FIGURE 13-1 Firefighter using an infrared camera.
© Jones and Bartlett Publishers. Courtesy of MIEMSS.

such issues as cardiovascular disease, hearing loss, and fatigue caused by shift work.

In addition to advancements in technology, with the ability to collect better data and the use of quality improvement programs to evaluate the data, long-standing medical protocols are now being called into question by advances in research. Evidence-based medicine, which is informed by accurate, definitive studies of the effectiveness of a product, drug, or treatment protocol, drives decisions on diagnostic and therapeutic equipment, ultimately improving the quality of patient care.

Although many of these new technologies and research may be very helpful for the FES organization, the process for research, experimentation, and acceptance tends to be sporadic and decentralized. The U.S. Fire Service is made up of more than 30,000 departments that characteristically serve small numbers of people and have minimal resources, especially for research. There is no large, well-funded, independent FES research agency to investigate and study new technologies, methods, or equipment. However, the National Institute of Science and Technology (NIST) Fire Research Division and NFPA Research Foundation have achieved some notable research projects that have added to the profession's knowledge. As stated by former Federal Emergency Management Agency (FEMA) administrator Dave Paulison, "It's just a start [the present level of fire research]. We must continue funding this type of research if we are truly committed to 'Everyone goes home'" (Data from Grant, 2009). Although most of this current research has been funded through the Department of Homeland Security, pressure to reduce the national deficit may threaten this funding in the future.

In the emergency medical services (EMS) area of FES service, research has a large and robust participation by the medical profession. For example, every research-based university has many students and faculty engaged in research. Funding by the medical industry and government is substantial. Techniques and medicines are constantly being reviewed and updated.

Furthermore, although many organizations label themselves as progressive, state-of-the-art agencies, they may be reluctant to accept all that is "new" and innovative, particularly in terms of equipment. As noted by NFPA research director Casey Grant, this issue seems to stem from the fact that the fire service is deeply rooted in tradition. "While this brings a certain stability to the profession, it can also mean that the fire service is sometimes slow to embrace improvements" (Data from Grant, 2009).

However, just because equipment or a technique is old does not automatically mean it is inferior. For example, at this time, plain water is still used to extinguish most structural fires. Fire protection experts have observed the inefficiencies of water because it is relatively heavy and not available everywhere **FIGURE 13-2**. Each new tool, idea, or technique needs to be judged on its own merits through research techniques that are based on standards and broad-based beta testing. This research is often best conducted independently. It is important to question information, observations, and studies cited by manufacturers. It is the manufacturer's job to tell the customer the best reasons to buy its product, and there is always the chance that the study process was biased.

The future is always difficult, if not impossible, to predict. In many cases, the implications of a product, new technology, or research result are not even apparent until many years after purchase or implementation. Although it is not realistic to expect 100% certainty before making a decision or implementing a change, it is necessary to be cautious and take the time to gain as much information as is needed to be reasonably certain. Flexibility and a willingness to adapt are key.

FIGURE 13-2 Plain water is used to extinguish most building and wildlands fires.
© a katz/Shutterstock.

Economic Impacts

Local governments primarily fund FES and are required to have a balanced budget. Unlike the federal government, local governments do not have the luxury of borrowing huge sums of money in times of economic downturn. In addition, elected officials seem less and less likely to try to raise taxes. Therefore, when tax revenues fall, services must be cut.

When this occurs, departments that are required to cut their budgets generally target nonpersonnel purchases first. Eventually, however, staffing also needs to be reduced. Often, this reduction begins with nonoperations staff, such as training and prevention personnel, but may also include fire-suppression personnel. For safety and efficiency issues, it is better to close stations than run companies with inadequate and unsafe crew sizes. Closing fire stations may be a more difficult decision for elected officials. When closed, these empty stations send a reminder to the public that their local government is in an economic struggle; reducing staffing levels, however, is rarely noticed by the public.

One idea to justify and acquire new funding can be found in the EMS field. Recently, community paramedicine has been replacing traditional in-hospital service in some regions. Community paramedicine is an effort to provide primary and preventive health care by having paramedics make house calls, especially in rural communities where access to health care is limited. These paramedics perform limited physical assessments, confirm prescription adherence, and evaluate an older person's home for fall hazards. Fees can be used to cover these services.

Another way to battle the issue of reduced funding is to consolidate fire departments. Consolidation has some substantial cost savings, such as reductions in mid- and upper-level management, staff functions, stations, and special services. These studies are very complex and time-consuming undertakings (Ref: https://www.fireengineering.com/2014/09/04/272718 /fire-department-mergers-consolidations-and -annexations/). In many cases, consolidation involves a county fire department absorbing a city fire department; however, there are also cases of a city absorbing a county department, such as the case of Jacksonville, Florida, which consolidated with Duval County.

Higher Education and Training

The results of an informal survey conducted for an article in Firehouse.com in 2014 indicated that more than 92% of fire chiefs at the time had a bachelor's degree or higher (Ref: https://www.firehouse.com/careers-education/article/12009765/higher-education-and-the-fire-service). It is now common for municipalities advertising for a vacant fire chief's position to list a minimum college degree requirement. At the national level, the NFA Executive Fire Officer program requires a minimum education level of a bachelor's degree; however, the NFPA Fire Officer professional standards still do not require any college education to comply with their essential career and technical education criteria, which are commonly used for certification in many areas. This discrepancy may send a mixed message to future officers.

Fire service college degrees for fire officers are readily available throughout the United States. They typically fall into two categories. The first are Associate two-year degrees commonly titled Fire Science. Many serve only the geographic areas near their campuses; however, there is a growing trend to offer courses at some higher education institutions via the Internet. In addition, Bachelor's degrees in Fire and Emergency Services are offered at a smaller number of four-year colleges and universities" **FIGURE 13-3**.

Many of the structural parts for a professional FES training and education system already exist. For example, state training agencies and other regional training centers are in place to provide firefighter and officer training. The problem is that most have been developed independently and many have different requirements. Dr. Denis Onieal, Superintendent of the NFA, summarizes the issue of training and education in the present fire service (Data from Onieal, 2004, p. 1):

It varies from place to place, depending upon the organization, the structure of the department

FIGURE 13-3 Distance learning is perfect for many members because they cannot get to most brick and mortar colleges. Also, students can access the courses any time, allowing complete flexibility to complete assignments. This is a great feature for members working shift schedules.
© Alexander_Safonov/Shutterstock.

and the governing agency. The process isn't the same wherever you go; frequently it is a slow and uneven process, or one solely based on popularity.... Right now, there is no one universally recognized and reciprocal system to acquire the knowledge and skills required in the Fire and Emergency Services.

Although most training programs use the NFPA professional qualifications standards, there are differences in the length of these training programs and the testing used for certification. In addition, some states maintain separate educational systems that provide officer and other specialized fire training and education. For example, a community college may offer a fire science program, but that does not mean the courses are recognized by the state training and certification agency. To become certified, applicants would then have to take the courses offered by the state training agency, even when the fire science program covers the same knowledge material.

In contrast, the EMS field has a model education, training, and certification system in place. The National Registry of Emergency Medical Technicians is a national certification system that was created in 1970 by the National Highway Traffic Safety Administration. This system is used by 92% of all states, with the other 8% still using the same training and skills but doing their own testing.

To be on the safe side, members who strive to become future leaders in the emergency services profession should, at a minimum, pursue a bachelor's degree. Fire services degrees that can be obtained by distance education, such as the NFA Degrees at a Distance programs, are often a good choice for students. However, if the student is fortunate enough to be in driving distance from a college that offers these degrees, face-to-face interaction with the instructor and other FES members is a significant advantage.

This is an area of emergency services that needs work and some type of consensus in the future. The key to leadership and change management is higher education. Without change, there cannot be progress. Education and training can give the chief officer the confidence and knowledge to deal with elected officials and the public, including union presidents, and to successfully lead the organization through change.

Risk Management

"America at Risk," a study that was concluded in 2002 by FEMA, contained the first discussion of total risk management at the federal level. According to the report, the future of emergency management involves more than just the risk of fire and may include other disasters such as hurricanes, floods, earthquakes, etc. "The establishment of FEMA, the growth of the emergency management community as a profession, the increase in disaster losses in America, and other factors, had dictated a different context for the fire service [in the future]" (Data from Federal Emergency Management Agency, 2002, p. 6).

The following is a list of the major topics discussed in this report:

- Implementation of loss prevention strategies (multihazard)
- The application and use of sprinkler technology
- Loss prevention education for the public
- The acquisition and analysis of data
- Improvements through research
- Codes and standards for fire loss reduction in the built environment
- Public education and awareness
- National accrediting and certification
- Firefighter health and safety
- Emergency medical services
- Diversity

"America at Risk" changed the expectations of many elected officials and the public, including the fire services perception of itself. FES agencies should focus on these topics when formulating goals to manage risk in the future.

Another theme that is present in this report is the emphasis on the effectiveness of prevention efforts. The United States has not experienced a major terrorist event since September 11, 2001; this could be the result of extensive prevention efforts by federal agencies such as the CIA, FBI, and U.S. military agencies **FIGURE 13-4**. September 11 was a series of four coordinated radical Islamic terrorist attacks using airplanes to crash into two 110-story high-rises (that completely collapsed) and the Pentagon; the last plane was forced by the passengers to crash in rural Pennsylvania. The attacks resulted in 2,977 fatalities (including 343 firefighters), over 25,000 injuries, and substantial long-term health consequences, in addition to at least $10 billion in infrastructure and property damage. For national security reasons, the federal government is not commenting at this time on any prevention successes—they just confirm that security is being pursued through prevention. Although it is also difficult to document in the FES field, prevention should continue to be a goal for the future.

FIGURE 13-4 Terrorist attacks of September 11, 2001.
© Anthony Correia/Shutterstock.

Homeland Security

The war on terrorism seems to be driving contemporary changes in FES administration and operations. On March 1, 2004, the U.S. Department of Homeland Security adopted NIMS (U.S. Department of Homeland Security, 2017). Previously, the emergency services had several incident command systems to choose from, each a little different from the others. "NIMS guides all levels of government, nongovernmental organizations (NGO), and the private sector to work together to prevent, protect against, mitigate, respond to, and recover from incidents. NIMS provides stakeholders across the whole community with the shared vocabulary, systems, and processes to successfully deliver the capabilities described in the National Preparedness System. NIMS defines operational systems, including the Incident Command System (ICS), Emergency Operations Center (EOC) structures, and Multiagency Coordination Groups (MAC Groups) that guide how personnel work together during incidents. NIMS applies to all incidents, from traffic accidents to major disasters." (*National Incident Management System*, Third Edition, October 2017; Retrieved from https://www.fema.gov/media-library-data/1508151197225-ced8c60378c3936adb92c1a3ee6f6564/FINAL_NIMS_2017.pdf).

The International Association of Fire Chiefs (2001, p. 3) has offered the following suggestions for NIMS compliance:

- Every responding agency (local, state, national) must use a standardized incident command system.
- Continual coordination between local, state, and national responders is essential for success

in protecting our personnel and the citizens we serve.

- National agencies need to involve local first responders in every level of discussions for resource allocation, especially training and equipment.

Many of the details of the NIMS are scheduled for study and adoption at a later date. These are very complex subjects, especially for the fire service, which has traditionally developed standards of service at the local level. As this project proceeds, expect to see efforts to define and implement the following for responses to a federal declared emergency:

- Mutual or automatic aid agreements
- Credentialing emergency responders for different competencies
- Resource typing
- Training for NIMS
- Integrated communications systems

For a look into the future, the federal system for responses to wildfires is a good model. This system can direct and coordinate hundreds of resources and thousands of firefighters operating at major wildfires. Wildfire resources and firefighters are brought in from hundreds and even thousands of miles away and from numerous individual departments and can work smoothly side by side. This type of effort to inspire and motivate others to provide efficient emergency services will be needed at the next major catastrophic hurricane, earthquake, wildfire, building fire, or terrorist event.

Advice for the Future

Investor's Business Daily (2004) has spent years analyzing leaders and other successful individuals. Most of these leaders have 10 traits that, when combined, can turn dreams and high goals into reality. The FES chief officer can refer to these traits when leading the organization toward change in the future (https://www.inc.com/peter-economy/this-study-of-300000-businesspeople-revealed-top-10-leader-traits-for-success.html):

1. Inspires and motivates others.
2. Displays high integrity and honesty.
3. Solves problems and analyzes issues.
4. Drives for results.
5. Communicates powerfully and prolifically.
6. Builds relationships.

Words of Wisdom

In addition, adhering to the following advice will help you in both your profession and private life:

- No one will ever get out of this world alive. Resolve therefore to maintain a reasonable sense of values.
- Take care of yourself. Good health is everyone's major source of wealth. Without it, happiness is almost impossible.
- Resolve to be cheerful and helpful. People will repay you in kind.
- Avoid angry, abrasive persons. They are generally vengeful.
- Avoid zealots. They are generally humorless.
- Resolve to listen more and to talk less. No one ever learns anything by talking.
- Be cautious of giving advice. Wise men don't need it and fools won't heed it.
- Resolve to be tender with the young, compassionate with the aged, sympathetic with the striving, and tolerant of the weak and the wrong. Sometimes in life you will have been all of the these.
- Do not equate money with success. There are many successful money makers who are miserable failures as human beings. What counts most about success is how a person achieves it.

By Lloyd Shearer, copyright 1989

Source: https://truthfulgrace.blogspot.com/2008/01/guide-to-happy-life-lloyd-shearer.html

7. Displays technical or professional expertise.
8. Displays a strategic perspective.
9. Develops others.
10. Innovates.

And finally, nothing substitutes for hard work and perseverance. In addition, whatever your personal or professional goals in life are today, they usually cannot be obtained without a focus on the future. These goals may change over your career and need to be constantly reviewed. Constantly look for opportunities to further your career, especially higher education. This means to be a professional your administrative competency must be based on a formal education. To close: stay safe because FES is a dangerous profession.

Wrap-Up

CHAPTER SUMMARY

- Explanation of innovated change in the past and future.
- Understanding technology and research and its influence on administration.
- Explaining the importance of higher education and training on achieving professional competency.
- At the national level, describing the place of risk management and homeland security in the present and future.
- Describing expectations of personal and organizational advice for the future.

KEY TERMS

Community paramedicine An effort to provide primary and preventive health care by having paramedics make house calls.

Consolidation The unification of two or more organizations by dissolution of existing ones and creation of a new single organization.

Evidence-based medicine Accurate, definitive studies of the effectiveness of a product, drug, or treatment protocol used to improve the quality of patient care.

Measurable outcomes Outcomes that physically can be counted, such as number of fire deaths, pickup trucks sold, etc.

Technology The knowledge of techniques and processes. It can also be incorporated in machines to allow for a process without detailed knowledge or actions of human staff.

REVIEW QUESTIONS

1. What do you feel is the greatest future threat FES personnel will have to face? Write a brief 500-word essay defending your choice and outlining at least three strategies the FES can use to overcome it.

2. Name a recent technology or protocol based on research that has been implemented in your workplace. What were the advantages of this change? Disadvantages? How was the implementation received, and what did leadership do to predict this change and bring it to fruition?

3. What are your personal goals for future education and training? Do you feel that these attainments should be standard for those in your ideal role(s) moving forward?

4. Did the COVID-19 virus change your FES department's policies or procedures? Were the changes justified?

5. What traits, education, and experience would you like to see in the next administrator for your organization's continued progress in providing great professional service to the public?

REFERENCES

Federal Emergency Management Agency. 2002. *America at risk, America burning recommissioned.* Retrieved from http://www.usfa.fema.gov/applications/publications

Grant C. 2009, March/April. The future of fire: Research renaissance? *NFPA Journal.*

International Association of Fire Chiefs. 2001. *Leading the way: Homeland security in your community.* Retrieved from http://www.iafc.org/downloads/chckweb.pdf

Investor's Business Daily. 2004. IBD's 10 secrets to success. *Leaders and Success.* Retrieved from http://www.stuartlevine.com/wp-content/uploads/2010/12/Investors-Business-Daily.pdf

Kotter J. P. 1996. *Leading change.* Boston, MA: Harvard Business School Press.

Onieal D. 2004. *Professional status: The future of fire service training and education.* Retrieved from http://www.usfa.fema.gov/downloads/pdf/nfa/higher-ed/ProfStatus Article.pdf

U.S. Department of Homeland Security. 2017. *National Incident Management System. Third Edition.* Retrieved from https://www.fema.gov/media-library-data/1508151197225-ced8c60378c3936adb92c1a3ee6f6564/FINAL_NIMS_2017.pdf

Appendix

Fire and Emergency Service Higher Education (FESHE) Correlation Grid for Fire and Emergency Services Administration Course Bachelor's Core (C0271)

Fire and Emergency Services Administration (C0271) Course Outcome	Chapter
1. Define and discuss the elements of effective departmental organization.	2, 3, 5, 6, 7, 11, 12, 13
2. Classify what training and skills are needed to establish departmental organization.	2, 3, 5, 6, 7, 8, 11, 12, 13
3. Analyze the value of a community-related approach to risk reduction.	2, 4, 5, 7, 9, 11, 12
4. Outline the priorities of a budget planning document while anticipating the diverse needs of a community.	5, 11, 12
5. Assess the importance of positively influencing community leaders by demonstrating effective leadership.	2, 3, 4, 5, 7, 9, 10, 11, 12, 13
6. Analyze the concept of change and the need to be aware of future trends in fire management.	1, 2, 3, 4, 6, 8, 9, 10, 12, 13
7. Report on the importance of communications technology, fire service networks, and the Internet, when conducting problem-solving analysis and managing trends.	3, 8, 10, 11, 12, 13
8. Develop a clear understanding of the national assessment models and their respective approaches to certification.	2, 6, 8, 9, 10, 12, 13

Fire and Emergency Services Administration (C0271) Course Objective	Chapter	Page
Module I: Leading and Managing Purposefully with a Community Approach		
1. Describe the role of the fire/emergency medical services department as a part of the community, government, and comprehensive plan.	1, 4, 7, 12	pp 1–14, 61–63, 64–67, 134–138, 219–238
2. Explain the importance of a good working relationship with public officials and the community as a whole.	2, 7, 10	pp 25–33, 134–142, 189–193
3. Assess ways to develop a good working relationship with public officials and the community.	2, 7, 12	pp 27–32, 134–142, 226–231
4. Identify local, state, and national organizations that will be beneficial to your department.	1, 6	pp 3–10, 124–126

Fire and Emergency Services Administration (C0271) Course Objective	Chapter	Page
5. Describe how to take a proactive role in local, state, and national organizations.	6	pp 124–127
6. Identify effective skills for developing a cooperative relationship with fire and emergency services personnel as well as public officials and the general public.	2, 6, 7	pp 25–33, 121–124, 134–142
Module II: Core Administrative Skills		
1. Identify the core skills essential to administrative success.	2	pp 18–22
2. Describe the integrated management of financial, human, facilities, and equipment and information resources.	3, 5, 6	pp 38–53, 74–97, 104–128
3. Explain the importance of public access to government operations.	7	pp 134–138
4. Describe the key elements of successful communication.	3, 10	pp 39–42, 189–190
5. Recognize the basic management theory in use in your agency.	3	pp 49–52
6. Recognize the formal and informal dynamics of public organizations and describe strategies to ensure success.	2, 6, 12	pp 25–27, 124–126, 226–238
7. Discuss the components and styles of leadership.	2	p 19
8. Identify and discuss a practical agency evaluation process.	3, 12	pp 47–49, 219–237
Module III: Planning and Implementation		
1. Describe the process of consensus building.	3, 12	pp 38–42, 237–238
2. Describe the components of project planning.	12	pp 219–237
3. Identify the steps of the planning cycle.	12	pp 219–225
4. Discuss how an environmental assessment determines the strategic issues and direction of an organization.	12	pp 224–225
5. Assess the interrelationship between budgeting, operational plans, and strategic plans.	12	p 219
6. Analyze the importance of an organizational culture and mission in the development of a strategic plan.	3, 12	pp 38–42, 219–226
7. Describe the purpose, function, and current and future security concerns of working document publication, storage, and integrity.	10	pp 189–191
8. Explain how a fire and emergency service administrator creates a vision of the future for his or her organization.	13	pp 243–250

Fire and Emergency Services Administration (C0271) Course Objective	Chapter	Page
Module IV: Leading Change		
1. Describe the importance of accepting and managing change within the fire and emergency service department.	1, 3, 4, 8, 9, 13	pp 1–14, 38–53, 151–164, 168–181, 189–190, 243–250
2. Identify models of change commonly used in organizations.	4	pp 61–70
3. Summarize the steps of the change management process.	3, 4	pp 53–61
4. Assess ways to create a positive climate for change and introduce new ideas within the organization.	3, 4, 8	pp 37–52, 49–50, 151–164
5. Describe how an organization can respond to current or emerging events or trends.	4, 5, 8, 9, 13	pp 61–70, 84–85, 151–164, 172–181, 243–250
6. Explain the benefits of employee involvement in departmental decisions.	3, 4, 9	pp 37, 63–65, 176–178
7. Demonstrate innovative ways to address traditional problems within the organization.	4, 10	pp 68–70, 189–190, 200–201
8. Describe ways to increase and reward professional development efforts.	3, 4, 8, 9	pp 37–38, 67–68, 159–164, 175–178
Module V: Customer Relationship Management (CRM)—A 21st Century FESA Responsibility		
1. Assess the importance of integrating fire and emergency services into a community's comprehensive plan.	7, 13	pp 134–142, 244–250
2. Assess your organization's capabilities and needs based on risk analysis probabilities.	12	pp 221–225
3. Describe the relationship between community risk analysis and strategic and operational planning.	12	pp 221–225
4. Identify the major steps of a community risk assessment.	12	pp 221–225
5. Identify direct and indirect costs associated with fire.	1, 5, 7	pp 3–6, 10–12, 90–97, 135–137, 138–141
6. Analyze economic incentives that encourage and discourage fire prevention.	1, 5, 7, 12	pp 3–6, 81–90, 138–141
7. Describe the role of fire and emergency services in the economic development and neighborhood preservation programs of the community.	5, 7	pp 76–80, 134–142

Glossary

360-degree performance review A type of performance evaluation that uses surveys sent to employees at all levels to obtain a more complete picture of a manager's performance.

A

Accidental Death and Disability The Neglected Disease of Modern Society 1966 publication that serves as a blueprint for the national effort to improve emergency medical care.

Accreditation To certify (a school, college, or the like) as meeting all formal official requirements of academic excellence, curriculum, facilities, faculty, etc.

Administration The practice of supervising and leading an organization to achieve its goals.

Affirmative action A hiring program used to recruit minorities and women in many FES organizations.

America Burning Report A compilation of 90 recommendations issued in 1973 by the National Commission on Fire Prevention and Control that began the national effort to professionalize the fire service.

Americans with Disabilities Act Federal act that provides civil rights protection to people with disabilities and guarantees equal opportunity in employment, public accommodations, transportation, government services, and telecommunications.

Analogy A comparison between something very familiar and a new concept, idea, situation, etc. This helps the reader quickly and accurately understand the new idea.

Apprentice phase Typically, 1-year period in which a new FES member learns additional knowledge and skills.

B

Banding This is a technique that replaces the traditional hiring or promotional lists used by many FES organizations. For example, for promotion a written exam is given and a list is created using the highest score first. This person is promoted first. However, in "banding" the list is separated and the top 15% are placed in the top band. Then either random choice or a diversity goal is used to choose the person for promotion.

Barriers Obstacles to change.

Battalion chief A chief officer who supervises a number of fire companies or a subdivision of the fire department.

Blind loyalty This behavior is seen in close acquaintances' of a leader, manager, or administrator. They are often seen supporting all actions even when strong evidence indicates the decision or information is wrong. Sometimes these individuals are known as "yes men" and can be recognized as those that agree with everything that is said by the supervisor.

Body mass index A high BMI can be an indicator of high body fatness. However, percent body fat computations tend to be more accurate.

Brainstorming A technique to generate ideas in which group members may suggest any solutions without initial judgments.

Broad-based empowerment Empowering low-level supervisors to alter or change policies or procedures to achieve a better outcome such as better quality or less time to complete the task. Once this is determined to be successful, it is assumed that the new or improved policies and/or procedures will be distributed to the entire organization.

Brownouts A cost-cutting strategy in which fire companies go unstaffed on a rotating basis.

Budget reports Periodic documents created by the administrator to monitor budget expenditures and detect any deviation.

Building and fire codes Government policies that regulate the construction and maintenance of structures.

Burnout A psychological state that results from physical or mental stress that affects a member's ability to think, behave, or physically perform effectively and safely.

C

Candidate Physical Ability Test Test comprising eight sequential events developed by the IAFC and the IAFF in 1999 to provide fire departments with a tool to select physically capable individuals to train as firefighters.

Capital item An operating cost that has a long life expectancy, such as a fire station.

Case studies A research process that tells a story about an administrative issue. After it is completed and verified, it is used many times as a lesson in public policy curriculums. Simply, these case studies describe either the failure or success of an administrative program.

Change The process of altering, modifying, or making different an idea, device, or policy.

Checkpoints Intermediate, short-term goals implemented to periodically monitor progress toward a long-term goal.

Chief officer The principal administrator of a fire department.

Civil rights acts The 13th and 14th Amendments to the U.S. Constitution guarantee equal treatment or protection against discrimination based on race, religion, gender, or national origin.

Code of ethics Ethical codes are adopted by organizations to assist members in understanding the difference between right and wrong and in applying that understanding to their administration actions.

Commission on Fire Accreditation International A self-assessment process that measures fire department capabilities, response times, and compliance with nationally recognized standards.

Community paramedicine An effort to provide primary and preventive health care by having paramedics make house calls.

Community risk assessment Community risk assessment is the identification of fire hazards and planning an appropriate suppression response force to mitigate medical and other emergencies. This all-hazards approach provides the opportunity to have a much bigger impact on life safety and property loss.

Compliance audit Documentation review that checks for adherence to a legal or policy mandate, which can include elements of fiscal management.

Compromise To compromise is to make a deal through negotiations between different parties where each party gives up part of their demand.

Confidence A sense or consciousness of reliance on circumstances, faith, or belief that one will act in a right, proper, or effective way.

Consensus Agreement reached by unanimous decision-making, a simple majority, or a specified majority.

Consensus building Usually used in a formal committee as the process to determine the conclusion of a discussion on an issue. The voting can be anywhere from a simple majority to complete agreement.

Consolidation The unification of two or more organizations by dissolution of existing ones and creation of a new single organization.

Constructive discharge A legal term indicating that a member was somehow convinced to resign because of intolerable working conditions.

Controlling Monitoring performance, measuring outcomes and progress toward goals, and making any adjustments necessary to keep the agency moving toward those goals.

Cooperative purchasing Joint buying arrangement that consolidates purchasing power across several departments, resulting in lower total costs.

Coping mechanism The strategies people often use in the face of stress and/or trauma to help manage painful or difficult emotions. Some good, some bad, such as overeating or using intoxicating drugs or alcohol.

Core of believers 10% to 15% of the members of a given group who believe that a proposed change is a good idea.

Cost-benefit analysis A cost-benefit analysis is used to measure the benefits of a decision or taking action versus the costs. It is commonly used in the business world because the benefits and profits can be more easily identified. A simple FES example would be the justification of a new fire station. The benefit would be reduced response time to a known number of citizens. This benefit in and of itself is not easily defined. Studies that have tried to correlate the relationship of reduced response time to either life or property loss have not been successful. Although, on a common sense level, reduced response time is always beneficial.

Critical Incident Stress Management A comprehensive crisis intervention system that includes components to manage all phases of a crisis.

Critique of operations A post-incident evaluation of the actions taken after a large-scale emergency or death of a firefighter. In a committee format, participants involved in the incident and experienced FES members are gathered to investigate and discuss the incident, resulting in a formal report containing recommendations for a better outcome in the future. Details of a comprehensive policy and process can be found at *Special Report: The After-Action Critique: Training Through Lessons Learned* (2008) (https://www.usfa.fema.gov/downloads/pdf/publications/tr_159.pdf).

Culture A set of shared attitudes, values, goals, and practices that are found in an institution, community, or organization generally unconsciously accepted by the members. Many times identified as "tradition."

Curriculum A set of courses constituting an area of educational specialization or degree criteria.

Customer service More commonly found in for-profit sales, it is the person-to-person interaction to assure the customer is completely satisfied with the purchase and company.

D

Debt ceiling In the federal government, this is the maximum national debt that cannot be exceeded by law. It is typically increased by Congress as needed with some posturing and even some compromising because the two political parties have opposing views on public policies.

Delegation Delegation is the process of handing decision-making authority to lower-level employees and to be successful, an employee must be able to obtain the resources and cooperation needed for successful completion of the delegated task.

Directing The process of a manager instructing the members under his or her authority to accomplish an organizational objective.

Discipline The action taken by a supervisor, government official, or parent for conduct that does not follow a rule or system of rules governing generally accepted appropriate behavior, conduct, or activity. This may take the form of a verbal or written reprimand and may include or some type of punishment like a fine, time off, even termination. Generally it is designed to change the behavior so the individual conforms to the rules and regulations, including generally accepted conduct of the culture.

Diversity A mix of employees that reflects a blend of sex, race, age, education, country of origin, and other characteristics, such as personal values.

E

Effectiveness The manager's success in attaining goals.

Efficiency A determination that a process has an advantage to others resulting in a better outcome for the amount of input. For example, my car is more efficient than yours because it gets better gas mileage. Many times this is determined by subjective determinations.

Emergency medical technician supervisors A supervisor who has the responsibly of several EMS units or a section of the organization.

Empowerment Allowing of the authority, right, or ability to perform various acts or duties without a supervisor's approval. More commonly found in the retail or service industries where immediate satisfaction of the customer is important.

EMS Workforce Agenda for the Future A document prepared by the University of California San Francisco (UCSF), with funding from the National Highway Traffic Safety Administration (NHTSA) and the Health Resources and Services

Administration (HRSA), that investigates developing and retaining a professional, well-trained EMS workforce and identifies four key research components considered critical to developing such a workforce.

Encumbrance Accounting procedure that sets aside designated funds for a particular expenditure until an invoice is received and payment is issued.

Environmental Protection Agency The mission of the U.S. agency, EPA, is to protect human health and the environment.

Equal Employment Opportunity Commission A federal agency that administers and enforces civil rights laws against workplace discrimination.

Ethics A principle of right or good conduct or a system of moral principles/values.

Everyone Goes Home program The result of a 2004 NFFF conference that established the 16 Firefighter Life Safety Initiatives.

Evidence-based medicine Accurate, definitive studies of the effectiveness of a product, drug, or treatment protocol used to improve the quality of patient care.

Expenditures This is the cost in dollars used to acquire some merchandise or service.

F

Facilitating Changing items or processes in interdependent systems to support the original change.

Fair Labor Standards Act The Fair Labor Standards Act of 1938 is a federal statute of the United States. The FLSA introduced the 40-hour work week, established a national minimum wage, guaranteed "time-and-a-half" for overtime in certain jobs, and prohibited most employment of minors in "oppressive child labor."

Fairness Treating people or ideas equally or in a way that is right or reasonable. However, this is a very subjective concept and can sometimes be clarified by an ethical evaluation of the person or idea by considering how one ought to act, morally. One father explained to his children, who wanted to be fairly judged (the "that's not fair" declaration), that if life were fair to him, he would have won the lotto several times. Life is not always fair but administrators should do all they can to be as fair as possible in their supervisory responsibilities.

Family and Medical Leave Act Federal act that mandates employers with more than 50 employees must offer employees up to 12 weeks of unpaid leave per year for family responsibilities.

Federal Emergency Management Agency (FEMA) The nation's incident manager for all hazards and major disasters, run by the Department of Homeland Security.

Federal Register The U.S. government's legal newspaper.

Financial audits Accounting procedure usually performed on an annual basis to ensure that financial records are accurate and complete.

Financial management The process of translating financial resources into services while providing fiscal accountability.

Financial manager A person (s) responsible for creating, justifying, and monitoring the department's budget.

Fire Research and Safety Act 1968 legislation that established the National Commission on Fire Prevention and Control.

Fire service tax A government tax that can be a set value for all residences or can vary depending on the square footage of the dwelling or the fire flow formula.

FirstNet The nationwide public safety communications system dedicated to America's first responders.

Fiscal year A full year budgetary period starting on July 1 for local and state governments and October 1 for the federal government.

G

Government bonds A form of money lending that has a low interest rate at a set interval for up to 30 years with tax-free interest.

Government fee for service These are fees charged by local, state, and federal governments for various services such as admission to national or state parks, drivers licenses, and toll roads.

Group decision-making Occurs when individuals, typically those that are affected by the decision, collectively study and make a choice from the potential realistic options.

Group think When some members of a group go along with majority opinion to avoid conflict.

Guiding coalition A group of individuals who represent the total organization and are appointed to define and devise a plan for implementation of a major change effort.

H

Heart Lung diseases This is a category of medical conditions that have been linked with the occupation of firefighting. Most firefighters have the benefit of Heart Lung presumption that assumes these medical conditions are job related and eligible for medical treatment, including justification for job-related disability and retirement.

Home (fire safety) inspection programs A voluntary service offer to the homeowner for fire department personnel to visit and inspect the home for fire prevention problems. Fire safety education tips are also typically offered to the homeowner.

Hostile environment Repeated workplace conduct of an inappropriate and/or sexual nature that may be offensive or intimidating.

Human resources The division responsible for recruiting, hiring, training, evaluating, and paying employees.

I

Imperfect information The lack of knowledge required to make an informed decision.

Incentive Positive (e.g., a rule or system of rules governing conduct or activity, money) and negative (e.g., discipline) reinforcement provided to make change more attractive.

Incident Command System A subcomponent of NIMS and the initial model of command for local incidents.

Income tax This is a tax based on a person's income and can be either progressive or flat. Most are based on the federal income tax that is progressive, requiring a larger percentage the more income is earned. Many earners at the low end of the pay scale are not taxed.

Incremental adjustment A method of implementation that works by making a small change and assessing the results.

In-service training An in-service program is a professional training or staff development effort where skills and knowledge to be proficient are discussed and demonstrated. In many cases, this includes new knowledge or newly generated SOPs.

Insubordination The act of willfully disobeying an order of one's superior.

Insurance Service Office (ISO)Grading Schedule A document that grades municipalities from Class 1 to 10 based with 1 being the best and 10 indicating long response times and no water supply.

International Association of Fire Chiefs (IAFC) "The International Association of Fire Chiefs (IAFC) represents the leadership of firefighters and emergency responders worldwide; our members are the world's leading experts in firefighting, emergency medical services, terrorism response, hazardous materials spills, natural disasters, search and rescue, and public safety policy" (https://www.iafc.org/about-iafc).

International Association of Firefighters (IAFF) A labor union representing nearly 313,000 firefighters and paramedics.

International Code Council (ICC) An organization created in 1994 by the three leading building code groups in the United States to produce one building code to be used throughout the country.

Intuition Intuition is commonly thought of as instinctual or even magical phenomenon, but is formed in our subconscious from the influence of past experience and cumulative knowledge.

J

Job analysis A common method of defining job requirements, including hiring criteria, recruit training and testing, and incumbent evaluation.

L

Leader A person generally in authority that guides and champions an organization or group through significant and often controversial changes.

Leadership Inspiring trained personnel to safely and professionally complete their mission using the resources provided by the FES organization.

Lease-purchase contracts An agreement to lease a capital item with the option to purchase it later.

Liability A liability is something a person or company owes, usually a sum of money.

Light duty A temporary assignment for workers who have been injured on the job in such areas as fire inspection, fire safety education, or 911 dispatching.

Line-item budgets A statement of used and available funds broken down into categories.

Live-fire drill Deliberately starting a structural fire for the purpose of safe successful extinguishment by firefighters. See NFPA 1403 *Standard on Live Fire Training Evolutions*, which provides minimum requirements for conducting live fire training to ensure they are conducted in safe facilities and a safe manner for participants.

Lying Deliberately presenting a false statement or action as true or hiding factual information with the intent to deceive.

M

Malfeasance Wrongdoing, including illegal actions or misconduct, especially by a public official.

Management by walking around Refers to administrators spending some part of each day listening to problems and ideas of their staff, while physically wandering around an organization.

Management Ensures that the organization is prepared and able to accomplish its goals by establishing that sufficient personnel and equipment are available for the organization to perform its duties.

Mandate To officially require (something). To make (something) mandatory.

Market failure This occurs when there is an inefficient allocation of resources and/or services in a free market. Market failure can occur due to a variety of reasons, such as monopoly (higher prices and less output), negative externalities (costs to third parties), and public goods that are usually not provided to everyone in a free market. As a public policy issue, the safety and wellbeing of the individual is always a chief concern.

Measurable outcomes Outcomes that physically can be counted, such as number of fire deaths, pickup trucks sold, etc.

Minority domination When a charismatic group member or several members representing a particular group disproportionately influence decision-making.

Misfeasance The performance of an administrative action in an illegal, immoral, or improper manner.

Monopoly Exclusive ownership through legal privilege, command of supply, or concerted action of a product or service.

Multimedia A technique using a combination of sound, video, and written text for conveying ideas or information.

Multitasking Performing multiple management functions at the same time.

N

National Board of Fire Underwriters "The National Board of Fire Underwriters was a group established in 1886 with the aim of preventing loss from fire and to protect property and lives from fire-related harms. It gathered statistics from fire incidents across the United States to provide a template for fire insurance" (https://www.insuranceopedia.com/definition/3060/national-board-of-fire-underwriters).

National Fallen Firefighters Foundation Since 1990 the National Fallen Firefighters Foundation has been officially designated by the U.S. Congress as the financial support for the National Memorial to career and volunteer fallen (death) firefighters. Located in Emmitsburg, Maryland, it was conceived as a tribute to American fire service.

National Fire Protection Association An international non-profit organization devoted to eliminating death, injury, property and economic loss due to fire, electrical and related hazards.

National Highway Traffic Safety Administration The NHTSA is responsible for keeping people safe on America's roadways. Through enforcing vehicle performance standards and partnerships with state and local governments, NHTSA reduces deaths, injuries, and economic losses from motor vehicle crashes (https://www.nhtsa.gov/about-nhtsa).

National Incident Management System Establishes a uniform set of processes, protocols, and procedures with which all responders must be compliant to command and control operations at catastrophic emergency incidents.

National Institute for Occupational Safety and Health (NIOSH) The U.S. federal agency responsible for conducting research and making recommendations for the prevention of work-related injury and illness.

National Labor Relations Act Enacted by Congress in 1935 to protect the rights of employees and employers, encourage collective bargaining, and curtail private sector labor and management practices that could harm the welfare of workers.

National Professional Development Model Primarily a visual presentation of the process of formal education, self-education, and experience needed to be a "professional" fire officer.

Nonfeasance Failure to do what ought to be done.

O

Obedience The act or an instance of obeying; dutiful or submissive behavior such as following orders of a supervisor at the scene of an emergency.

Occupational hazard Occupational hazards, experiences in the workplace, can encompass many types of hazards, including chemical hazards, biological hazards, psychosocial hazards, and physical hazards.

Occupational Safety and Health Administration (OSHA) A U.S. agency that is responsible for providing safe and healthful working conditions by setting and enforcing standards, and by providing training, outreach, education, and assistance.

Organizing The process of determining what tasks need to be done and who is to perform these tasks.

Outcome A measured benefit to stakeholders, created as a direct result of dollars spent by the organization.

Oversight Review by entities such as the government or professional organizations to regulate ethical behavior.

P

Performance measures Generally, are measures that are quantifiable and are essential to manage anything. Used as a planning input or to assess a person's contribution to the organization or their meeting some performance measure of job accomplishment.

Personal ethics May include complying with laws and societal norms, telling the truth, showing concern for others' well-being, and respecting others' freedom and independence.

Pilot program A small-scale preliminary study conducted to evaluate the feasibility, time, cost, and adverse consequences, many times unintended, prior to the implementation of a change in policy or equipment.

Planning Creating and defining goals and making important decisions as well as establishing strategic plans to achieve these goals.

Policy analysis A method used in government and nonprofit organizations to examine and evaluate the available options to implement the goals of laws, regulations, and agency policies. It identifies problems and the need for change, assesses the urgency, and recommends potential solutions. Generally, there is no profit involved, just service to the public, such as FES organizations.

Posttraumatic stress disorder A psychological disorder that can develop after a person is exposed to a traumatic event, such as sexual assault, warfare, traffic collisions, child abuse, or other threats on a person's life.

Preventive medicine A policy or program whose goal is the prevention of a disease, either by preventing the occurrence of a disease or by halting a disease and averting resulting complexions after its onset. Preventive medicine can be practised by governmental agencies, primary care physicians, and individuals themselves. Wellness programs, providing safety equipment and policies are a big part of these efforts.

Probationary period A specified length of time after hiring (say, 1 year) used to evaluate a new hire, during which the employee can be separated without cause.

Procedural audit A review designed to evaluate management policies and procedures, which also can include financial policies and procedures.

Professional ethics May include complying with company policies and procedures, telling the truth to the public and elected officials, protecting confidentiality, and ensuring the safety of FES workers and the public.

Program analysis A systematic process that examines alternative means, methods, or policies to accomplish the goals and objectives of the FES organization.

Program budgeting A budgeting system that describes measurable outcomes and gives detailed costs of every activity to be carried out in a budget.

Progressive consequences A disciplinary system requiring more severe repercussions for repeat violations or serious offenses.

Property interests A legal term indicating that the employer is required to use due process in disciplinary procedures to separate an employee.

Property tax Government tax assessed as a percentage of the total value of property and improvements, or in some cases based on income for a commercial business.

Public relations An agenda of persuading the public to have understanding for and support toward a person, idea(s), philosophy, or organization.

Public survey Used to recognize how the public perceives the department's present service, identify areas in which they would like improvement or expansion of service, and determine how much the public is willing to pay for improvements in service.

Punitive The imposition of an undesirable or unpleasant outcome upon a group or individual, imparted by an authority as a response and deterrent to a particular action or behavior that is deemed undesirable or unacceptable.

Q

Quid pro quo Harassment that occurs when a person in authority takes advantage of their control over an employee's future in the organization.

R

Reductions in force A method to identify employees to be laid off when budget reductions dictate reducing the total budget.

Refresher training See in-service training.

Regulation A rule or order issued by an executive authority or regulatory agency of a government and having the force of law.

Revenue This is the money received by the political jurisdiction from taxes, fees, or borrowing.

S

Safety committee A group of stakeholders with varied experience that plans and implements changes needed to comply with safety standards.

Sales tax Government tax applied to consumer purchases, such as electronics and vehicles.

Seminar A seminar is a form of academic instruction, either at an academic institution or offered by a commercial or professional organization.

Sense of urgency This is the process to persuade others that the change proposed needs to be at the top of this list to start and complete in an expeditious fashion.

Sensitivity training Education offered to employees about personal differences to achieve social harmony.

Simulation drills Simulation drills are used to practice in a nonemergency setup the actions and process needed to manage a large-scale disaster situation. Used both as a training exercise and to identify areas needing improvement.

Sleep deprivation Sleep deprivation is the condition of not having enough or good quality of sleep. It can be either chronic or acute and may vary widely in severity. A chronic sleep-restricted state adversely affects the brain and cognitive function.

Special interest group A group with a shared interest in advancing a specific area of knowledge, learning, political view, financial interest, or technology where members cooperate to affect or to produce justifications and solutions to support their point of view.

Standard operating procedures Rules that clarify consistent actions that result in proven outcomes and safe practices.

Statistics The field of scientific study that concerns the collection, organization, analysis, and presentation of data as opposed to opinions. One simple example would be the percentage of people who have blue eyes. If you could count everyone, this would be a hard number. However, many times statistical inferences are used, such as polling 1,000 people. The study may report 10% have blue eyes with a plus or minus of 2%. This is a mathematical technique.

Strategic planning A type of planning that identifies the department's vision, mission, and strategies.

Stress Stress is a fact of life in which psychological pressures from the inside or outside world affect the individual, either one's emotional or physical well-being, or both. Typically, it can be helpful in short-term situations but harmful when it is chronic long term.

Supreme Court The highest court in the federal judiciary of the United States of America. It has ultimate (and largely discretionary) appellate jurisdiction over all federal and state court cases that involve a point of federal law.

T

Tactical plans Tactical plans are the very actions that are necessary to carry out the strategy. For example, the strategy may be to travel to a distant city. Some of the tactics could be fill the car with gas, check all fluids, get directions including rest and food stops, etc.

Task force A group of employees assembled to discuss an important decision before implementation and to make recommendations.

Technology The knowledge of techniques and processes. It can also be incorporated in machines to allow for a process without detailed knowledge or actions of human staff.

Tiered response Also known as "priority dispatching," this practice allows for efficient use of resources by making response decisions after dispatch call screening and medical evaluation.

Total cost purchasing A purchasing strategy that considers both the cost of purchase and any other associated costs, such as maintenance.

Total quality management Constant process of detecting and reducing or eliminating errors in manufacturing, streamlining supply chain management, improving the customer experience, and ensuring that employees are up to speed with training.

Tradition An inherited, established, or customary pattern of thought, action, or behavior handed down in an organization or society consisting of information, beliefs, and customs. It is passed down without written instruction or formal instructions of a culture of social attitudes, beliefs, customs, and institutions.

U

Unfunded mandates Regulations requiring expenditures that are not combined with any government funding.

Unions Groups of workers who have formed an association to discuss issues with management.

Unity of command An organizing principle based on the idea that each individual reports to only one manager.

User tax A government tax that applies only to the use of specific goods or services, such as a road tax.

U.S. Firefighter Fatalities and Injuries report Annual report issued by the NFPA providing statistics on American firefighter fatalities and injuries.

V

Vision A thought or concept formed by the perception of a need for change from the study of data, experiences of others or events, common sense, or intuition. In FES, it is frequently the result of analyzing and understanding the requirements of national recognized professional standards.

W

Wellness The characteristic of being in good health physically and mentally.

Workers' compensation A form of insurance providing wage replacement and medical benefits to employees injured in the course of employment in exchange for mandatory relinquishment of the employee's right to sue his or her employer for the tort of negligence.

Z

Zero-based budgeting (ZBB) Financial decision-making technique in which every line item must be approved as if it was the first time it was requested.

Zoning Regulations created to prevent noncompatible uses of buildings and land.

Index